NMR OF PROTEINS

TOPICS IN MOLECULAR AND
STRUCTURAL BIOLOGY

Series Editors:

Stephen Neidle
Institute of Cancer Research
Sutton, Surrey, UK

Watson Fuller
Department of Physics
University of Keele, UK

Jack Cohen
Georgetown University
USA

Recent titles:

Protein–Nucleic Acid Interaction
Edited by Wolfram Saenger and Udo Heinemann (1989)

Calcified Tissue
Edited by David W. L. Hukins (1989)

Oligodeoxynucleotides: Antisense Inhibitors of Gene Expression
Edited by Jack S. Cohen (1989)

Molecular Mechanisms in Muscular Contraction
Edited by John M. Squire (1990)

Connective Tissue Matrix, Part 2
Edited by David W. L. Hukins (1990)

New Techniques of Optical Microscopy and Microspectroscopy
Edited by Richard J. Cherry (1990)

Molecular Dynamics: Applications in Molecular Biology
Edited by Julia M. Goodfellow (1990)

Water and Biological Macromolecules
Edited by Eric Westhof (1993)

Molecular Aspects of Anticancer Drug–DNA Interactions, Volume 1
Edited by Stephen Neidle and Michael Waring (1993)

Molecular Aspects of Anticancer Drug–DNA Interactions, Volume 2
Edited by Stephen Neidle and Michael Waring (1994)

Molecular Modelling and Drug Design
Edited by Andrew Vinter and Mark Gardner (1994)

NMR of Proteins

Edited by

G. M. Clore
A. M. Gronenborn

Laboratory of Chemical Physics
National Institutes for Diabetes and Digestive and Kidney Diseases
National Institutes of Health
Bethesda, MD 20892
USA

CRC Press
Boca Raton   Ann Arbor   Tokyo

First published 1993

Published in the USA, its Dependencies, and Canada by
CRC Press Inc.
2000 Corporate Blvd, N.W.
Boca Raton, FL33431, USA

Library of Congress Cataloging-in-Publication Data
NMR of proteins / edited by G. M. Clore, A. M. Gronenborn.
p.   cm. — (Topics in molecular and structural biology)
Includes index.
ISBN 0–8493–7771–4
1. Proteins—Analysis.   2. Nuclear magnetic resonance
spectroscopy.   I. Clore, G. M.   II. Gronenborn, A. M.   III. Series:
Topics in molecular and structural biology (Boca Raton, Fla.)
QP551.N545   1993
574.19'245—dc20   93–25445
                                                             CIP

ISBN 0–8493–7771–4
Catalog #Z7771

Printed in Great Britain

Contents

Preface

The last few years have witnessed great advances in the application of nuclear magnetic resonance techniques in molecular biology. Significant strides have been made with respect to studying both structural and dynamical properties of proteins. Methods are now in place to determine three-dimensional structures of proteins up to molecular weights of 30–40 kDa in solution. Thus, NMR now stands side by side with X-ray crystallography as a technique providing the atomic details of important proteins. NMR, however, is not solely limited to structure determination; indeed, it can be used to effectively study important aspects of protein dynamics over a wide range of time-scales, both in solution and in the solid state, and may also be applied to investigate questions pertinent to the protein folding problem.

The present book is not intended to provide a comprehensive overview of all the NMR studies that have been carried out over the last few years, but rather serves to highlight some of the recent advances with respect to both methodology and applications. The different chapters collected here reflect some of the variety and different emphasis found among the NMR laboratories engaged in biological NMR, and as such constitute only a few examples of the application of NMR to proteins. We nevertheless hope that, in addition to providing up-to-date results, this book may stimulate the more biochemically inclined reader to follow the field more closely with respect to possible applications of these techniques in his or her area of research.

National Institutes of Health, Bethesda,   G. M. C.
September 1992   A. M. G.

The Contributors

D. Eric Anderson
Department of Chemistry and
  Institute of Molecular Biology
University of Oregon
Eugene
Oregon 97403
USA

Ad Bax
Laboratory of Chemical Physics
National Institutes of Diabetes and
  Digestive and Kidney Diseases
National Institutes of Health
Bethesda
MD 20892
USA

Iain D. Campbell
Department of Biochemistry
University of Oxford
South Parks Road
Oxford
OX1 3QU
UK

David A. Case
Department of Molecular Biology
The Scripps Research Institute
La Jolla
CA 92037
USA

G. Marius Clore
Laboratory of Chemical Physics
Building 2
National Institutes of Diabetes and
  Digestive and Kidney Diseases
National Institutes of Health
Bethesda
MD 20892
USA

Holly B. R. Cole
Bone Research Branch
National Institute of Dental
  Research
NIH
Bethesda
MD 20892
USA

Frederick W. Dahlquist
Department of Chemistry and
  Institute of Molecular Biology
University of Oregon
Eugene
Oregon 97403
USA

Paul C. Driscoll
Department of Biochemistry
University of Oxford
South Parks Road
Oxford
OX1 3QU
UK

Stephen W. Fesik
NMR Research
Department 47G
Building AP9
Pharmaceutical Discovery Division
Abbott Laboratories
Abbott Park
IL 60064–3500
USA

Angela M. Gronenborn
Laboratory of Chemical Physics
Building 2
National Institutes of Diabetes and
    Digestive and Kidney Diseases
National Institutes of Health
Bethesda
MD 20892
USA

Stephen Grzesiek
Laboratory of Chemical Physics
National Institutes of Diabetes
    and Digestive and Kidney
    Diseases
National Institutes of Health
Bethesda
MD 20892
USA

Sven Hyberts
Harvard Medical School
Dept of Biological Chemistry and
    Molecular Pharmacology
240 Longwood Avenue
Boston
MA 02115
USA

Lewis E. Kay
Department of Chemistry
University of Toronto
Toronto
Ontario
M5S 1A8
Canada

Jirong Lu
Department of Chemistry
    and Institute of Molecular
    Biology
University of Oregon
Eugene
Oregon 97403
USA

P. A. McDonnell
Department of Chemistry
University of Pennsylvania
Philadelphia
Pennsylvania 19104
USA

Lawrence McIntosh
Departments of Chemistry and
    Biochemistry
University of British Columbia
Vancouver
Canada

Placido Neri
NMR Research
Department 47G
Building AP9
Pharmaceutical Discovery Division
Abbott Laboratories
Abbott Park
IL 60064–3500
USA

Linda K. Nicholson
Bone Research Branch
National Institute of Dental Research
NIH
Bethesda
MD 20892
USA

S. J. Opella
Department of Chemistry
University of Pennsylvania
Philadelphia
Pennsylvania 19104
USA

Jeffrey W. Peng
Harvard Medical School
Dept of Biological Chemistry and
    Molecular Pharmacology
240 Longwood Avenue
Boston
MA 02115
USA

Dennis A. Torchia
Bone Research Branch
National Institute of Dental Research
NIH
Bethesda
MD 20892
USA

Gerhard Wagner
Harvard Medical School
Dept of Biological Chemistry and
  Molecular Pharmacology
240 Longwood Avenue
Boston
MA 02115
USA

Peter E. Wright
Department of Molecular Biology
The Scripps Research Institute
La Jolla
CA 92037
USA

1

Determination of Structures of Larger Proteins in Solution by Three- and Four-dimensional Heteronuclear Magnetic Resonance Spectroscopy*

G. Marius Clore and Angela M. Gronenborn

1 Introduction

A complete understanding of protein function and mechanism of action can only be accomplished with a knowledge of its three-dimensional structure at atomic resolution. At the present time there are two methods available for determining such structures. The first method, which has been established for many years, is X-ray diffraction of protein single crystals. The second method has only blossomed in the last 5–10 years and is based on the application of nuclear magnetic resonance (NMR[†]) spectroscopy of proteins in solution. The driving force for the development of an alternative to X-ray crystallography was threefold. First, many proteins do not crystallize; and even when they do, the crystals may diffract poorly or difficulties in solving the phase problem (e.g. finding suitable heavy atom derivatives) may be encountered. Second, there may be significant and possibly important functional differences between structures in the crystal state and in solution. Third, dynamic processes ranging from the picosecond to second time-scales are amenable to study by NMR. Despite these attractive features, it should be borne in mind that, just like crystallography, NMR also has a number

---

*Adapted from a review that appeared in *Science*, **252**, 1390–1399 (1991).
[†]Acronyms: NMR, nuclear magnetic resonance; NOE, nuclear Overhauser enhancement; NOESY, nuclear Overhauser enhancement spectroscopy; ROESY, rotating frame Overhauser spectroscopy; HOHAHA, homonuclear Hartmann–Hahn spectroscopy; TOCSY, total correlation spectroscopy; HMQC, heteronuclear multiple quantum coherence.

of limitations. In particular, the protein under investigation must be soluble and should not aggregate up to a concentration of at least 1 mM. Further, the dependence of linewidth on rotational correlation time probably sets an intrinsic upper molecular weight limit of about 50–60 kDa for the applicability of current solution NMR technology.

The advent of two-dimensional (2D) NMR (Jeener, 1971; Aue *et al.*, 1976; Jeener *et al.*, 1979; Bax and Lerner, 1986; Ernst *et al.*, 1987) set the stage for the determination of the first low-resolution structures of small proteins in the mid-1980s (Williamson *et al.*, 1985; Clore *et al.*, 1985, 1986b; Kaptein *et al.*, 1985). Subsequent improvements over the next few years led to a tremendous increase in the precision and accuracy of such protein structure determination., such that it is now possible, using 2D NMR methods, to determine structures of proteins up to about 100 residues which are comparable in quality to 2–2.5 Å resolution X-ray structures (Clore and Gronenborn, 1991b). This progress has been summarized in a number of reviews (Wüthrich, 1986, 1989, 1990; Clore and Gronenborn, 1987, 1989, 1991a). As proteins get larger than about 100 residues, however, conventional 2D methods can no longer be applied successfully. This chapter focuses on the novel methodological developments of heteronuclear 3D and 4D NMR, which have been designed to overcome the limitations imposed by the increased molecular weight and spectral complexity inherent to these proteins, and describes their application, which has recently culminated in the determination of the first high-resolution NMR structure of a protein greater than 150 residues (Clore *et al.*, 1991c).

The concept of increasing spectral dimensionality to extract information can perhaps most easily be understood by analogy. Consider, for example, the *Encyclopedia Britannica*. In a one-dimensional representation, all the information (i.e. words and sentences arranged in a particular set order) present in the encyclopedia would be condensed into a single line. If this line were expanded to two dimensions in the form of a page, the odd word may be resolved but the vast majority would still be superimposed on each other. When this page is expanded into a book (i.e. three dimensions) comprising a set number of lines and words per page, as well as a fixed number of pages, some pages may become intelligible, but many words will still lie on top of each other. The final expansion to the multivolume book (i.e. four dimensions) then makes it possible to extract in full all the information present in the individual entries of the encyclopedia.

2 The Nature of the Structural Data Derived from NMR Measurements

The principal source of geometric information used in NMR protein structure determination lies in short approximate interproton distance restraints derived from nuclear Overhauser enhancement (NOE) measurements (Noggle and Schirmer, 1971). The physical basis for the NOE effect is relatively simple and is based on the fact that each proton spin possesses a property known as magnetization. Magnetization is exchanged between the spins by a process termed cross-relaxation, and the rate constant for this process is directly related to r^{-6}. The chemical analogy to such a system is one with a large number of interconverting species in equilibrium with each other. The phenomenon of cross-relaxation is observed by perturbing the magnetization of a particular spin and observing the resulting change in magnetization, known as the NOE, of the other spins as the equilibrium is re-established. Thus, the time dependence of the NOE is governed by a set of coupled first-order differential equations. The experiment is therefore similar in spirit to chemical relaxation kinetics (e.g. a temperature jump experiment), except that the initial perturbation in the chemical system involves rate constants rather than species concentration. If the NOE is observed only a short time after the perturbation, the size of the NOE is proportional to the cross-relaxation rate and, hence, to r^{-6}. Because of the r^{-6} dependence, the magnitude of these effects decreases rapidly as the interproton distances increase, so that effects are generally not observable beyond 5 Å. The interproton distance restraints derived from the NOE measurements may also be supplemented by backbone and side-chain torsion angle restraints derived from three-bond coupling constants and appropriate NOEs (Wagner *et al.*, 1987; Güntert *et al.*, 1989; Kraulis *et al.*, 1989; Nilges *et al.*, 1990).

With the approximate interproton distance and torsion angle restraints in hand, a number of computational strategies can be applied to locate the minimum of a target function comprising terms for the experimental restraints, covalent geometry (i.e. bonds, angles, planes and chirality) and non-bonded contacts (e.g. a van der Waals repulsion term to prevent atoms from coming too close together). The types of algorithms employed operate either in *n*-dimensional distance space followed by projection into real space (e.g. metric matrix distance geometry—Havel *et al.*, 1983; Crippen and Havel, 1988) or directly in real space (e.g. minimization in torsion angle space with a variable target function—Braun and Go, 1985; dynamical simulated annealing—Nilges *et al.*, 1988a,b,c; and restrained molecular dynamics—Brünger *et al.*, 1986; Clore *et al.*, 1985, 1986a; Kaptein *et al.*, 1985). All real-space methods require initial structures which can be random structures with correct covalent geometry, structures very far from the final structure (e.g. an extended strand), structures made up of a completely random array of atoms or structures generated by distance space methods. The key requirements of all these methods is that they have large radii of convergence and that they fully sample in an unbiased fashion the conformational space consistent with the experimental, geometrical

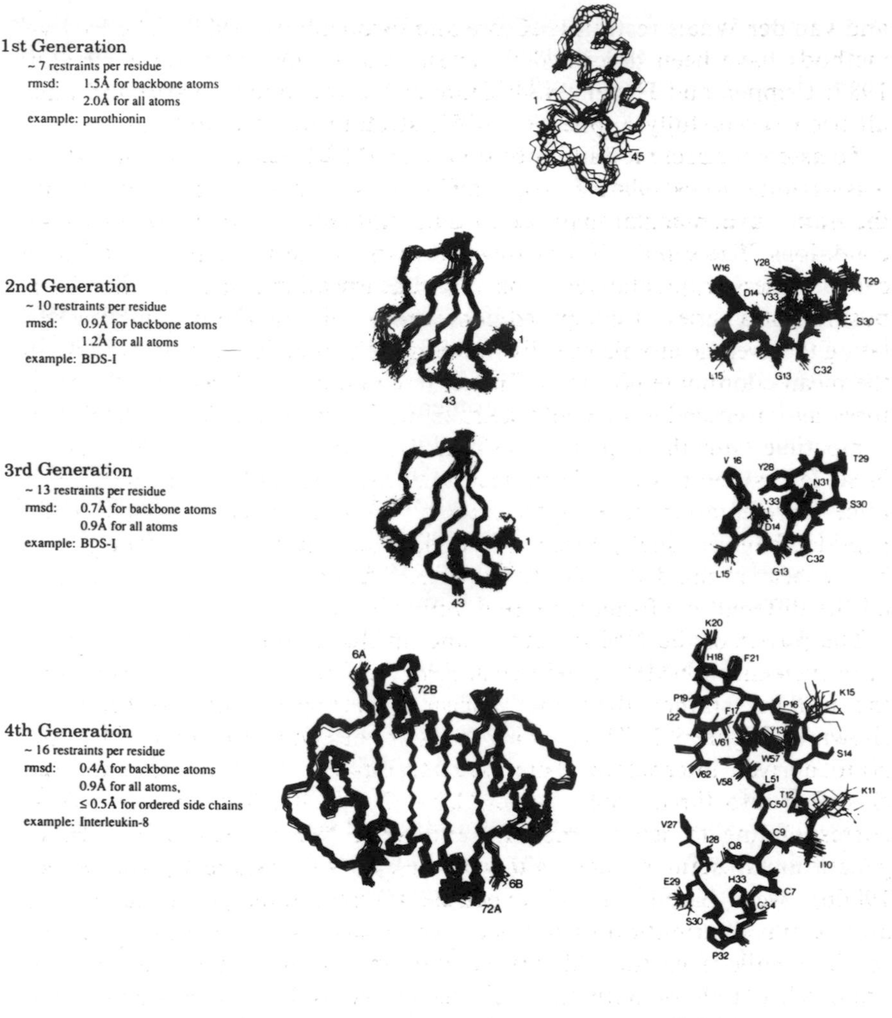

1st Generation
~ 7 restraints per residue
rmsd: 1.5Å for backbone atoms
        2.0Å for all atoms
example: purothionin

2nd Generation
~ 10 restraints per residue
rmsd: 0.9Å for backbone atoms
        1.2Å for all atoms
example: BDS-I

3rd Generation
~ 13 restraints per residue
rmsd: 0.7Å for backbone atoms
        0.9Å for all atoms
example: BDS-I

4th Generation
~ 16 restraints per residue
rmsd: 0.4Å for backbone atoms
        0.9Å for all atoms,
        ≤ 0.5Å for ordered side chains
example: Interleukin-8

Figure 1.1 Illustration of the progressive improvement in the precision and accuracy of NMR structure determinations with increasing number of experimental restraints. The fourth-generation structures are equivalent in quality to ~ 2–2.5 Å resolution X-ray structures. Each protein is represented by a best-fit superposition of a number of independently computed conformers: there are 9 for α1-purothionin, 31 for the second-generation version of BDS-1, 42 for the third generation of BDS-I, 30 for the backbone atoms of interleukin-8 and 15 for the side-chains of interleukin-8. In all the examples shown, the NOE-derived inter-proton distance restraints are only classified into three broad ranges, 1.8–2.7 Å, 1.8–3.3 Å and 1.8–5.0 Å, corresponding to strong, medium and weak NOEs. The structure of α1-purothionin (Clore *et al.*, 1986b) was calculated by metric matrix distance geometry followed by restrained molecular dynamics (Clore *et al.*, 1985), while the structures of BDS-I (Driscoll *et al.*, 1989a,b) and inter-

(*Cont. opposite page...*)

and van der Waals restraints (Clore and Gronenborn, 1989). The various methods have been described in detail in a number of reviews (Braun, 1987; Crippen and Havel, 1988; Clore and Gronenborn, 1989) and have all been successfully applied to NMR structure determinations.

To assess the accuracy and precision of an NMR structure determination, it is essential to calculate a large number of structures independently with the same experimental data set, using different starting structures or conditions. The spread of structures consistent with the experimental data can be assessed qualitatively from a visual inspection of a best-fit super-position of a series of computed conformers, and quantitatively by calcu-lating the average atomic rms distribution of the individual structures about the mean coordinate positions. The representation of NMR solution struc-tures as an ensemble of conformers in which each individual member is compatible with the experimental data may still be regarded as unusual, in so far as one has long been accustomed to the traditional single-chain trace representation of X-ray structures. In the latter case, a single poly-peptide chain is usually fitted to the electron density map, although one has to bear in mind that the density arises from a linear superposition of all the different conformers present within the crystal.

The power of the NMR method and an illustration of the progressive improvement in NMR protein structures as a function of increasing *only* the number, rather than the precision, of experimental restraints is shown in Figure 1.1. Thus, in all the examples shown in Figure 1.1, the NOE derived interproton distance restraints are only approximate and classified into three broad ranges, 1.8–2.7 Å, 1.8–3.3 Å and 1.8–5.0 Å, corresponding to strong, medium and weak NOEs, respectively. First-generation structures, such as α1-purothionin (45 residues) (Clore *et al.*, 1986b), were based on 5–7 experimental restraints per residue. The atomic rms distribution of the backbone atoms is rather large ($\sim$1.5 Å), so that while a cartoon-like representation of the polypeptide fold is obtained, details of both the local backbone and side-chain conforma-tions are essentially obscured by the large variations between the struc-tures. As the experimental restraints are increased to $\sim$10 per residue, including some backbone ϕ torsion angle restraints derived from a qualitative interpretation of the $^3J_{HN\alpha}$ couplings, a significant improve-ment is observed and the resulting second-generation structures have a backbone atomic rms distribution of $\sim$0.9 Å (Driscoll *et al.*, 1989b). The side-chain conformations, however, are still rather blurred. Additional improvement requires stereospecific assignments of β-methylene protons

leukin-8 (Clore *et al.*, 1990a) were calculated by the hybrid distance geometry–simulated annealing method (Nilges *et al.*, 1988c). The rmsd is the average atomic rms distribution of the individual calculated structures about the mean co-ordinate positions

and χ_1 side-chain torsion angle restraints. In a simplified approach these are obtained from a qualitative interpretation of $^3J_{\alpha\beta}$ coupling constants and intraresidue NOEs (Wagner *et al.*, 1987). The resulting third-generation structures are based on ~13 experimental restraints per residue and display a marked improvement not only in the definition of the backbone atoms, but more importantly in the side-chains as well (Driscoll *et al.*, 1989a). This is readily apparent on comparing the second- (Driscoll *et al.*, 1989b) and third-generation (Driscoll *et al.*, 1989a) structures of the same protein BDS-I (43 residues). Finally, in the fourth-generation structures, a much larger number of stereospecific assignments and loose torsion angle restraints are obtained by carrying out conformational grid searches of ϕ, ψ, χ_1 torsion angle space on the basis of the $^3J_{HN\alpha}$ and $^3J_{\alpha\beta}$ coupling constants and intraresidue and sequential interresidue NOEs involving the NH, $C^\alpha H$ and $C^\beta H$ protons (Nilges *et al.*, 1990). This leads to 16–20 experimental restraints per residue, a backbone atomic rms distribution of ≤ 0.4 Å and an atomic rms distribution of ≤ 0.5 Å for ordered side-chains. The errors in the atomic coordinates of such fourth-generation structures, like that of interleukin-8 (a dimer with 72 residues per subunit) (Clore *et al.*, 1990a) shown in Figure 1.1, are similar to those of 2–2.5 Å resolution X-ray structures (Clore and Gronenborn, 1991b). Indeed, the solution structure of interleukin-8 was used by necessity to solve the X-ray structure by molecular replacement, as traditional methods based on heavy atom derivatives had proved unsuccessful, despite several years of effort (Baldwin *et al.*, 1991).

While there are many examples of first- and second-generation structures in the literature, relatively few third- and fourth-generation structures have been published to date. In addition to the two structures shown in Figure 1.1, examples of third-generation structures are tendamistat (Kline *et al.*, 1988), hirudin (Folkers *et al.*, 1989; Haruyama and Wütrich, 1989) and *E. coli* thioredoxin (Dyson *et al.*, 1990), while examples of fourth-generation structures are the C-terminal domain of cellobiohydrolase (Kraulis *et al.*, 1989), the homeodomain of the Antennapedia protein (Billeter *et al.*, 1990), a zinc finger domain from a human enhancer binding protein (Omichinski *et al.*, 1990), human thioredoxin (Forman-Kay *et al.*, 1991), the IgG binding domain of protein G (Gronenborn *et al.*, 1991), plastocyanin (Moore *et al.* 1991) and interleukin-1β (Clore *et al.*, 1991c).

3   General Principles Involved in Deriving Experimental Restraints from NMR Data

The experimental restraints derived from the NMR data require the identification of specific interactions between proton pairs which may be either through-space (i.e. via the NOE) or through-bonds (i.e. via coupling constants). The power of the NMR method, compared with other spectroscopic techniques, lies in the fact that each proton gives rise to a specific resonance in the spectrum. Thus, a key aspect of any NMR structure determination is the requirement to assign each resonance to an individual proton and then to uniquely identify each pairwise through-space NOE interaction. In principle, this can be accomplished in a relatively straightforward manner, using correlation experiments to identify resonances belonging to different amino acid types via through-bond connectivities, and NOE experiments to subsequently link these residues in a sequential manner along the polypeptide chain on the basis of sequential and short-range interresidue NOEs involving the NH, $C^{\alpha}H$ and $C^{\beta}H$ protons (Wüthrich, 1986; Clore and Gronenborn, 1987, 1989). In particular, the types of NOE interactions that are most instructive for this purpose involve $NH(i)-NH(i + 1,2)$, $C^{\alpha}H(i)-NH(i + 1,2,3,4)$, $C^{\beta}H(i)-NH(i + 1)$ and $C^{\alpha}H(i)-C^{\beta}H(i + 3)$ connectivities, and the pattern of observed NOEs provides a very good indication of the different secondary structure elements along the polypeptide chain. With the resonance assignments in hand, one can then proceed to identify long-range NOE interactions between protons belonging to residues far apart in the sequence but close together in space which yield the crucial information for determining the tertiary structure of the protein.

While the principles of sequential resonance assignment are simple, the practice is difficult. Even for a small protein of, say, 50 residues, there are likely to be some 300–400 protons for which resonances have to be uniquely assigned. This number goes up linearly with the number of residues, so that for a 150 residue protein there will be 900–1200 protons. Because of the large number of protons, there is an extensive degree of resonance overlap and chemical shift degeneracy. As a result, one-dimensional NOE and decoupling experiments can only be applied with any degree of confidence for peptides up to about 10 residues, and even then there may be serious difficulties. The major conceptual advance in the application of NMR as a method of protein structure determination was the introduction of 2D NMR (reviewed by Ernst *et al.*, 1987). By spreading out the correlations in two ^1H frequency dimensions, each interaction is labelled by two chemical shifts, namely the frequencies of the originating and destination protons. This not only results in a tremendous increase in spectral resolution but equally importantly enables one to detect and interpret effects that would not have been possible in one dimension.

All 2D experiments can be reduced to the same basic conceptual scheme shown in Figure 1.2. This comprises a preparation pulse, an evolution period (t_1) during which the spins are labelled according to

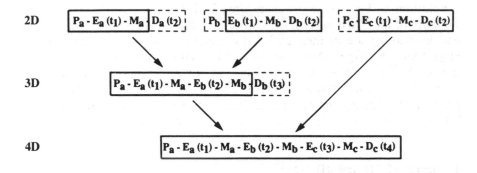

Figure 1.2 General representation of pulse sequences used in multidimensional NMR, illustrating the relationship between the basic schemes used to record 2D, 3D and 4D NMR spectra. Note how 3D and 4D experiments are constructed by the appropriate linear combination of 2D ones. Abbreviations: P, preparation; E, evolution; M, mixing; and D, detection. In 3D and 4D NMR, the evolution periods are incremented independently

their chemical shifts, a mixing period (M_1) during which the spins are correlated with one another and finally a detection period (t_2). The experiment is repeated several times with successively linearly incremented values of the evolution period t_1 to yield a data matrix $s(t_1, t_2)$. Fourier transformation in the t_2 dimension yields a set of n 1D spectra in which the intensities of the resonances are sinusoidally modulated as a function of the t_1 duration. Subsequent Fourier transformation in the t_1 dimension yields the desired 2D spectrum $S(\omega_1, \omega_2)$.

2D methods have proved extremely powerful for the structure determination of small proteins. The largest proteins (in terms of number of residues) where this approach has been successfully applied to achieve a complete 3D structure determination are *E. coli* (Dyson *et al.*, 1990) and human thioredoxin (Forman-Kay *et al.*, 1991), which have 108 and 105 residues, respectively. Beyond this limit of about 100 residues, 2D methods soon break down, owing to two fundamental problems. The first is associated with extensive spectral overlap due to the larger number of resonances. Consequently, the contour plot of a 2D spectrum of a 150 residue protein assumes the appearance of an intangible network of intersecting circles as an ever larger number of cross-peaks merge into one another. The result is that the 2D spectra of such a protein can no longer be interpreted. The second is a sharp decrease in the efficiency with which magnetization can be transferred through the small three-bond ^1H–^1H J couplings (3–12 Hz) as the linewidths become larger than the couplings, owing to increasing rotational correlation time. This leads to incomplete delineation of spin systems (i.e. amino acid types) in through-bond correlation experiments.

Solutions to both these problems are obtained by extending the dimensionality of the NMR spectra to remove resonance overlap and degeneracy, and by making use of through-bond correlations via heteronuclear couplings that are larger than the linewidths. This necessitates the use of uniformly ^{15}N- and/or ^{13}C-labelled protein. In proteins that can be over-expressed in bacterial systems, such labelling can be readily achieved by growing the organism in minimal medium supplemented by ^{15}NH$_4$Cl and/or [^{13}C$_6$]-glucose as the sole nitrogen and carbon sources, respectively.

4 Basics of 3D and 4D NMR

The design and implementation of higher-dimensionality NMR experiments can be carried out by the appropriate combination of 2D NMR experiments, as illustrated schematically in Figure 1.2 (Oschkinat *et al.*, 1988). A 3D experiment is constructed from two 2D pulse schemes by leaving out the detection period of the first experiment and the preparation pulse of the second. This results in a pulse train comprising two independently incremented evolution periods t_1 and t_2, two corresponding mixing periods M_1 and M_2, and a detection period t_3. Similarly, a 4D experiment is obtained by combining three 2D experiments in an analogous fashion. Thus, conceptually *n*-dimensional NMR can be conceived as a straightforward extension of 2D NMR. The real challenge, however, of 3D and 4D NMR is twofold: first, to ascertain which 2D experiments should be combined to best advantage; and second, to design the pulse sequences in such a way that undesired artefacts, which may severely interefere with the interpretation of the spectra, are removed. This task is far from trivial.

The first application of 3D NMR to a small protein, namely α1-purothionin (45 residues), was presented in 1988 (Oschkinat *et al.*, 1988). The experiment was of the proton homonuclear variety, in which a through-bond correlation experiment (HOHAHA) was combined with a through-space one (NOESY). While this experiment demonstrated the potential of the methodology, it suffered from a number of drawbacks which severely limited its application to larger proteins. First, the correlation portion of the experiment relied on small ^{1}H–^{1}H couplings. Second, all homonuclear 3D spectra are substantially more difficult to interpret than the equivalent 2D versions, as the number of cross-peaks present in the former far exceeds that in the latter.

Fortunately, heteronuclear 3D and 4D NMR experiments do not suffer from any of these disadvantages and yield important additional information in the form of ^{15}N and ^{13}C chemical shifts. They exploit a series of large one-bond heteronuclear couplings for magnetization transfer through-bonds which are summarized in Figure 1.3. This, together with the fact that the ¹H nucleus is always detected, renders these experiments very sensitive. Indeed, high-quality 3D and 4D hetero-nuclear-edited spectra can easily be obtained on samples of 1–2 mM uniformly labelled protein in a time-frame that is limited solely by the number of increments that have to be collected for appropriate digitiza-tion and the number of phase cycling steps that have to be used to reduce artefacts to an acceptably low level. Typical measurement times are 1.5–3 days for 3D experiments and 2.5–5 days for 4D ones. A detailed technical review of heteronuclear multidimensional NMR has been provided by Clore and Gronenborn (1991c).

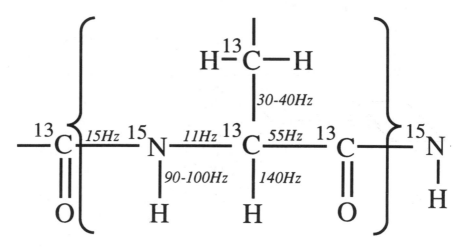

Figure 1.3 Summary of the one-bond heteronuclear couplings along the poly-peptide chain utilized in 3D and 4D NMR experiments. The backbone torsion angles ϕ and ψ involve rotations about the N_i–C_i^α and C_i^α–C_i bonds, respectively, while the side-chain torsion angle χ_1 involves a rotation about the C_i^α–C_i^β bond

Many of the 3D and 4D experiments are based on heteronuclear-editing of ¹H–¹H experiments, so that the general appearance of con-ventional 2D experiments is preserved and the total number of cross-peaks present is the same as that in the 2D equivalents. The progression from a 2D spectrum to 3D and 4D heteronuclear-edited spectra is depicted schematically in Figure 1.4. Consider, for example, the cross-peaks involving a particular ¹H frequency in a 2D NOESY spectrum, a 3D ¹⁵N- or ¹³C-edited NOESY spectrum, and finally a 4D ¹⁵N/¹³C- or ¹³C/¹³C-edited NOESY spectrum. In the 2D spectrum a series of cross-peaks will be seen from the originating proton frequencies in the F_1 dimension to the single destination ¹H frequency along the F_2 dimension. From the 2D experiment it is impossible to ascertain whether these NOEs involve only a single destination proton or several destination protons with identical chemical shifts. By spreading the spectrum into a third dimension according to the chemical shift of the heteronucleus attached to the destination proton(s), NOEs involving different destination protons will appear in distinct ^{1}H–^{1}H planes of the 3D spectrum. Thus, each interaction is simultaneously labelled by three chemical shift

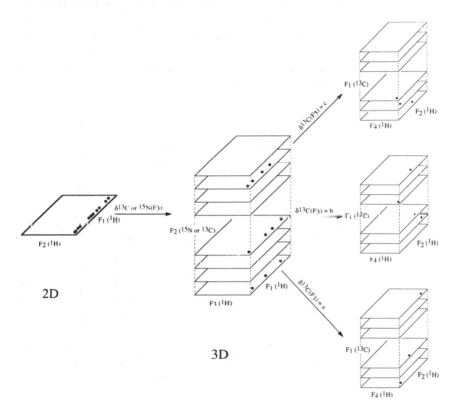

Figure 1.4 Schematic illustration of the progression and relationship between 2D, 3D and 4D heteronuclear NMR experiments. The closed circles represent NOE cross-peaks. In the example shown, there are 11 NOEs originating from 11 different protons in the F_1 dimension to a single frequency position in the F_2 dimension. In the 2D spectrum, it is impossible to ascertain whether there is only one destination proton or several in the F_2 dimension. By spreading the spectrum into a third dimension (labelled F_2), according to the chemical shift of the heteronucleus attached to the destination proton, it can be seen that the NOEs now lie in three distinct ^{1}H(F_1)–^{1}H(F_3) planes, indicating that three different destination protons are involved. However, the ^{1}H chemical shifts still provide the only means of identifying the originating protons. Hence, the problem of spectral overlap still prevents the unambiguous assignment of these NOEs. By extending the dimensionality of the spectrum to four, each NOE interaction is labelled by four chemical shifts along four orthogonal axes. Thus, the NOEs in each plane of the 3D spectrum are now spread over a cube in the 4D spectrum according to the chemical shift of the heteronucleus directly attached to the originating protons. Adapted from Clore *et al.* (1991b)

coordinates along three orthogonal axes of the spectrum. The projection of all these planes onto a single plane yields the corresponding 2D spectrum. For the purposes of sequential assignment, heteronuclear-edited 3D spectra are often sufficient for analysis. However, when the goal of the analysis is to assign NOEs between protons far apart in the sequence, a 3D ^{15}N- or ^{13}C-edited NOESY spectrum will often prove inadequate. This is because the originating protons are only specified by their ^{1}H chemical shifts, and, more often than not, there are several protons which resonate at the same frequencies. For example, in the case of the 153 residue protein interleukin-1β, there are about 60 protons which resonate in a 0.4 ppm interval between 0.8 and 1.2 ppm (Clore *et al.*, 1990b). Such ambiguities can then be resolved by spreading out the 3D spectrum still further into a fourth dimension according to the chemical shift of the heteronucleus attached to the originating protons, so that each NOE interaction is simultaneously labelled by four chemical shift coordinates along four orthogonal axes, namely those of the originating and destinations protons and those of the corresponding heteronuclei directly bonded to these protons (Kay *et al.*, 1990; Clore *et al.*, 1991b; Zuiderweg *et al.*, 1991). The result is a 4D spectrum in which each plane of the 3D spectrum constitutes a cube in the 4D spectrum.

For illustration purposes it is also useful to compare the type of information that can be extracted from a very simple system, using 2D, 3D and 4D NMR. Consider a molecule with only two NH and two aliphatic protons in which only one NH proton is close to an aliphatic proton. In addition, the chemical shifts of the NH protons are degenerate, as are those of the aliphatic protons, so that only two resonances are seen in the one-dimensional spectrum. In the 2D NOESY spectrum, an NOE will be observed between the resonance position of the NH protons and the resonance position of the aliphatic protons, but it will be impossible to ascertain which one of the four possible NH-aliphatic proton combinations gives rise to the NOE. By spreading the spectrum into a third dimension—for example, by the chemical shift of the ^{15}N atoms attached to the NH protons—the number of possibilities will be reduced to two, provided, of course, that the chemical shifts of the two nitrogen atoms are different. Finally, when the fourth dimension corresponding to the chemical shift of the ^{13}C atoms attached to the aliphatic protons is introduced, a unique assignment of the NH-aliphatic proton pair giving rise to the NOE can be made.

5   Heteronuclear 3D and 4D NMR in Practice

Figure 1.5(A) presents a portion of the 2D ^{15}N-edited NOESY spectrum of interleukin-1β (153 residues) illustrating NOE interactions between

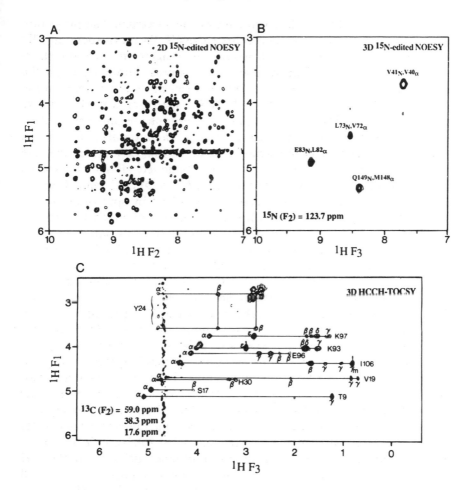

Figure 1.5 Example of 2D and 3D spectra of interleukin-1β recorded at 600 MHz (Driscoll *et al.*, 1990a; Clore *et al.*, 1990a). The 2D spectrum in panel A shows the NH(F_2 axis)–C$^\alpha$H(F_1 axis) region of a 2D ^{15}N-edited NOESY spectrum. The same region of a single NH(F_3)–^{1}H(F_1) plane of the 3D ^{15}N-edited NOESY at δ^{15}N(F_2) = 123.7 ppm is shown in panel B. The actual 3D spectrum comprises 64 such planes and projection of these on a single plane would yield the same spectrum as in (A). Panel C shows a single ^{1}H(F_3)–^{1}H(F_1) plane of the 3D HCCH-TOCSY spectrum at δ^{13}C(F_2) = 38.3 ± nSW (where SW is the spectral width of 20.71 ppm in the ^{13}C dimension), illustrating both direct and relayed connectivities originating from the C$^\alpha$H protons. Note how easy it is to delineate complete spin systems of long side-chains such as Lys (i.e. cross-peaks to the C$^\beta$H, C$^\gamma$H, C$^\delta$H and C$^\epsilon$H protons are observed) owing to the fact that magnetization along the side-chain is transferred via large $^1J_{CC}$ couplings. Several features of the HCCH-TOCSY spectrum should be pointed out. First, extensive folding is employed which does not obscure analysis as ^{13}C chemical shifts for different carbon types are located in characteristic regions of the ^{13}C spectrum

(*Cont. next page...*)

the NH protons along the F_2 axis and the $C^\alpha H$ protons along the F_1 dimension. Despite the fact that a large number of cross-peaks can be resolved, it can be seen that many of the cross-peaks have identical chemical shifts in one or other dimensions. For example, there are 15 cross-peaks involving NH protons at an $F_2(^1H)$ chemical shift of ~9.2 ppm. A single $^1H(F_1)-^1H(F_3)$ plane of the 3D ^{15}N-edited NOESY spectrum of interleukin-1β at $\delta^{15}N(F_2) = 123.7$ ppm is shown in Figure 1.5(B). Not only is the number of cross-peaks in this slice small, but also at $\delta\ ^1H(F_3) \sim 9.2$ ppm there is only a single cross-peak involving one NH proton. The correlations observed in the ^{15}N-edited NOESY spectrum (Fesik and Zuiderweg, 1988; Marion *et al.*, 1989) are through-space ones. Intraresidue correlations from the NH protons to the $C^\alpha H$ and $C^\beta H$ protons can similarly be resolved, using a 3D ^{15}N-edited HOHAHA spectrum (Marion *et al.*, 1989; Clore *et al.*, 1991a) in which efficient isotropic mixing sequences are used to transfer magnetization between protons via three-bond $^1H-^1H$ couplings.

The 3D ^{15}N-edited NOESY and HOHAHA spectra constitute only one of several versions of a 3D heteronuclear-edited spectrum. Many alternative through-bond pathways can be utilized to great effect. Consider, for example, the delineation of amino acid spin systems which involves grouping those resonances which belong to the same residue. In 2D NMR, correlation experiments are used to delineate either direct or relayed connectivities via small three-bond $^1H-^1H$ couplings. Even for proteins of 50–60 residues, it can be difficult to delineate long-chain amino acids such as Lys and Arg in this manner. In heteronuclear 3D NMR an alternative pathway can be employed which involves transferring magnetization first from a proton to its directly attached carbon atom via the large $^1J_{CH}$ coupling ~130 Hz, followed by either direct or relayed transfer of magnetization along the carbon chain via the $^1J_{CC}$ coupling (~30–40 Hz), before transferring the magnetization back to protons (Bax *et al.*, 1990a,b; Fesik *et al.*, 1990). An example of such a spectrum is the so called HCCH-TOCSY (Bax *et al.*, 1990b; Clore *et al.*, 1990b) shown in Figure 1.5(C). The $^1H(F_1)-^1H(F_3)$ plane at $\delta^{13}C(F_2) = 59$ ppm illustrates both direct and relayed connectivities along various side-chains originating from $C^\alpha H$ protons. As expected, the resolution of the spectrum is excellent and there is no spectral overlap. Just as importantly, however, the sensitivity of the experiment is extremely high and complete spin systems are readily identified in interleukin-1β even for

with little overlap. Second, the spectrum is edited according to the chemical shift of the heteronucleus attached to the originating proton rather than the destination one. Third, multiple cross-checks on the assignments are readily made by looking for the symmetry-related peaks in the planes corresponding to the ^{13}C chemical shifts of the destination protons in the original slice

long side-chains, such as those of two lysine residues shown in the figure. Indeed, analysing spectra of this kind, it was possible to obtain complete ^{1}H and ^{13}C assignments for the side-chains of interleukin-1β (Clore *et al.*, 1990b).

3D NMR also permits one to devise experiments for sequential assignment which are based solely on through-bond connectivities via heteronuclear couplings (Ikura *et al.*, 1990; Powers *et al.*, 1991) and thus do not rely on the NOESY experiment. This becomes increasingly important for larger proteins, as the types of connectivites observed in these correlation experiments are entirely predictable, whereas in the NOESY spectrum, which relies solely on close proximity of protons, it may be possible to confuse sequential connectivities with long-range ones. These 3D heteronuclear correlation experiments are of the triple resonance variety and make use of one-bond ^{13}CO(i − 1)–^{15}N(i), ^{15}N(i)–^{13}C$^\alpha$(i) and ^{13}C$^\alpha$(i)–^{13}CO(i) couplings, as well as two-bond ^{13}C$^\alpha$(i − 1)–^{15}N(i) couplings. In this manner multiple independent pathways for linking the resonances of one residue with those of its adjacent neighbour are available, thereby avoiding ambiguities in the sequential assignment. A summary of some of the useful triple resonance experiments and the connectivities observed in them is provided in Figure 1.6 and Table 1.1, and an example of an HNCA, HN(CO)CA and HNCO spectrum which provide C$^\alpha$(i − 1, i)–N(i)–NH(i), C$^\alpha$(i − 1)–N(i)–NH(i) and CO(i − 1)–N(i)–NH(i) connectivities, respectively, is shown in Figure 1.7. A more detailed discussion of the various triple resonance experiments is provided in Chapter 2 of this volume.

The power of 4D heteronuclear NMR spectroscopy for unravelling interactions that would not have been possible in lower-dimensional spectra is illustrated in Figure 1.8 by the ^{13}C/^{13}C-edited NOESY spectrum of interleukin-1β (Clore *et al.*, 1991b). Figure 1.8(A) shows a small portion of the aliphatic region between 1 and 2 ppm of a conventional 2D NOESY spectrum of interleukin-1β. The overlap is so great that no single individual cross-peak can be resolved. One might therefore wonder just how many NOE interactions are actually superimposed, for example, at the ^{1}H chemical shift coordinates of the letter X at 1.39 (F_1) and 1.67 (F_2) ppm. A ^{1}H(F_2)–^{1}H(F_4) plane of the 4D spectrum at δ^{13}C(F_1), δ^{13}C(F_3) = 44.3, 34.6 ppm is shown in panel B and the square box at the top right-hand side of this panel encloses the region between 1 and 2 ppm. Only two cross-peaks are present in this region, and the arrow points to a single NOE between the C$^\gamma$H and C$^\beta$H protons of Lys-77 with the same ^{1}H chemical shift coordinates as the letter X in panel A. All the other NOE interactions at the same ^{1}H chemical shift coordinates can be determined by inspection of a single ^{13}C(F_1)–^{13}C(F_3) plane taken at δ^1H(F_2), δ^1H(F_4) = 1.39, 1.67 ppm. This reveals a total of 7 NOE interactions superimposed at the ^{1}H

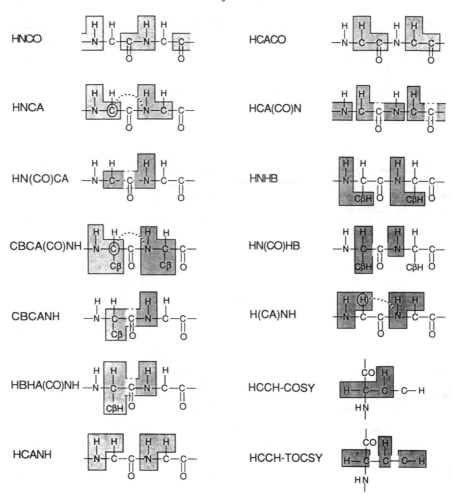

Figure 1.6 Summary of some useful 3D double- and triple-resonance experiments and the correlations and connectivities observed in them. Direct and relayed scalar correlations from the $C^\alpha H$ and $C^\beta H$ protons to the NH protons via three-bond ^1H–^1H couplings are detected in the 3D ^{15}N-edited HOHAHA spectrum; direct and relayed scalar correlations between aliphatic protons via the large $^1J_{CC}$ couplings are observed in the 3D HCCH-COSY and HCCH-TOCSY experiments, respectively. $^{13}C^\alpha(i-1,i)$–$^{15}N(i)$–$^{15}N(i)$, $^{13}CO(i-1)$–$^{15}N(i)$–NH(i) and $C^\alpha H(i)$–$^{13}C^\alpha(i)$–$^{13}CO(i)$ correlations via direct through-bond connectivities are observed in the 3D HNCA, HNCO and HCACO experiments, respectively. $^{13}C^\alpha(i-1)$–$^{15}N(i)$–NH(i) and $C^\alpha H(i-1)$–$^{13}C^\alpha(i-1)$–$^{15}N(i)$ correlations are observed in the HN(CO)CA and HCA(CO)N relayed experiments, respectively; while $C^\alpha H$ $(i-1,i)$–$^{15}N(i)$–NH(i) correlations can be observed in the H(CA)NH experiment which relays magnetization via the $^{13}C^\alpha$ atom. $^{13}C^\alpha/C^\beta(i-1,i)$–$^{15}N(i)$–$^{15}N(i)$, $^{13}C^\alpha/C^\beta(i-1)$–$^{15}N(i)$–$^{15}N(i)$ and $H^\alpha/H^\beta(i-1,i)$–$^{15}N(i)$–$^{15}N(i)$ connectivities are observed in the 3D CBCANH, CBCA(CO)NH and HBHA(CO)NH experiments, respectively. Finally, $C^\beta H(i)$–N(i)–NH(i) and $C^\beta H/C^\alpha H(i-1)$–N$(i)$–NH$(i)$ correlations are observed in the HNHB and HN(CO)HB experiments, respectively

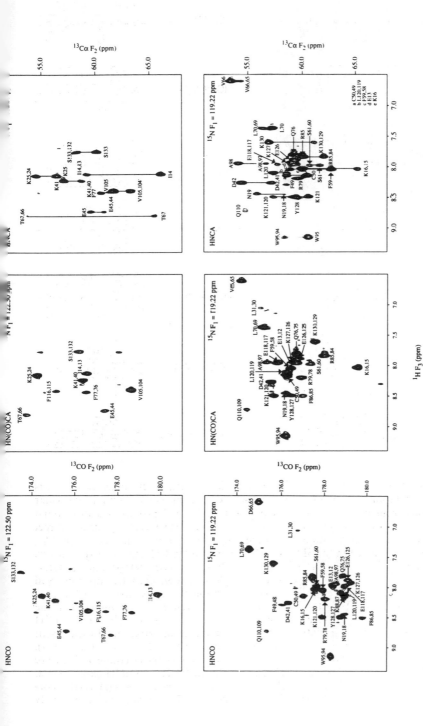

Figure 1.7 An example of selected $^{13}C^\alpha(F_2)-{}^1H(F_3)$ planes of the 3D HNCA and HN(CO)CA spectra and selected $^{13}CO(F_2)-{}^1H(F_3)$ planes of the HNCO spectrum of interleukin-4 at two $^{15}H(F_2)$ frequencies (Powers *et al.*, 1992a), illustrating $^{13}C^\alpha(i-1, i)-{}^{15}N(i)-NH(i)$, $^{13}C^\alpha(i-1)-{}^{15}N(i)-NH(i)$ and $^{13}CO(i-1)-{}^{15}N(i)-NH(i)$ correlations, respectively

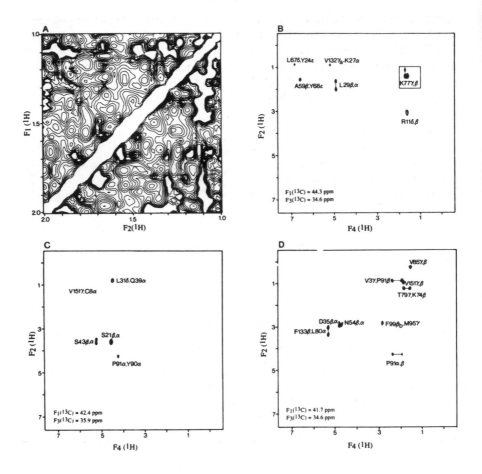

Figure 1.8 Comparison of 2D and 4D NMR spectra of interleukin-1β recorded at 600 MHz (Clore *et al.*, 1991b). The region between 1 and 2 ppm of the 2D NOESY spectrum is shown in (A). ^{1}H(F_2)–^{1}H(F_4) planes at several ^{13}C(F_1) and ^{13}C(F_3) frequencies of the 4D ^{13}C/^{13}C NOESY spectrum are shown in panels B–D. No individual cross-peaks can be observed in the 2D spectrum and the letter X has ^{1}H coordinates of 1.39 and 1.67 ppm. In contrast, only two cross-peaks are observed in the boxed region in panel B between 1 and 2 ppm, one of which (indicated by an arrow) has the same ^{1}H coordinates as the letter X. Further analysis of the complete 4D spectrum reveals the presence of 7 NOE cross-peaks superimposed at the ^{1}H coordinates of the letter X. This can be ascertained by looking at a ^{13}C(F_1)–^{13}C(F_3) plane taken at the ^{1}H coordinates of X. True diagonal peaks corresponding to magnetization that has not been transferred from one proton to another, as well as intense NOE peaks involving protons attached to the same carbon atom (i.e. methylene protons), appear in only a single ^{1}H(F_2)–^{1}H(F_4) plane of each ^{13}C(F_1), ^{1}H(F_2), ^{1}H(F_4) cube at the carbon frequency where the originating and destination carbon atoms coincide (i.e at $F_1 = F_3$). Thus, these intense resonances no longer obscure NOEs between

(*Cont. opposite page...*)

chemical shift coordinates of the letter X. Another feature of the 4D spectrum is illustrated by the two $^1H(F_2)$–$^1H(F_4)$ planes at different F_1 and F_3 ^{13}C frequencies shown in panels C and B. In both cases, there are cross-peaks involving protons with identical or near chemical shifts, namely that between Pro-91($C^{\alpha}H$) and Tyr-90($C^{\alpha}H$), diagnostic of a *cis*-proline, in panel C, and between Phe-99($C^{\beta b}H$) and Met-95($C^{\gamma}H$) in panel D. These interactions could not be resolved in either a 2D spectrum or a 3D ^{13}C-edited spectrum, as they would lie on the spectral diagonal (i.e. the region of the spectrum corresponding to magnetization that has not been transferred from one proton to another). In the 4D spectrum, however, they are easy to observe, provided, of course, that the ^{13}C chemical shifts of the directly bonded ^{13}C nuclei are different.

Because the number of NOE interactions present in each $^1H(F_4)$–$^1H(F_2)$ plane of 4D $^{13}C/^{15}N$- or $^{13}C/^{13}C$-edited NOESY spectra is so small, the inherent resolution in a 4D spectrum is extremely high, despite the low level of digitization. Indeed, spectra with equivalent resolution can

---

protons with similar or degenerate chemical shifts. Two examples of such NOEs can be seen in panels C (between the $C^{\alpha}H$ protons of Pro-91 and Tyr-90) and D (between one of the $C^{\beta}H$ protons of Phe-77 and the methyl protons of Met-95). These various planes of the 4D spectrum also illustrate another key aspect of 3D and 4D NMR, namely the importance of designing the pulse scheme to optimally remove undesired artefacts which may severely interfere with the interpretation of the spectra. Thus, while the 4D $^{13}C/^{13}C$-edited NOESY experiment is conceptually analogous to that of a 4D $^{13}C/^{15}N$-edited one, the design of a suitable pulse scheme is actually much more complex in the $^{13}C/^{13}C$ case. This is due to the fact that there are a large number of spurious magnetization transfer pathways that can lead to observable signals in the homonuclear $^{13}C/^{13}C$ case. For example, in the 4D $^{15}N/^{13}C$-edited case there are no 'diagonal peaks' which would correspond to magnetization that has not been transferred from one hydrogen to another, as the double heteronuclear filtering (i.e. ^{13}C and ^{15}N) is extremely efficient at completely removing these normally very intense and uninformative resonances. Such a double filter is not available in the $^{13}C/^{13}C$ case, so that both additional pulses and phase cycling are required to suppress magnetization transfer through these pathways. This task is far from trivial, as the number of phase cycling steps in 4D experiments is severely limited by the need to keep the measurement time down to practical levels (i.e. less than 1 week). The results of such care in pulse design can be clearly appreciated from the artefact-free planes shown in panels B–D. However, when a 4D $^{13}C/^{13}C$-edited NOESY spectrum is recorded with the same pulse scheme as that used in the 4D $^{15}N/^{13}C$ experiment (with the obvious replacement of ^{15}N pulses by ^{13}C pulses), a large number of spurious peaks are observed along a pseudo-diagonal at $\delta^1H(F_2) = \delta^1H(F_4)$ in planes where the carbon frequencies of the originating and destination protons do *not* coincide. As a result, it become virtually impossible under these circumstances to distinguish artefacts from NOEs between protons with the same ^1H chemical shifts, as was possible with complete confidence in panels C and D

be recorded at magnetic field strengths considerably lower than 600 MHz, although this would obviously lead to a reduction in sensitivity. Further, it can be calculated that 4D spectra with virtual lack of resonance overlap and good sensitivity can be obtained on proteins with as many as 400 residues. Thus, once complete ^{1}N, ^{15}N and ^{13}C assignments are obtained, analysis of 4D spectra should permit the automated assignment of almost all NOE interactions.

Table 1.1 Summary of correlations observed in the 3D double and triple resonance experiments used for sequential and side-chain assignments.

Experiment	*Correlation*	*J Coupling*[a]
^{15}N-edited HOHAHA	$C^{\alpha}H(i)-^{15}N(i)-NH(i)$	$^3J_{HN\alpha}$ (~3–11 Hz)
	$C^{\beta}(i)-^{15}N(i)-NH(i)$	$^3J_{HN\alpha}$ (~3–11 Hz) and $^3J_{\alpha\beta}$(~3–11 Hz)
H(CA)NH	$C^{\alpha}H(i)-^{15}N(i)-NH(i)$	$^1J_{NC\alpha}$(~9–13 Hz)
	$C^{\alpha}H(i-1)-^{15}N(i)-NH(i)$	$^2J_{NC\alpha}$(~5–10 Hz)
HN(CO)HB	$C^{\alpha}H(i-1)-^{15}N(i)-NH(i)$	$^2J_{COH\alpha}$ (~4–7 Hz)
	$C^{\beta}H(i-1)-^{15}N(i)-NH(i)$	$^3J_{COH\beta}$ (~8 Hz for *trans* coupling)
HNHB	$C^{\beta}H(i)-^{15}N(i)-NH(i)$	$^3J_{NH\beta}$ (~5 Hz for *trans* coupling)
HNCA	$^{13}C^{\alpha}(i)-^{15}N(i)-NH(i)$	$^1J_{NC\alpha}$ (~9–13 Hz)
	$^{13}C^{\alpha}(i-1)-^{15}N(i)-NH(i)$	$^2J_{NC\alpha}$ (~5–10 Hz)
CBCANH	$^{13}C^{\beta}/C^{\alpha}(i)-^{15}N(i)-NH(i)$	$^1J_{CC}$(~30–40 Hz), $^1J_{NC\alpha}$ (~9–13 Hz)
	$^{13}C^{\beta}/C^{\alpha}(i-1)-^{15}N(i)-NH(i)$	$^1J_{CC}$(~30–40 Hz), $^2J_{NC\alpha}$ (~5–10 Hz)
HN(CO)CA	$^{13}C^{\alpha}(i-1)-^{15}N(i)-NH(i)$	$^1J_{NCO}$ (~15 Hz) and $^1J_{C\alpha CO}$ (~55 Hz)
CBCA(CO)NH	$^{13}C^{\beta}/C^{\alpha}(i-1)-^{15}N(i)-NH(i)$	$^1J_{CC}$ (~30–40 Hz), $^1J_{NCO}$ (~15 Hz) and $^1J_{C\alpha CO}$ (~55 Hz)
HBHA(CO)NH	$^{13}C^{\beta}/C^{\alpha}(i-1)-^{15}N(i)-NH(i)$	$^1J_{CC}$ (~30–40 Hz), $^1J_{NCO}$ (~15 Hz) and $^1J_{C\alpha CO}$ (~55 Hz)
HNCO	$^{13}CO(i-1)-^{15}N(i)-NH(i)$	$^1J_{NCO}$ (~15 Hz)
HCACO	$C^{\alpha}H(i)-^{13}C^{\alpha}(i)-^{13}CO(i)$	$^1J_{C\alpha CO}$ (~55 Hz)
HCA(CO)N	$C^{\alpha}H(i)-^{13}C^{\alpha}(i)-^{15}N(i+1)$	$^1J_{C\alpha CO}$ (~55 Hz) and $^1J_{NCO}$ (~15 Hz)
HCCH-COSY	$H(j)-^{13}C(j)-^{13}C(j\pm1)-H(j\pm1)$	$^1J_{CC}$ (~30–40 Hz)
HCCH-TOCSY	$H(j)-^{13}C(j) \ldots ^{13}C(j\pm n)-H(j\pm n)$	$^1J_{CC}$ (~30–40 Hz)

[a]In addition to the couplings indicated, all the experiments make use of the $^1J_{CH}$ (~140 Hz) and/or $^1J_{NH}$ (~95 Hz) couplings.

6 Application of 3D and 4D NMR to the Determination of Larger Protein Structures

While the potential of heteronuclear 3D and 4D NMR methods in resolving problems associated with both extensive resonance overlap and large linewidths is obvious, how does this new approach fare in practice? In this regard, it should be borne in mind that resonance assignments are only a means to an end, and the true test of multidimensional NMR lies in examining its success in solving the problem which it was originally designed to tackle, namely the determination of high-resolution three-dimensional structures of larger proteins in solution. This goal has now been attained in the case of interleukin-1β, a protein of 153 residues and 17.4 kDa, which plays a central role in the immune and inflammatory responses (Clore *et al.*, 1991c). This protein is 50% larger than any other protein whose three-dimensional structure had been previously determined by NMR at that time (Dyson *et al.*, 1990; Forman-Kay *et al.*, 1991). Subsequently, structures of two other larger proteins have been determined, namely the complex of calmodulin with a target peptide from light-chain myosin kinase (Ikura *et al.*, 1992) and the cytokine interleukin-4 (Powers *et al.*, 1992b; Smith *et al.*, 1992), albeit at medium resolution (equivalent to a second-generation structure; cf. Figure 1.1).

Despite extensive analysis of 2D spectra obtained at different pH values and temperatures, as well as examination of 2D spectra of mutant proteins, it did not prove feasible to obtain unambiguous ^1H assignment for more than about 30% of the residues of interleukin-1β (Driscoll *et al.*, 1990a). Thus, any further progress could only be made by resorting to higher-dimensionality heteronuclear NMR, and a summary of the strategy we employed for determining its structure is shown in Figure 1.9. The initial step involved the complete assignment of the ^1H, ^{15}N and ^{13}C resonances of the backbone and side-chains, using the entire gammut of double- and triple-resonance 3D experiments listed in the top left-hand panel of the figure (Clore *et al.*, 1990b; Driscoll *et al.*, 1990a,b). In the second step, backbone and side-chain torsion angle restraints, as well as stereospecific assignments for β-methylene protons, were obtained by means of a three-dimensional systematic grid search of ϕ, ψ, χ_1 space (Nilges *et al.*, 1990). In the third step, approximate interproton distance restraints between non-adjacent residues were derived from analysis of 3D and 4D heteronuclear-edited NOESY spectra. Analysis of the 3D heteronuclear-edited NOESY spectra alone was sufficient to derive a low-resolution structure on the basis of a small number of NOEs involving solely NH, $C^\alpha H$ and $C^\beta H$ protons (Clore *et al.*, 1990d). However, further progress using 3D NMR was severely hindered by the numerous ambiguities still present in these spectra—in particular, for NOEs arising from the large number of aliphatic protons.

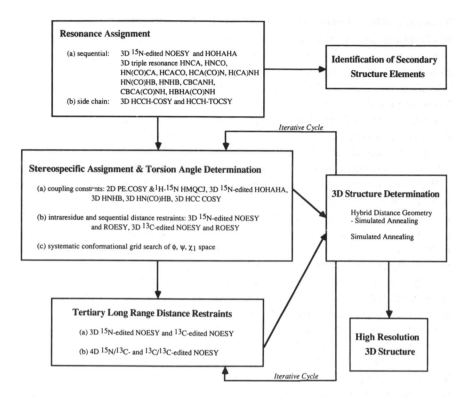

Figure 1.9 Outline of the general strategy employed in our laboratory to determine the three-dimensional structure of larger proteins such as interleukin-1β by 3D and 4D NMR. The various NMR experiments are as follows. Through-space interactions are detected in the heteronuclear-edited 3D and 4D NOESY and ROESY experiments. $^3J_{NH\alpha}$ and $^3J_{\alpha\beta}$ ^1H-^1H coupling constants, which are related to the torsion angles φ and χ_1 via empirical Karplus relationships (Pardi *et al.*, 1984), are measured from 2D heteronuclear ^1H-^{15}N HMQC-J (Kay and Bax, 1990; Forman-Kay *et al.*, 1990) and homonuclear ^1H-^1H PE. COSY (Mueller, 1987) correlation spectra, respectively. A semiquantitative measure of the $^3J_{\alpha\beta}$ couplings, which is sufficient for securing stereospecific assignments and χ_1 torsion angle restraints, can also be obtained from the relative magnitude of the NH–C$^\beta$H correlations observed in the 3D ^{15}N-edited HOHAHA spectrum (Clore *et al.*, 1991a). Similarly, a semiquantitative measure of the $^3J_{NH\beta}$ and $^3J_{COH\beta}$ couplings can be obtained from the relative magnitude of the cross-peaks observed in the 3D HNHB (Archer *et al.*, 1991) and HN(CO)HB (Grzesiek *et al.*, 1992) experiments, respectively. Stereospecific assignments and torsion angle restraints are obtained by comparing the relevant experimental data (i.e. coupling constants and intraresidue and sequential distance restraints involving the NH, C$^\alpha$H and C$^\beta$H protons) with the calculated values of these parameters present in two databases. The first is a systematic one covering the complete φ, ψ and χ_1 conformational space (in a three-dimensional grid spaced at 10° intervals) of a tripeptide fragment with idealized geometry; while the second comprises a library of tripeptide segments from high-resolution X-ray

(Cont. opposite page...)

Thus, the 4D heteronuclear-edited NOESY spectra proved to be absolutely essential for the successful completion of this task. In addition, the proximity of backbone NH protons to bound structural water molecules was ascertained from a 3D ^{15}N-separated ROESY spectrum which permits one to distinguish specific protein–water NOE interactions from chemical exchange with bulk solvent (Clore *et al.*, 1990c). In this regard, we should emphasize again that in our laboratory all the NOE data are interpreted in as conservative a manner as possible, and are simply classified into three distance ranges, 1.8–2.7 Å, 1.8–3.3 Å and 1.8–5.0 Å, corresponding to strong-, medium- and weak-intensity NOEs.

With an initial set of experimental restraints in hand, 3D structure calculations were initiated. Typically we use the hybrid distance geometry–dynamical simulated annealing method, in which an approximate polypeptide fold is obtained by projection of a subset of atoms from *n*-dimensional distance space into cartesian coordinate space followed by simulated annealing including all atoms (Nilges *et al.*, 1988c). Alternatively, we employ simulated annealing starting from either random structures with intact covalent geometry (Nilges *et al.*, 1988b) or from a completely random array of atoms (Nilges *et al.*, 1988a). All these simulated annealing protocols involve solving Newton's equations of motion subject to a simplified target function comprising terms for the experimental restraints, covalent geometry and non-bonded contacts. The underlying principle lies in raising the temperature of the system followed by slow cooling in order to overcome false local minima and large potential energy barriers along the path towards the global minimum region of the target function, and to sample efficiently and comprehensively the conformational space consistent with the experimental restraints. A key aspect of the overall strategy lies in the use of an iterative approach whereby the experimental data are re-examined in the light of the initial set of calculated structures in order to resolve ambiguities in NOE assignments, to obtain more stereospecific assignments (e.g. the α-methylene protons of glycine and the methyl groups of valine and leucine) and torsion angle restraints, and to assign backbone hydrogen bonds associated with slowly exchanging NH protons as well as with bound water molecules. The iterative cycle comes to an end when all the experimental data have been interpreted.

The final experimental data set for interleukin-1β comprised a total of 3146 approximate and loose experimental restraints made up of

---

structures (Nilges *et al.*, 1990). This procedure is carried out for both possible stereospecific assignments, and when the experimental data are only consistent with one of the two possibilities, the correct stereospecific assignements, as well as allowed ranges for ϕ, ψ and χ_1, are obtained. The minimum ranges that we employ for the torsion angle restraints are ±30°, ±50° and ±20°, respectively

2780 distance and 366 torsion angle restraints (Clore *et al.*, 1991c). This represents an average of ~21 experimental restraints per residue. If one takes into account the fact that interresidue NOEs affect two residues, while intraresidue NOE and torsion angle restraints only affect individual residues, the average number of restraints influencing the conformation of each residue is ~33. A superposition of 32 independently calculated structures is shown in Figure 1.10. All 32 structures satisfy the experimental restraints within their specified errors, display very small deviations from idealized covalent geometry and have good non-bonded contacts. It can be seen that both the backbone as well as ordered side-chains are exceptionally well-defined. Indeed, the atomic rms distribution about the mean coordinate positions is 0.4 Å for the backbone atoms, 0.8 Å for all atoms and 0.5 Å for side-chains with ≤ 40% of their surface (relative to that in a tripeptide Gly–X–Gly) accessible to solvent (Clore *et al.*, 1991c).

The structure of interleukin-1β itself resembles a tetrahedron and displays threefold internal pseudosymmetry. There are 12 β-strands arranged in an exclusively anti-parallel β-structure, and 6 of the strands form a β-barrel (seen in the front of Figure 1.10A) which is closed off at the back of the molecule by the other 6 strands. Each repeating topological unit is composed of 5 strands arranged in an antiparallel manner with respect to each other, and one of these units is shown in Figure 1.8(B). Water molecules occupy very similar positions in all three topological units, as well as at the interface of the three units, and are involved in bridging backbone hydrogen bonds. Thus, in the case of the topological unit shown in Figure 1.10(B), the water molecule labelled W5 accepts a hydrogen bond from the NH of Phe-112 in strand IX and donates two hydrogen bonds to the backbone carbonyls of Ile-122 in strand X and Thr-144 in strand XII. The packing of some internal residues with respect to one another, as well as the excellent definition of internal side-chains is illustrated in Figure 1.8(C). Because of the high resolution of the interleukin-1β structure, it was possible to analyse in detail side-chain–side-chain interactions involved in stabilizing the structure. In addition, examination of the structure in the light of mutational data permitted us to propose the presence of three distinct sites involved in the binding of interleukin-1β to its cell surface receptor (Clore *et al.*, 1991c).

7 Combining Experimental Information from Crystal and Solution Studies: Joint X-ray and NMR Refinement

It is clear from the preceding discussion that NMR is a valid method, alongside X-ray crystallography, for determining high-resolution structures

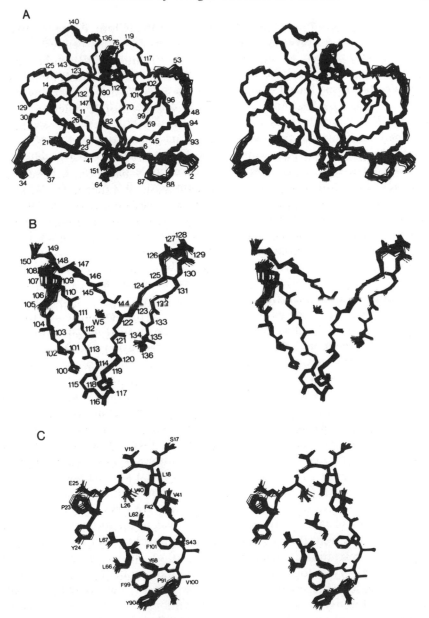

Figure 1.10 Stereoviews of the 3D structure of interleukin-1β determined by 3D and 4D heteronuclear NMR spectroscopy on the basis of a total of 3146 approximate and loose experimental NMR restraints (Clore *et al.*, 1991c). Best-fit superpositions of the backbone (N, Cᵅ, C) atoms of residues 2–151, the backbone (N, Cᵅ, C, O) atoms of one of the three repeating topological units including a water molecule (W5), and selected side-chains of 32 simulated annealing structures are shown in (A), (B) and (C), respectively. The N-terminal residue and the two C-terminal ones (residues 152–153) are ill-defined

of small to medium-sized proteins of less than about 35 kDa. Inter-leukin-1β offers an ideal system for comparing the results of NMR and X-ray crystallography as, in addition to the solution structure, there are three independently solved X-ray structures at 2 Å resolution of the same crystal form (Finzel *et al.*, 1989; Priestle *et al.*, 1989; Veerapandian *et al.*, 1992). The backbone atomic rms difference between the NMR and the X-ray structures is about 1 Å, with the largest differences being confined to some of the loops and turns connecting the β strands. Interestingly, however, the atomic rms distribution of the 32 calculated solution structures about their mean coordinate positions (~0.4 Å for the backbone atoms, ~0.8 Å for all atoms and ~0.5 Å for all atoms of internal residues) is approximately the same as the atomic rms differ-ences between the three X-ray structures, indicating that the positional errors in the atomic coordinates determined by the two methods are similar (Clore and Gronenborn, 1991b). Upon initial inspection, the X-ray structures appear to be incompatible with the NMR data, as manifested by a relatively large number of NOE and torsion angle violations, and, conversely, the NMR structure fits the X-ray data poorly with an *R* factor of 40–50%. Because of the very different nature of the two methods, it is not immediately apparent that these discrep-ancies reflect genuine differences between the solution and X-ray struc-tures or whether they reflect differences in the computational procedures employed. To analyse this in more detail, we have developed a new method of structure determination in which the NMR and X-ray data are combined and used simultaneously in the structure refinement (Shaanan *et al.*, 1992). Using this approach, we have shown that a model can readily be generated from a joint NMR/X-ray refinement which is compatible with the data from both techniques. Thus, there are only minimal violations of the NMR restraints (NOEs and torsion angles), the value of the crystallographic *R* factor is comparable to, if not better than, that derived from refinement against the crystallographic data alone, and the deviations from idealized covalent geometry are small. In addition, the *R* free (Brünger, 1992) for the model refined with the NMR and X-ray restraints is smaller than that of the model obtained by conventional crystallographic refinement, indicating that the crystallo-graphic phases obtained by the joint NMR/X-ray refinement are more accurate. Moreover, the few NMR observations that are still violated by the model serve as an indicator for genuine differences between the crystal and solution structures.

The implications of the joint NMR/X-ray refinement method to struc-tural biology are of considerable significance. In particular, the full potential and future use of the method will be for structure determina-tions of multidomain proteins, for which only low-resolution X-ray data for the entire protein are available but for which detailed structural information may be obtained by NMR on the individual domains. Using the joint X-ray/NMR refinement approach in such cases will open the way to the study of proteins which may otherwise never be structurally accessible by either of the two methods alone.

8 Concluding Remarks

In this chapter we have summarized the recent developments in hetero-nuclear 3D and 4D NMR which have been designed to extend the NMR methodology to medium-sized proteins in the 15–30 kDa range. The underlying principle of this approach consists of extending the dimensionality of the spectra to obtain dramatic improvements in spectral resolution while simultaneously exploiting large heteronuclear couplings to circumvent problems associated with larger linewidths. A key feature of all these experiments is that they do not result in any increase in the number of observed cross-peaks relative to their 2D counterparts. Hence, the improvement in resolution is achieved without raising the spectral complexity, rendering data interpretation straightforward. Thus, for example, in 4D heteronuclear-edited NOESY spectra, the NOE interactions between proton pairs are labelled not only by the ^{1}H chemical shifts but also by the corresponding chemical shifts of their directly bonded heteronuclei in four orthogonal axes of the spectrum. Also important in terms of practical applications is the high sensitivity of these experiments, which makes it feasible to obtain high-quality spectra in a relatively short time-frame on 1–2 mM protein samples uniformly labelled with ^{15}N and/or ^{13}C.

Just as 2D NMR opened the application of NMR to the structure determination of small proteins of less than about 100 residues, 3D and 4D heteronuclear NMR provide the means of extending the methodology to medium-sized proteins in the 150–300 residue range. Indeed, the recent determination of the high-resolution structure of interleukin-1β using 3D and 4D heteronuclear NMR (Clore *et al.*, 1991c) demonstrates beyond doubt that the technology is now available for obtaining the structures of such medium-sized proteins at a level of accuracy and precision that is comparable to the best results attainable for small proteins.

Acknowledgements

We thank Ad Bax for many stimulating discussions. This work was supported in part by the AIDS Targeted Anti-Viral Program of the Office of the Director of the National Institutes of Health.

References

Archer, S. J., Ikura, M., Torchia, D. A. and Bax, A. (1991). An alternative 3D NMR technique for correlating backbone ^{15}N with side chain Hβ resonances in larger proteins. *J. Magn. Reson.*, **95**, 636–641

Aue, W. P., Bartholdi, E. and Ernst, R. R. (1976). Two-dimensional spectroscopy: application to nuclear magnetic resonance. *J. Chem. Phys.*, **64**, 2229–2246

Baldwin, E. T., Weber, I. T., St. Charles, R., Zuan, J. C., Appella, E., Matsushima, K., Edwards, B. F. P., Clore, G. M., Gronenborn, A. M. and Wlodower, A. (1991). Crystal structure of interleukin-8: symbiosis of NMR and crystallography. *Proc. Natl Acad. Sci. USA*, **88**, 502–506

Bax, A., Clore, G. M., Driscoll, P. C., Gronenborn, A. M., Ikura, M. and Kay, L. E. (1990a). Practical aspects of proton–carbon–carbon–proton three-dimensional correlation spectroscopy of ^{13}C-labeled proteins. *J. Magn. Reson.*, **87**, 620–627

Bax, A., Clore, G. M. and Gronenborn, A. M. (1990b). ^{1}H–^{1}H correlation via isotropic mixing of ^{13}C magnetization: a new three-dimensional approach for assigning ^{1}H and ^{13}C spectra of ^{13}C-enriched proteins. *J. Magn. Reson.*, **88**, 425–431

Bax, A. and Lerner, L. (1986). Two-dimensional NMR spectroscopy. *Science*, **232**, 960–970

Billeter, M., Qian, Y., Otting, G., Müller, M., Gehring, W. J. and Wüthrich, K. (1990). Determination of the three-dimensional structure of the *Antennapedia* homeodomain from *Drosophila* in solution by ^{1}H nuclear magnetic resonance spectroscopy. *J. Mol. Biol.*, **214**, 183–197

Braun, W. (1987). Distance geometry and related methods for protein structure determination of NMR data. *Quart. Rev. Biophys.*, **19**, 115–157

Braun, W. and Go, N. (1985). Calculation of protein conformation by proton–proton distance constraints: a new efficient algorithm. *J. Mol. Biol.*, **186**, 611–626

Brünger, A. T. (1992). Free-R value: a novel statistical quantity for assessing the accuracy of crystal structures. *Nature*, **355**, 472–474

Brünger, A. T., Clore, G. M., Gronenborn, A. M. and Karplus, M. (1986). Three-dimensional structures of proteins determined by molecular dynamics with interproton distance restraints: application to crambin. *Proc. Natl Acad. Sci. USA*, **83**, 3801–3805

Clore, G. M. and Gronenborn, A. M. (1987). Determination of three-dimensional structures of proteins in solution by nuclear magnetic resonance spectroscopy. *Protein Eng.*, **1**, 275–288

Clore, G. M. and Gronenborn, A. M. (1989). Determination of three-dimensional structures of proteins and nucleic acids in solution by nuclear magnetic resonance spectroscopy. *CRC Crit. Rev. Biochem. Mol. Biol.*, **24**, 479–564

Clore, G. M. and Gronenborn, A. M. (1991a). Two, three and four dimensional NMR methods for obtaining larger and more precise three-dimensional structures of proteins in solution. *Ann. Rev. Biophys. Biophys. Chem.*, **21**, 29–63

Clore, G. M. and Gronenborn, A. M. (1991b). Comparison of the solution nuclear magnetic resonance and X-ray structures of human recombinant interleukin-1β. *J. Mol. Biol.*, **221**, 47–53

Clore, G. M. and Gronenborn, A. M. (1991c). Applications of three- and four-dimensional heteronuclear NMR spectroscopy to protein structure determination. *Prog. NMR Spectrosc.*, **23**, 43–92

Clore, G. M., Appella, E., Yamada, M., Matsushima, K. and Gronenborn, A. M. (1990a). The three-dimensional structure of interleukin-8 in solution. *Biochemistry*, **29**, 1689–1696

Clore, G. M., Bax, A., Driscoll, P. C., Wingfield, P. T. and Gronenborn, A. M. (1990b). Assignment of side chain ^{1}H and ^{13}C resonances of interleukin-1β using double and triple resonance heteronuclear three-dimensional NMR spectroscopy. *Biochemistry*, **29**, 8172–8184

Clore, G. M., Bax, A. and Gronenborn, A. M. (1991a). Stereospecific assignment of β-methylene protons in larger proteins using three-dimensional ^{15}N-separated Hartmann–Hahn and ^{13}C-separated rotating frame Overhauser spectroscopy. *J. Biomol. NMR*, **1**, 13–22

Clore, G. M., Bax, A., Wingfield, P. T. and Gronenborn, A. M. (1990c). Identification and localization of bound internal water in the solution structure of interleukin-1β by heteronuclear three-dimensional ^{1}H rotating frame Overhauser ^{15}N–^{1}H multiple quantum coherence NMR spectroscopy. *Biochemistry*, **29**, 5671–5676

Clore, G. M., Brünger, A. T., Karplus, M. and Gronenborn, A. M. (1986a). Application of molecular dynamics with interproton distance restraints to three-dimensional protein structure determination: a model study of crambin. *J. Mol. Biol.*, **191**, 523–551

Clore, G. M., Driscoll, P. C., Wingfield, P. T. and Gronenborn, A. M. (1990d). Low resolution structure of interleukin-1β in solution derived from ^{1}H–^{15}N heteronuclear three-dimensional NMR spectroscopy. *J. Mol. Biol.*, **214**, 811–817

Clore, G. M., Gronenborn, A. M., Brünger, A. T. and Karplus, M. (1985). The solution conformation of a heptadecapeptide comprising the DNA binding helix F of the cyclic AMP receptor protein of *Escherichia coli*: combined use of ^{1}N-nuclear magnetic resonance and restrained molecular dynamics. *J. Mol. Biol.*, **186**, 435–455

Clore, G. M., Kay, L. E., Bax, A. and Gronenborn, A. M. (1991b). Four dimensional ^{13}C/^{13}C-edited nuclear Overhauser enhancement spectroscopy of a protein in solution: application to interleukin-1β. *Biochemistry*, **30**, 12–18

Clore, G. M., Nilges, M., Sukuraman, D. K., Brünger, A. T., Karplus, M. and Gronenborn, A. M. (1986b). The three-dimensional structure of a1-purothionin in solution: combined use of nuclear magnetic resonance, distance geometry and restrained molecular dynamics. *EMBO Jl*, **5**, 2728–2735

Clore, G. M., Wingfield, P. T. and Gronenborn, A. M. (1991c). High resolution three dimensional structure of interleukin-1β in solution by three and four dimensional nuclear magnetic resonance spectroscopy. *Biochemistry*, **30**, 2315–2323

Crippen, G. M. and Havel, T. F. (1988). *Distance Geometry and Molecular Conformation*. Wiley, New York

Driscoll, P. C., Clore, G. M., Marion, D., Wingfield, P. T. and Gronenborn, A. M. (1990a). Complete resonance assignment of the polypeptide backbone of interleukin-1β using three-dimensional heteronuclear NMR spectroscopy. *Biochemistry*, **29**, 3542–3556

Driscoll, P. C., Gronenborn, A. M., Beress, L. and Clore, G. M. (1989a). Determination of the three-dimensional structure of the anti-hypertensive and anti-viral protein BDS-I from the sea anemone *Anemonia sulcata*: a study using nuclear magnetic resonance and hybrid distance geometry–dynamical simulated annealing. *Biochemistry*, **28**, 2188–2198

Driscoll, P. C., Gronenborn, A. M. and Clore, G. M. (1989b). The influence of stereospecific assignments on the determination of three-dimensional structures of proteins by nuclear magnetic resonance spectroscopy: application to the sea anemone protein BDS-1. *FEBS Lett.*, **243**, 223–233

Driscoll, P. C., Gronenborn, A. M., Wingfield, P. T. and Clore, G. M. (1990b). Determination of the secondary structure and molecular topology of inter-leukin-1β using two- and three-dimensional heteronuclear $^{15}N-^1H$ NMR spectroscopy. *Biochemistry*, **29**, 4468–4682

Dyson, H. J., Gippert, G. P., Case, D. A., Holmgren, A. and Wright, P. E. (1990). Three dimensional structure of reduced thioredoxin from *Escherichia coli* determined by nuclear magnetic resonance spectroscopy. *Biochemistry*, **29**, 4129–4136

Ernst, R. R., Bodenhausen, G. and Wokaun, A. (1987). *Principles of Nuclear Magnetic Resonance in One and Two Dimensions.* Clarendon Press, Oxford

Fesik, S. W., Eaton, H. L., Olejniczak, E. T., Zuiderweg, E. R. P., McIntosh, L. P. and Dahlquist, F. W. (1990). 2D and 3D NMR spectroscopy employing $^{13}C-^{13}C$ magnetization transfer by isotropic mixing: spin system identification in large proteins. *J. Am. Chem. Soc.*, **112**, 886–887

Fesik, S. W. and Zuiderweg, E. R. P. (1988). Heteronuclear three dimensional NMR spectroscopy: a strategy for the simplification of homonuclear two dimensional NMR spectra. *J. Magn. Reson.*, **78**, 588–593

Finzel, B. C., Clancy, L. L., Holland, D. R., Muchmore, S. W., Watenpaugh, K. D. and Einspahr, H. M. (1989). Crystal structure of recombinant human interleukin-1β at 2.0 Å resolution. *J. Mol. Biol.*, **209**, 779–791

Folkers, P. J. M., Clore, G. M., Driscoll, P. C., Dodt, J., Køohler, S. and Gronenborn, A. M. (1989). The solution structure of recombinant hirudin and the Lys-47→Glu mutant: a nuclear magnetic resonance and hybrid distance geometry–dynamical simulated annealing study. *Biochemistry*, **28**, 2601–2617

Forman-Kay, J. D., Clore, G. M., Wingfield, P. T. and Gronenborn, A. M. (1991). The high resolution three-dimensional structure of reduced recombin-ant human thioredoxin in solution. *Biochemistry*, **30**, 2685–2698

Forman-Kay, J. D., Gronenborn, A. M., Kay, L. E., Wingfield, P. T. and Clore, G. M. (1990). Studies on the solution conformation of human thioredoxin using heteronuclear $^{15}N-^1N$ nuclear magnetic resonance spectroscopy. *Biochemistry*, **29**, 1566–1572

Gronenborn, A. M., Filpula, D. R., Essig, N. Z., Achari, A., Whitlow, M., Wiungfield, P. T. and Clore, G. M. (1991). A novel highly stable fold of the immunoglobulin binding domain of streptococcal protein G. *Science*, **253**, 657–661

Grzesiek, S., Ikura, M., Clore, G. M., Gronenborn, A. M. and Bax, A. (1992). A 3D triple resonance NMR technique for qualitative measurement of carbonyl-Hβ couplings in isotopically enriched proteins. *J. Magn. Reson.*, **96**, 215–221

Güntert, P., Braun, W., Wider, W. and Wüthrich, K. (1989). Automated stereo-specific ^1H NMR assignments and their impact on the precision of protein structure determinations in solution. *J. Am. Chem. Soc.*, **111**, 3997–4004

Haruyama, H. and Wüthrich, K. (1989). Conformation of recombinant desulfato-hirudin in aqueous solution determined by nuclear magnetic resonance. *Biochemistry*, **28**, 4301–4312

Havel, T. F., Kurtz, I. D. and Crippen, G. M. (1983). Theory and practice of distance geometry. *Bull. Math. Biol.*, **45**, 665–720

Ikura, M., Clore, G. M., Gronenborn, A. M., Zhu, G., Klee, C. B. and Bax, A. (1992). Solution structure of a calmodulin–target peptide complex by multi-dimensional NMR. *Science*, **256**, 632–638

Ikura, M., Kay, L. E. and Bax, A. (1990). A novel approach for sequential assignment of ^1H, ^{13}C and ^{15}N spectra of larger proteins: heteronuclear triple-resonance NMR spectroscopy. Application to calmodulin. *Biochemistry*, **29**, 4659–4667

Jeener, J. (1971). Unpublished lecture, Ampere International Summer School, BaskoPolj, Yugoslavia

Jeener, J., Meier, B. H., Bachmann, P. and Ernst, R. R. (1979). Investigation of exchange processes by two-dimensional NMR spectroscopy. *J. Chem. Phys.*, **71**, 4546–4553

Kaptein, R., Zuiderweg, E. R. P., Scheek, R. M., Boelens, R. and van Gunsteren, W. F. (1985). A protein structure from nuclear magnetic resonance data: lac repressor headpiece. *J. Mol. Biol.*, **182**, 179–182

Kay, L. E. and Bax, A. (1990). New methods for the measurement of NH–C^αH J couplings in ^{15}N labeled proteins. *J. Magn. Reson.*, **86**, 110–126

Kay, L. E., Clore, G. M., Bax, A. and Gronenborn, A. M. (1990). Four-dimensional heteronuclear triple resonance NMR spectroscopy of interleukin-1β in solution. *Science*, **249**, 411–414

Kline, A. D., Braun, W. and Wüthrich, K. (1988). Determination of the complete three-dimensional structure of the α-amylase inhibitor tendamistat in aqueous solution by nuclear magnetic resonance and distance geometry. *J. Mol. Biol.*, **204**, 675–724

Kraulis, P. J., Clore, G. M., Nilges, M., Jones, A. T., Petterson, G., Knowles, J. and Gronenborn, A. M. (1989). Determination of the three-dimensional solution structure of the C-terminal domain of cellobiohydrolase I from *Trichoderma ressei*: a study using nuclear magnetic resonance and hybrid distance geometry–dynamical simulated annealing. *Biochemistry*, **28**, 7241–7257

Marion, D., Driscoll, P. C., Kay, L. E., Wingfield, P. T., Bax, A., Gronenborn, A. M. and Clore, G. M. (1989). Overcoming the overlap problem in the assignment of ^{1}H–NMR spectra of larger proteins using three-dimensional heteronuclear ^{1}H–^{15}N Hartmann–Hahn and nuclear Overhauser–multiple quantum coherence spectroscopy: application to interleukin-1β. *Biochemistry*, **29**, 6150–6156

Moore, J. M., Lepre, C., Gippert, G. P., Chazin, W. J., Case, D. A. and Wright, P. E. (1991). High resolution solution structure of reduced French bean plasticyanin and comparison with the crystal structure of popular plastocyanin. *J. Mol. Biol.*, **221**, 533–555

Müller, L. (1987). PE.COSY: a simple alternative to E.COSY. *J. Magn. Reson.*, **72**, 191–196

Nilges, M., Clore, G. M. and Gronenborn, A. M. (1988a). Determination of three-dimensional structures of proteins from interproton distance data by dynamical simulated annealing from a random array of atoms. *FEBS Lett.*, **239**, 129–136

Nilges, M., Clore, G. M. and Gronenborn, A. M. (1990). ^{1}H–NMR stereospecific assignments by conformation database searches. *Biopolymers*, **29**, 813–822

Nilges, M., Gronenborn, A. M., Brünger, A. T. and Clore, G. M. (1988b). Determination of three-dimensional structures of proteins by simulated annealing with interproton distance restraints: application to crambin, potatocarboxypeptidase inhibitor and barley serine proteinase inhibitor 2. *Protein Eng.*, **2**, 27–38

Nilges, M., Gronenborn, A. M. and Clore, G. M. (1988c). Determination of three-dimensional structures of proteins from interproton distance data by hybrid distance geometry–dynamical simulated annealing calculations. *FEBS Lett.*, **229**, 317–324

Noggle, J. H. and Schirmer, R. E. (1971). *The Nuclear Overhauser Effect— Chemical Applications*. Academic Press, New York

Omichinski, J. G., Clore, G. M., Appella, E., Sakaguchi, K. and Gronenborn, A. M. (1990). High resolution three-dimensional solution structure of a single zinc finger from a human enhancer binding protein in solution. *Biochemistry*, **29**, 9324–9334

Oschkinat, H., Griesinger, C., Kraulis, P. J., Sørensen, O. W., Ernst, R. R., Gronenborn, A. M. and Clore, G. M. (1988). Three-dimensional NMR spectroscopy of a protein in solution. *Nature*, **332**, 374–376

Pardi, A., Billetter, M. and Wüthrich, K. (1984). Calibration of the angular dependence of the amide proton–C^α proton coupling constants, $^3J_{HN\alpha}$, in a globular protein. *J. Mol. Biol.*, **180**, 741–751

Priestle, J. P., Scär, H. P. and Grütter, M. G. (1989). Crystallographic refinement of interleukin-1β at 2.0 Å resolution. *Proc. Natl Acad. Sci. USA*, **86**, 9667–9671

Powers, R., Garrett, D. S., March, C. J., Frieden, E. A., Gronenborn, A. M. and Clore, G. M. (1992a). ^1H, ^{15}N, ^{13}C and ^{13}CO assignments of interleukin-4 using three-dimensional double and triple resonance heteronuclear magnetic resonance spectroscopy. *Biochemistry*, **31**, 4334–4347

Powers, R., Garrett, D. S., March, C. J., Frieden, E. A., Gronenborn, A. M. and Clore, G. M. (1992b). Three-dimensional structure of interleukin-4 by multi-dimensional heteronuclear magnetic resonance spectroscopy. *Science*, **256**, 1673–1677

Powers, R., Gronenborn, A. M., Clore, G. M. and Bax, A. (1991). Three dimensional triple resonance NMR of $^{13}C/^{15}N$ enriched proteins using constant-time evolution. *J. Magn. Reson.*, **94**, 209–213

Shaanan, B., Gronenborn, A. M., Cohen, G. H., Gilliland, G. L., Veerapandian, B., Davies, D. R. and Clore, G. M. (1992). Combining experimental information from crystal and solution studies: joint X-ray and NMR refinement. *Science*, **257**, 961–964

Smith, L. J., Redfied, C., Boyd, J., Lawrence, G. M. P., Edwards, R. G., Smith, R. A. G. and Dobson, C. M. (1992). Human interleukin-4. The solution structure of a four-helix bundle protein. *J. Mol. Biol.*, **224**, 899–904

Veerapandian, B., Gilliland, G. L., Raag, R., Svensson, A. L., Masui, Y., Hirai, Y. and Poulos, T. L. (1992). Functional implications of interleukin-1b based on the three dimensional structure. *Protein Struct. Funct. Genet.*, **12**, 10–21

Wagner, G., Braun, W., Havel, T. F., Schaumann, T., Go, N. and Wüthrich, K. (1987). Protein structures in solution by nuclear magnetic resonance and distance geometry. The polypeptide fold of the basic pancreatic trypsin inhibitor determined using two different algorithms, DISGEO and DISMAN. *J. Mol. Biol.*, **196**, 611–639

Williamson, M. P., Havel, T. F. and Wüthrich, K. (1985). Solution conformation of proteinase inhibitor IIA from bull seminal plasma by ^1H nuclear magnetic resonance and distance geometry. *J. Mol. Biol.*, **182**, 295–315

Wüthrich, K. (1986). *NMR of Proteins*. Wiley, New York

Wüthrich, K. (1989). Protein structure determination in solution by nuclear magnetic resonance spectroscopy. *Science*, **243**, 45–50

Wüthrich, K. (1990). Protein structure determination in solution by NMR spectroscopy. *J. Biol. Chem.*, **265**, 22059–22062

Zuiderweg, E. R. P., Petros, A. M., Fesik, S. W. and Olejniczak, E. T. (1991). Four dimensional [^{13}C, ^1H, ^{13}C, ^1H] HNQC-NOE-HMQC NMR spectroscopy: resolving tertiary NOE distance constraints in the spectra of larger proteins. *J. Am. Chem. Soc.*, **113**, 370–371

2

Methodological Advances in Protein NMR*

Ad Bax and Stephan Grzesiek

1 Introduction

Since the first experimental observation of nuclear magnetic resonance (NMR) in bulk matter more than 45 years ago (Bloch *et al.*, 1946; Purcell *et al.*, 1946), its history has been punctuated by a series of revolutionary advances that have greatly expanded its horizons. Indeed, methodological and instrumental developments witnessed over the past two decades have turned NMR into the most diverse spectroscopic tool currently available. Applications vary from exploration of natural resources and medical imaging to determination of the three-dimensional structure of biologically important macromolecules (Wüthrich, 1986; Kaptein *et al.*, 1988; Bax, 1989; Clore and Gronenborn, 1989; Markley, 1989; Wüthrich, 1989; Wagner *et al.*, 1992). The present chapter focuses primarily on the methodological advances in this latter application, particularly as they relate to the study of proteins in solution.

After the development of Fourier transform NMR (Ernst, 1966), the introduction of a second frequency dimension in NMR spectroscopy by Jeener (1971) provided a critical trigger to the development of this field. An enormous variety of experimental schemes, all based on the two-dimensional (2D) concept and largely developed by the group of Ernst (for a review, see Ernst *et al.*, 1987) expanded the applicability of NMR to the characterization of quite complex molecules, including natural products, sugars, synthetic polymers and peptides. Nearly a decade ago, protein structure determination was added to the realm of applications by the introduction of new systematic procedures for spectral analysis,

*Adapted from Bax, A. and Grzesiek, S., *Accounts of Chemical Research* (in press).

primarily developed by Wüthrich and co-workers (Wüthrich, 1986). During the 1980s, development of new experimental pulse schemes continued to increase the power and applicability of 2D NMR to structural characterization of biopolymers. The most important development was undoubtedly the addition of a third frequency dimension to the NMR spectra (Oschkinat *et al.*, 1988; Vuister *et al.*, 1988). The concept of 3D NMR is so similar to 2D NMR that no new formalism for the description of such experiments is required. The main problem that had to be solved for the development of such techniques was a way to record and process the enormous data matrices associated with such experiments. The second, as will be discussed later, was that sensitivity of the 3D experiments frequently is much lower than for analogous 2D experiments unless the third dimension corresponds to the chemical shift of a ^{13}C or ^{15}N nucleus and isotopic enrichment is used (Fesik and Zuiderweg, 1988; Marion *et al.*, 1989b). The advances in genetic engineering techniques that have occurred in the last decade enable many proteins to be overproduced and labelled in micro-organisms with the NMR-observable stable isotopes. With isotopic enrichment, sensitivity of many of the heteronuclear 3D experiments is sufficiently high to add yet another frequency dimension to the NMR spectrum, dispersing resonance frequencies in four orthogonal dimensions (Kay *et al.*, 1990; Clore *et al.*, 1991; Zuiderweg *et al.*, 1991). Below, we shall discuss the advantages and problems associated with extending the dimensionality of the NMR spectrum, and attempt to provide an answer to the question 'How many dimensions do we really need?'

2 Principles of Multidimensional NMR

Although the principles of 2D NMR have been reviewed many times, we shall briefly reiterate some of these in order to clarify the basis of 3D and 4D NMR. Most of the useful *n*D NMR experiments are of the so-called 'correlated' type, in which the chemical shift of a nucleus is correlated with the chemical shifts of other nuclei on the basis of an interaction between them. For example, in the important 2D NOESY experiment, protons are correlated on the basis of the dipole–dipole coupling between their magnetic moments, giving rise to magnetization transfer via the nuclear Overhauser effect (NOE). The pulse scheme used for the NOESY experiment is sketched in Figure 2.1(a). In this scheme, three RF pulses are applied to the proton spins, and the scheme is repeated many times for systematically incremented durations of time t_1. The signals detected during time t_2 are modulated by the frequencies present during time t_1, and a two-dimensional Fourier transformation of the acquired data matrix results in the 2D NMR spectrum.

In the NOESY spectrum, correlations between a resonance of proton A and proton B will be observed if A and B are sufficiently close in space (less than ~5 Å). However, before the distance information in the NOESY spectrum can be fully interpreted, it is necessary to assign each of the resonances in the ^{1}H NMR spectrum to its site in the chemical structure. To accomplish this, it is also necessary to record so-called J-correlated experiments, in which magnetization is transferred between chemically bonded nuclei via the J coupling mechanism. The oldest 2D NMR pulse scheme, in which magnetization is transferred from one proton to another via ^{1}H–^{1}H J coupling, is known as the COSY experiment and is probably the most popular experiment in the NMR analysis of small molecules. This experiment requires that J_{HH} is not much smaller than the ^{1}H resonance linewidth. This linewidth is approximately proportional to the inverse of the molecular tumbling rate and therefore increases approximately linearly with the size of the protein. For larger proteins ^{1}H–^{1}H J couplings are frequently smaller than the linewidth, making the COSY experiment ineffective. Other J correlation techniques, such as the one depicted in Figure 2.1(b), can correlate the frequency of a proton with that of its directly attached heteroatom (^{13}C or ^{15}N). The heteronuclear one-bond couplings, $^1J_{CH}$ (125–160 Hz) and $^1J_{NH}$ (~92 Hz), are much larger than $^3J_{HH}$, and frequently as much as 50–90% of the magnetization can be transferred back and forth between protons and their directly coupled heteronuclei. Consequently, the 2D heteronuclear shift correlation techniques, based on such magnetization transfers, are highly sensitive and can often be carried out even without isotopic enrichment.

The concept of 2D NMR is easily extended to higher dimensionality. For example, the two pulse schemes of Figure 2.1(a, b) can be concatenated in a manner depicted in Figure 2.1(c), yielding a 3D experiment. The signals, acquired during time t_3, are now obtained for many different t_1 and t_2 durations. As was the case in the 2D NOESY experiment, the data are modulated in the t_1 dimension by the frequencies of other nearby protons; however, in the t_2 dimension the modulation frequency is that of the ^{15}N nucleus that is directly attached to the observed proton. Consequently, a 3D Fourier transformation (with respect to the time variables t_1, t_2 and t_3) yields a 3D frequency domain NMR spectrum. For practical reasons, such a spectrum is usually displayed and analysed as a series of adjacent cross-sections through the 3D data matrix.

Figure 2.2 illustrates the practical advantage of 3D over 2D NMR. The region of the conventional 3D spectrum that displays the NOE interactions involving amide protons of the protein staphylococcal nuclease (156 residues) is shown in Figure 2.2(b), and shows a high degree of resonance overlap. Figure 2.2(a) represents a slice taken through the 3D ^{15}N-separated NOESY spectrum, and displays NOE interactions only

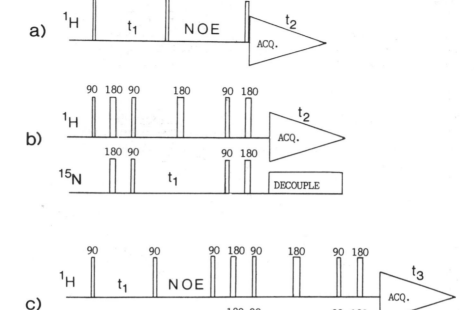

Figure 2.1 Examples of timing diagrams of 2D and 3D NMR pulse sequences: (a) 2D NOESY experiment; (b) 2D ^{1}H-detected ^{1}H–^{15}N HSQC correlation experiment; (c) 3D pulse scheme for 3D ^{15}N-separated NOESY-HSQC experiment, obtained by concatenating schemes (a) and (b). Radiofrequency pulses are marked by vertical bars and have typical durations of tens of microseconds. Signal is acquired during the time t_2 (schemes a and b) and t_3 (c); the scheme is repeated many times while the duration of t_1 (and t_2, for scheme c) is systematically incremented from 0 to ~30 ms

for amide protons attached to ^{15}N nuclei with a chemical shift near 121.4 ppm. The entire 3D spectrum consists of 64 such slices, representing interactions to amides with ^{15}N chemical shifts ranging from 130 to 105 ppm.

Although resonance overlap in the 3D spectrum is dramatically reduced compared with 2D, interpretation of the 3D NMR spectrum frequently is not necessarily straightforward. First, even if the chemical shift frequency were known for each proton in the protein, these shifts are often insufficiently unique to identify the proton. For example, if in the 3D NOESY spectrum we observe that the amide proton (at 6.7 ppm) of Ala-69 (^{15}N at 121.4 ppm) interacts with a proton at 4.20 ppm,

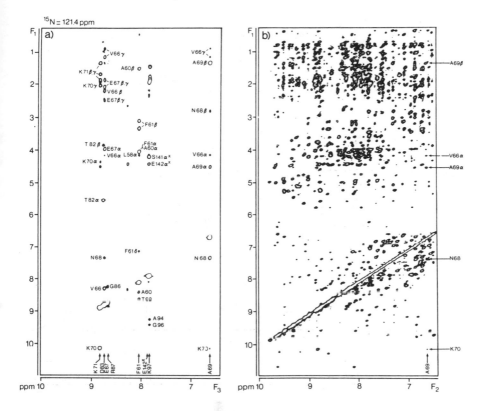

Figure 2.2 (a) One out of 64 parallel cross-sections through the 3D ^{15}N-separated NOESY spectrum of the protein staphylococcal nuclease, displaying NOE interactions involving amide protons that are attached to a ^{15}N with 121.4 ± 0.3 ppm chemical shift. (b) Corresponding region of a regular 2D NOESY spectrum, displaying NOE interactions involving all amide protons. From Marion *et al.* (1989b)

this does not identify uniquely the second proton, because nearly a dozen protons resonate in the 4.20 ± 0.02 ppm region. Therefore, it is useful to disperse the 3D spectrum in yet another dimension in order to reveal the frequency of the ^{13}C directly attached to the proton at 4.2 ppm. The pair of ^{1}H and ^{13}C shifts associated with the second proton frequently identifies it in a unique or nearly unique manner, greatly facilitating the identification of NOE interactions. Extending the 3D experiment into four dimensions is straightforward and involves inserting a 2D ^{1}H–^{13}C correlation scheme in the 3D experiment of Figure 2.2(c) (Kay *et al.*, 1990). The power of 4D NMR is illustrated in Figure 2.3, for the amide proton of Asn-17 in the protein interferon-γ, a homodimer with total molecular weight of 31.4 kDa. Figure 2.3(a) shows a 'strip' taken through the 3D spectrum at the F_1 and F_2

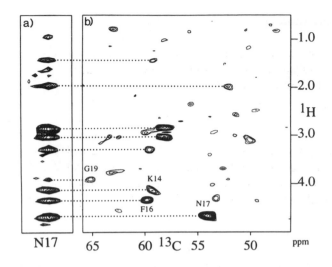

Figure 2.3 (a) Narrow 'strip' region of a cross-section, taken at a ^{15}N shift of 117.9 ppm, through the ^{15}N-separated NOESY spectrum of interferon-γ. The strip shown is centred at a ^{1}H frequency of 8.1 ppm and shows the NOE correlations from aliphatic protons to the backbone amide of Asn-17. The strip corresponds to a narrow band taken from a 2D cross-section such as the one shown in Figure 2.2(a). (b) Cross-section through the 4D ^{15}N/^{13}C-separated NOESY spectrum, displaying the chemical shifts of the protons that have an NOE interaction to the amide of Asn-17, together with the shifts of the ^{13}C nuclei directly attached to these protons. Broken contours correspond to ^{13}C nuclei in the 46–26 ppm chemical shift range, which have been aliased once in the ^{13}C dimension. Adapted from Grzesiek *et al.* (1992)

coordinates of the Asn-17 amide, and shows the chemical shifts of all protons that have an NOE interaction with this amide proton. Figure 2.3(b) shows the corresponding slice through the 4D spectrum, and, as can be seen from this figure, it yields both the ^{1}H and ^{13}C chemical shifts.

Clearly, 4D NMR presents a desirable and logical approach for identification of NOE interactions in larger proteins. The method is more symmetrical than the 3D ^{15}N- or ^{13}C-separated NOESY experiments, where the chemical shift of the attached heteronucleus is obtained for only one of the two interacting protons. The 4D method mentioned above is only suitable for separation of NOEs between ^{15}N- and ^{13}C attached protons. A conceptually similar, but technically more demanding, 4D experiment can separate NOE interactions between ^{13}C-attached protons (Clore *et al.*, 1991; Zuiderweg *et al.*, 1991). This is crucial for the unambiguous identification of interresidue side-chain–side-chain NOE contacts.

Assignment Approach

For small proteins, assignment of the ^{1}H spectrum can be made by combined analysis of NOE- and ^{1}H–^{1}H J-correlated 2D spectra. However, as indicated earlier, ^{1}H–^{1}H J correlation techniques frequently fail for larger proteins because of the increased ^{1}H resonance linewidth. This problem is even worse for proteins enriched with ^{13}C, because the ^{13}C–^{1}H dipolar interaction causes additional proton line broadening. Because assignments cannot be made on the basis of NOE interactions alone, other experiments for obtaining through-bond J correlations are essential. Moreover, in order to utilize the additional ^{15}N and ^{13}C chemical shift information available from the 3D and 4D ^{15}N- and ^{13}C-separated NOESY spectra, it is necessary to assign all protonated ^{15}N and ^{13}C nuclei.

In recent years, a novel assignment procedure has been developed that is applicable to uniformly isotopically enriched proteins (Oh *et al.*, 1988; Ikura *et al.*, 1990; Bax *et al.*, 1991; Clore and Gronenborn, 1991). This procedure is quite different from the traditional approach and is based primarily on one-bond J couplings between adjacent atoms. The one-bond J couplings are relatively uniform and depend only weakly on conformation. Typical values for the relevant coupling constants are indicated in Figure 2.4. Equally important are the magnitudes of the transverse relaxation times, T_2, which determine the resonance linewidths. T_2 depends approximately linearly on the molecular tumbling rate, i.e. on molecular size and inversely on the viscosity, but also on the degree of internal mobility and on local conformation. As a rough guide, residues that do not have a high degree of internal mobility in a globular protein of 20 kDa at 35 °C have linewidths of ~12 Hz for the amide proton, ~7 Hz for the amide nitrogen in the ^{1}H-coupled mode and ~4 Hz in the ^{1}H-decoupled mode, ~15 Hz for ^{13}C$_\alpha$ and ~25 Hz for a ^{13}C attached ^{1}H$_\alpha$. The linewidth of the carbonyl carbon is dominated by chemical shift anisotropy and therefore proportional to the square of the applied magnetic field; for the 20 kDa protein, values of ~5 Hz are observed at 500 MHz ^{1}H frequency. Comparison of these numbers with the J values shown in Figure 2.4 indicates that for a protein of 20 kDa most one-bond J couplings are significantly larger than the linewidths. This means that magnetization can be transferred with high efficiency (>~50%) from one nucleus to its directly coupled neighbour. In this way a number of highly efficient 3D J-correlated NMR experiments have been constructed, correlating the backbone and side-chain resonances in a manner schematically indicated in Figure 2.4. For example, the HBHA(CO)NH experiment (Grzesiek and Bax, 1993) correlates the amide ^{1}H and ^{15}N resonances of one residue with the H$_\alpha$ and H$_\beta$ chemical shifts of its preceding residue. Thus, for each amide two or three (in

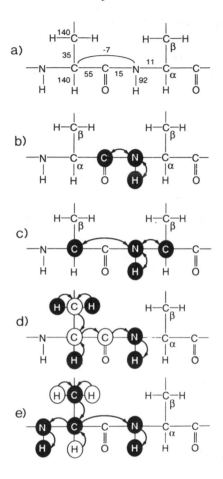

Figure 2.4 (a) A dipeptide segment of a protein backbone with the approximate values for the *J* couplings which are essential for the assignment procedure in isotopically enriched proteins. (b–e) Schematic diagrams of the nuclei that are correlated in the (b) HNCO, (c) HNCA, (d) HBHA(CBCACO)NH and (e) CBCANH experiments. Nuclei for which the chemical shift is measured in the 3D experiment are marked by solid circles. Nuclei involved in the magnetization transfer pathway, but not observed, are marked by open circles. Magnetization transfer in these experiments is marked by curved solid lines and the direction of the transfer is marked by arrows

the case of non-equivalent H_β protons) resonances are observed in the 3D HBHA(CO)NH spectrum, and their coordinates in the F_1, F_2 and F_3 dimensions of the 3D spectrum correspond to the $^1H_\alpha$ and $^1H_\beta$ (F_1), ^{15}N (F_2) and ^1H_N (F_3) chemical shift frequencies.

Table 2.1 lists the experiments that are needed to make the backbone ^1H, ^{13}C and ^{15}N assignments, and to determine the secondary structure.

Also indicated in this table is the approximate measuring time needed for acquiring each of these spectra. This measuring time should be considered to be only a rough estimate for proteins in the 15–25 kDa range at concentrations of *c.* 1 mM in a volume of ~0.2 ml. Longer measuring times may be needed for proteins that require high digital resolution because of particularly severe resonance overlap, for more

Table 2.1 Experiments for determining backbone assignments and secondary structure of isotopically labelled proteins

| | | | Labelling | | | | |
| | | | ^{15}N | ^{13}C | | | |
Experiment	Purpose	Solvent	^{15}N	^{13}C	S/N	Time[a]	Ref.
^{15}N-separated HOHAHA	$H_N(i)$, $N(i)$, $H_\alpha(i)$	H_2O	+	–	±	3	Marion *et al.* (1989a)
^{15}N-separated NOESY	Short-range NOE	H_2O	+	–	±	3	Fesik and Zuiderweg (1988); Marion *et al.* (1989a)
HNCO	$H_N(i)$, $N(i)$, $CO(i-1)$	H_2O	+	+	++	2	Ikura *et al.* (1990); Grzesiek and Bax (1992a)
HNCA[b]	$H_N(i)$, $N(i)$, $C_\alpha(i)/C_\alpha(i-1)$	H_2O	+	+	+	2	Ikura *et al.* (1990); Grzesiek and Bax (1992a)
HCACO[b]	$H_\alpha(i)$, $C_\alpha(i)$, $CO(i)$	D_2O	+/–	+	++	2	Ikura *et al.* (1990); Palmer *et al.* (1992a)
CBCA(CO)NH	$C_\beta(i-1)/C_\alpha(i-1)$, $N(i)$, $H_N(i)$	H_2O	+	+	+	2.5	Grzesiek and Bax (1992b)
HBHA(CO)NH	$H_\beta(i-1)/H_\alpha(i-1)$, $N(i)$, $H_N(i)$	H_2O	+	+	+	2.5	Grzesiek and Bax (1993)
CBCANH	$H_\beta(i-1)/H_\alpha(i-1)/$ $H_\beta(i)/H_\alpha(i)$, $N(i)$, $H_N(i)$	H_2O	+	+	±	2.5	Grzesiek and Bax (1992c)
HN(CA)CO[b]	$N(i)$, $H_N(i)$, $CO(i)$	H_2O	+	+	±	2.5	Clubb *et al.* (1992)

[a]Minimum measuring time in days, determined by the required digital resolution and the minimum number of phase cycling steps. ++, + and ± refer to the inherent sensitivity of the experiment. Experiments labelled ++ and + may be shortened significantly by the use of pulsed field gradient methodology.
[b]For proteins with favourable resonance dispersion and sensitivity, these three experiments may not be necessary.

dilute samples or for proteins approaching the molecular size limit of a particular experiment, in which case the resonance linewidth seriously degrades the sensitivity of the experiment. On the other hand, shorter measuring times could be afforded for spectra that are relatively well resolved and for samples of concentrations significantly higher than 1 mM. In favourable cases, complete backbone assignments may be obtained, using only a subset of the experiments listed in Table 2.1. However, for proteins with substantial overlap in the H_α–C_α correlation, as typically encountered for larger proteins rich in α helices, the entire arsenal, including additional experiments not listed here, may be required.

For determining the tertiary structure of a protein, additional experiments need to be carried out for obtaining side-chain resonance assignments and for collecting the required NOE data. For obtaining a 'high-resolution' structure, even more experiments are typically needed to measure the numerous homonuclear and heteronuclear J couplings and to determine stereospecific assignments of non-equivalent methylene protons and methyl groups in valine and leucine residues. Consequently, approximately 45–90 days of measuring time is at present required to gather all spectral information needed for determining a high-resolution protein structure. For very soluble proteins ($>\sim 2$ mM) smaller than ~ 20 kDa, it may be anticipated that recent advances in pulsed field gradient methodology may shorten this minimum (Vuister *et al.*, 1991; Bax and Pochapsky, 1992; Davis *et al.*, 1992). However, for experiments involving NOE or J coupling measurements, which are usually limited by low signal-to-noise ratios, no dramatic improvement is expected. At present, analysis of the multitude of 3D and 4D spectra is even more time-consuming than the data acquisition itself, particularly as software for completely automated analysis of the NMR spectra is not yet available. Development of such software, although conceptually quite straightforward, is fraught with many practical problems that have not yet been solved in a general manner.

3   Sensitivity of Multidimensional NMR

The sensitivity of 3D NMR experiments has been treated in a rigorous manner by Griesinger *et al.* (1989). Here we shall briefly discuss some of the most critical factors. The sensitivity of multidimensional NMR experiments is determined primarily by the efficiency of the magnetization transfer steps. For example, if in the NOESY experiment, sketched in Figure 2.1(a), only 0.1% of the nuclear spin magnetization of proton A is transferred to proton B, this immediately results in a 1000-fold reduction compared with the resonance intensity of proton A in a conventional single-pulse one-dimensional spectrum. Other factors that affect the sensitivity of multidimensional NMR are the decay caused by transverse relaxation during the evolution periods of the experiment. For example, if, for the NOESY experiment mentioned above, high resolution in the final 2D spectrum is critical, it will be necessary to use relatively long acquisition times in the t_1 dimension, i.e. the pulse scheme must be repeated for a range of t_1 durations that stretches from zero to several times the transverse relaxation time, T_2, of the protons. Note that the experiments with long t_1 durations carry little signal, and therefore decrease the signal-to-noise ratio obtainable per unit of time. Another important but frequently overlooked detail is the fact that each time the spectral dimensionality is increased, the sensitivity drops by $\sqrt{2}$, because both the real and imaginary component of the signal must be sampled in separate experiments. Consequently, 4D experiments, for example, are inherently less sensitive by a factor of 2 than 2D experiments. An ingenious approach, applicable to a number of J-correlated experiments, which reduces this loss at the expense of an increase in the complexity of the experiment, has recently been proposed by Palmer *et al.* (1992a).

As indicated above, the efficiency of the magnetization transfer processes is the main factor determining the sensitivity. Consequently, many of the original homonuclear 3D experiments, which contain two relatively inefficient homonuclear magnetization transfer steps, require concentrated samples to overcome their low inherent sensitivity. Nevertheless, such 3D experiments can be extremely powerful in resolving ambiguities that are invariably present in 2D spectra of proteins in the 5–15 kDa molecular weight range.

How Many Dimensions Do We Need?

With the recent introduction of a fourth dimension in NMR spectroscopy, the logical question to ask is what the optimal dimensionality of an NMR spectrum is. There is no single answer to this question, but the following discussion is intended to clarify some of the issues. As pointed out by Sørensen (1990), 3D or 4D NMR spectra can be considered as mathematical products of their corresponding 2D building blocks. Although the 3D and 4D experiments can also be used to delineate the pathways through which magnetization transfer takes place, their main use is to resolve overlap problems present in the 2D spectra. Alternatively, resonance overlap in the 2D spectrum may be reduced by using longer acquisition times in the orthogonal time dimensions. However, as indicated above, extending the acquisition time much beyond the applicable transverse relaxation time, T_2, rapidly decreases the inherent sensitivity of the experiment, and the pay-off in increased resolution drops sharply. Therefore, if acceptable spectral resolution cannot be obtained with t_1 acquisition times in the indirectly detected dimensions of the order of the applicable T_2, increasing the spectral dimensionality will be useful, provided that an efficient additional magnetization transfer step is available. For proteins isotopically enriched with ^{13}C and/or ^{15}N, such an additional efficient transfer frequently can be generated.

Once the spectrum is adequately resolved, it usually does not pay to increase the spectral dimensionality any further. Consider, for example, a J-correlated experiment in which we are trying to correlate the $^{13}C_\alpha$ and $^1H_\alpha$ resonances of one residue with the ^{15}N and ^1H_N backbone amide resonances of the next residue and suppose, for convenience, that the 2D $^{15}N-^1H$ correlation spectrum does not yield any spectral overlap. We now have the choice of correlating resonances of all four nuclei simultaneously in a 4D experiment, or to conduct two 3D experiments, one which correlates the amide resonances with $^{13}C_\alpha$ and one which correlates them with $^1H_\alpha$. To make the example less abstract, let us assume identical acquisition times in the ^{15}N and ^1H_N dimensions for the 3D and 4D experiments, and let us also assume that 8 complex increments are taken in the $^{13}C_\alpha$ and in the $^1H_\alpha$ dimension of the 4D experiment, i.e. the 2D $^{15}N-^1H$ correlation experiment must be repeated $(2 \times 8) \times (2 \times 8) = 256$ times. In the same amount of time, two 3D equally sensitive experiments could be recorded with 64 complex increments each in their respective $^1H_\alpha$ and $^{13}C_\alpha$ dimensions. In this case, the two 3D experiments determine the peak position of the $^1H_\alpha/^{13}C_\alpha$ pair up to 64 times more precisely than does the 4D experiment. The information not available from the pair of 3D experiments but readily observed in the 4D spectrum concerns the question as to which of the indirectly observed protons (in one of the 3D experiments) is coupled to which of the ^{13}C nuclei, indirectly observed in the other 3D experiment. For many of the triple-resonance J-correlated experiments this question rarely poses a problem. This discussion merely serves to illustrate that the highest possible dimensionality is not necessarily always the best choice. On the other hand, as argued by Boucher et al. (1992), there may be other practical reasons that make it preferable to obtain information from a single 4D experiment instead of from two separate 3D spectra that have been acquired at different times, possibly under different conditions. Clearly, there are advantages and disadvantages associated with increasing the dimensionality of the NMR spectrum. In practice, an increase in the spectral dimensionality requires that an additional, efficient magnetization transfer step be available. For example, in the 4D NOESY experiment, the two transfer steps, added to the 2D NOESY scheme, involve correlating the two protons with their directly attached heteroatoms. These additional steps are efficient because they utilize the large one-bond J couplings. With few exceptions, however, increasing the dimensionality in non-NOESY-type experiments is necessary and beneficial only if resolution in the 3D spectrum is limited by the natural resonance linewidth, and not by the use of acquisition times that are shorter than the transverse relaxation time, T_2. If resolution is not limited by T_2, a combination of two or three experiments of the lower dimensionality frequently is more efficient at yielding the desired information than is an increase in dimensionality.

4 Structural Parameters

To date, NMR protein structures have been calculated almost exclusively on the basis of interproton distances (derived from NOE measurements) and dihedral angles (derived from ^{1}H–^{1}H J couplings). In addition, slow exchange of amide protons with water is usually interpreted as an indication of hydrogen bonding.

Over the past decade, complete ^{1}H resonance assignments have been made for a large number of proteins for which the 3D structure is accurately known from X-ray crystallographic studies. Careful analysis of the ^{1}H chemical shifts indicates that they can be predicted on the basis of the protein structure to within a few tenths of a ppm (Ösapay and Case, 1991). Their study reveals that these chemical shifts are exquisitely sensitive to very small changes in the protein conformation. Therefore, there is hope that the chemical shift data may become useful for obtaining more accurate protein structures.

The introduction of uniform isotopic enrichment yields access to additional structural parameters that may increase further the level of detail at which both the structure and the internal dynamics of a protein can be studied. These parameters are briefly discussed below.

^{15}N and ^{13}C Chemical Shifts

The new isotope-assisted methodology (Table 2.1) relies on ^{13}C and ^{15}N chemical shifts to resolve the very crowded ^{1}H spectra of larger proteins. The ^{13}C and ^{15}N resonances themselves also contain valuable information, however. For example, a clear correlation has been found between the protein backbone angles ϕ and ψ and the $^{13}C_\alpha$ and $^{13}C_\beta$ chemical shifts (Spera and Bax, 1991; Wishart *et al.*, 1991). The $^{13}C_\beta$ resonance follows the same trend as the H_α proton resonance: for extended structures, with ϕ, ψ ~130°, on average a downfield shift from the random coil position is observed, whereas for α-helical structures (ϕ, ψ ~ $- 50°$) a small upfield shift is observed. The $^{13}C_\alpha$ resonance follows the opposite trend, with significant downfield shifts (~3 ppm) for helical structures and an upfield shift (~1–2 ppm) for extended structures.

Significant variations in the ^{13}C chemical shifts of side-chain carbons beyond C_β are observed for many residues. For example, the $^{13}C_\delta$ shifts of leucine residues may vary by as much as 6 ppm, without any concomitant large shift changes for the methyl protons. At present, no good correlation between protein structure and these variations in ^{13}C chemical shift has been determined, but it may be anticipated that, once the main contributor to the ^{13}C secondary shift has been identified, they may become important structural parameters.

The ^{15}N chemical shifts of backbone amides in proteins can also deviate substantially from their random coil values. For example, the ^{15}N shifts of Val-39 and Val-99 in staphylococcal nuclease differ by 32 ppm. Clearly, such deviations must be related to local structural differences. On average, amides in a β sheet resonate downfield by about 5 ppm compared with α-helical amides, but large exceptions to this rule occur. Attempts to correlate structural features with ^{15}N chemical shifts at a more detailed level so far have remained unsuccessful. Likely, there are a number of important contributors, including the φ, ψ backbone angles, the planarity of the peptide bond, hydrogen bonding of the amide proton and its adjacent carbonyl, and the environment of the lone pair N electrons. Consequently, no adequate correlation of ^{15}N shifts with protein structure is currently available and a large database of chemical shift assignments for proteins for which a highly accurate structure is available will be needed before such a correlation can be established in a reliable manner.

J Couplings

To date, use of J couplings in protein structure determination has been restricted mainly to the three-bond H_N–H_α and H_α–H_β couplings. These couplings are correlated with dihedral angles following well-known Karplus equations and have been used to restrain the φ and χ_1 angles, and for making stereospecific assignments of non-equivalent H_β methylene resonances. In larger proteins, where these couplings typically cannot be obtained from the ^{1}H–^{1}H multiplet structures, they may be measured, using a variety of different methods that utilize the presence of stable isotopes. These isotopes also provide access to a large number of heteronuclear J couplings, and the three-bond ^{1}H–^{15}N and ^{1}H–^{13}C couplings carry particularly important conformational information that can be interpreted readily, using Karplus equations (Bystrov, 1976).

Although little used to date, one-bond ^{1}H–^{13}C J couplings are also related to structure in a simple manner, and these couplings frequently can readily be measured in isotopically enriched proteins (Vuister *et al.*, 1992). Finally, ^{13}C–^{13}C and ^{13}C–^{15}N J couplings, accessible in labelled proteins, provide additional sources of structural information. Proteins are probably the best 'model compounds' for establishing quantitatively how these parameters relate to structure. A large database of J couplings for proteins with accurately known structures may yield such a relationship and increase the value of these couplings for structural characterization.

Recently, a technique has been described for measuring multibond ^{13}C–^{13}C J couplings (Bax et al., 1992). For proteins in the 15–20 kDa range, these relatively small couplings ($< \sim 4$ Hz) can be measured only for methyl carbons, which, because of their non-exponential transverse relaxation, have a narrow component to their resonance line shape. More than 100 of these two- and three-bond couplings were measured for a complex of the protein calmodulin and a 26-residue peptide. These $^{3}J_{CC}$ couplings contain valuable structural information regarding the side-chain torsion angles in Leu, Ile, Val and Thr residues, and can make it possible to determine stereospecific assignments of the C_γ and C_δ carbons of Val and Leu residues.

^{15}N and ^{13}C Relaxation Times

The degree of internal protein flexibility not only is of fundamental interest, but also is critical for understanding protein recognition. For proteins at natural isotopic abundance, the dynamic behaviour of a protein is not easily quantified. In contrast, for proteins isotopically enriched with ^{15}N and/or ^{13}C, detailed information regarding internal dynamics becomes readily accessible. With well-established procedures (Lipari and Szabo, 1982; Kay et al., 1989), this information can be retrieved from ^{15}N and ^{13}C longitudinal and transverse relaxation times and from the heteronuclear $^{15}N\{-^{1}H\}$ and $^{13}C\{-^{1}H\}$ NOEs. The strength of this approach for the characterization of dynamics stems from the fact that the ^{15}N and ^{13}C relaxation times are dominated by the time dependence of the strong heteronuclear dipolar interaction to their attached proton(s). For most practical purposes, this time dependence is determined solely by reorientation of the internuclear bond vector. The relaxation measurements provide information on both the amplitudes of individual angular bond vector fluctuations and the time-scale on which they occur. For practical reasons, most of the detailed protein relaxation measurements so far have focused on backbone amides. For example, we used this methodology to show that the so-called 'central helix' of the protein calmodulin, which separates its two globular domains in the crystalline state, functions as a flexible linker in solution (Barbato et al., 1992). Specifically, the amide N–H bond vectors of four adjacent residues near the middle of this central helix show large angular fluctuations on a 100 ps time-scale, and the anisotropy of the overall molecular tumbling, expected for a rigid 'central helix', was not observed. In contrast, each of the globular domains reoriented in a nearly isotropic manner, with the time-scale of the smaller of the two domains being faster than for the larger domain.

Extending the relaxation measurements to aliphatic side-chain carbons in principle is straightforward. However, in practice extra care needs to be paid to technical details, related to $^{13}C-^{13}C$ J couplings, which can affect measurements of ^{13}C transverse relaxation times, and to analysis of the data when more than one proton is coupled to a particular ^{13}C nucleus. In the latter case, dipolar cross-correlation can affect the quantitative interpretation of measured parameters. Although the theoretical formalism for describing this effect is well established, for proteins cross-correlation can seriously affect the measured parameters themselves unless specific precautions are taken (Kay *et al.*, 1992).

5 Conclusion and Outlook

It is clear that the introduction of 3D and 4D NMR, combined with uniform ^{13}C and ^{15}N isotopic enrichment, significantly extends the molecular weight limit of proteins for which a solution structure can be determined by NMR. However, it is equally clear that such structural studies of larger proteins are time-consuming and costly. The minimum amount of data acquisition needed for detailed characterization of a single protein in the 25–35 kDa range is expected to remain at least several months. Tens of milligrams of isotopically enriched protein, which must remain stable over such long measuring periods, are required for this process. Depending on the efficiency and the type of expression system, this can require large amounts of expensive isotopically enriched precursors. Analysis of the multitude of 3D and 4D NMR spectra needed for a detailed structural characterization also is at present a labour-intensive and time-consuming process. However, development of suitable support software is expected to alleviate this tedious burden in the near future, significantly reducing the time needed for structure determination.

The present well-established 2D methodology for determining protein structure starts to fail when the ^1H linewidth, which is proportional to the protein molecular weight and inversely proportional to viscosity, becomes significantly larger than the homonuclear $^1H-^1H$ J couplings. For such larger proteins, resonance overlap in the 2D NOESY spectrum presents an additional serious barrier for structure determination. However, this latter limitation is less fundamental in nature, since it may be resolved by going to higher field strengths or by recording homonuclear 3D experiments (Oschkinat *et al.*, 1988; Vuister *et al.*, 1988).

For proteins enriched uniformly with ^{13}C and ^{15}N, the multinuclear methodology described in this chapter works very well for intermediate-size proteins but it starts to fail when the ^{13}C linewidth becomes significantly larger than the ^{13}C–^{13}C couplings. At room temperature this occurs for proteins of ~30 kDa. Indeed, for the protein interferon-γ (31.4 kDa) linewidths of ~45 Hz for the ^{13}C$_\alpha$ resonances are observed at 27 °C, making complete assignments of all side-chain resonances extremely difficult (Grzesiek *et al.*, 1992). Procedures for assignment of the backbone resonances are, in our experience, more robust and function quite well, even for proteins as large as interferon-γ. Recording of the 4D NOESY spectra also becomes problematic for these larger proteins, because the sensitivity loss occurring during the ^{1}H–^{13}C correlation step(s) in these experiments is significant. For example, in interferon-γ the ^{1}H linewidths of many of the ^{13}C attached protons are in excess of 50 Hz, resulting in a threefold loss in NOE cross-peak sensitivity from relaxation during the ^{1}H–^{13}C correlation step. Altogether, it therefore appears that 30 kDa presents an upper molecular weight limit for proteins that can be studied in detail with the technology outlined herein. Slightly larger proteins may be accessible if solvent viscosity can be decreased by raising the temperature or if the protein can be studied at concentrations significantly higher than 1 mM.

For completeness, it should be mentioned that the ^{13}C- and ^{15}N-based multidimensional approach described here is not the only possible approach for increasing the molecular weight limit of proteins that can be studied in detail by solution NMR. One powerful alternative utilizes either random or residue-specific deuteration, thus simplifying the ^{1}H spectrum and narrowing ^{1}H resonance linewidths (LeMaster and Richards, 1988; Arrowsmith *et al.*, 1990). This approach has the advantage that the required NMR experiments are all relatively simple and of the homonuclear ^{1}H type. However, when residue-specific labelling is called for, a large number of samples with different deuterated amino acid residues are needed, and obtaining a comprehensive set of resonance assignments remains a difficult and labour-intensive task. It is conceivable that partial ^{2}H labelling in combination with ^{13}C- and ^{15}N-based multidimensional experiments may further increase the molecular weight limit of proteins whose solution structure can be determined by NMR. Alternatively, a combination of the ^{13}C- and ^{15}N-based multidimensional experiments and residue-specific ^{13}C- and/or ^{15}N-labelling may prove to be effective for this purpose.

Acknowledgments

We thank Ted Becker, Marius Clore, Angela Gronenborn, David Live, John Schwab, Attila Szabo and Dennis Torchia for stimulating discussions and useful suggestions during the preparation of this manuscript.

The work in the authors' laboratory is supported in part by a grant from the Intramural AIDS Targeted Anti-Viral Program of the Office of the Director of the National Institutes of Health.

References

Arrowsmith, C. H., Pachter, R., Altman, R. B., Iyer, S. B. and Jardetzky, O. (1990). Sequence-specific ^{1}H NMR assignments and secondary structure in solution of *Escherichia coli trp* repressor. *Biochemistry*, **29**, 6332–6341

Barbato, G., Ikura, M., Kay, L. E., Pastor, R. W. and Bax, A. (1992). Backbone dynamics of calmodulin studied by ^{15}N relaxation using inverse detected two-dimensional NMR spectroscopy: The central helix is flexible. *Biochemistry*, **31**, 5269–5278

Bax A. (1989). Two-dimensional NMR and protein structure. *Ann. Rev. Biochem.*, **58**, 223–256

Bax, A., Ikura, M., Kay, L. E., Barbato, G. and Spera, S. (1991). Multi-dimensional triple resonance NMR spectroscopy of isotopically uniformly enriched proteins: a powerful new strategy for structure determination. In *Protein Conformation 1991*. Wiley, New York (Ciba Foundation Symposium 161), pp. 108–135

Bax, A., Max, D. and Zax, D. (1992). Measurement of long-range [13]C–[13]C J couplings in a 20-kDa protein–peptide complex. *J. Am. Chem. Soc.*, **114**, 6923–6925

Bax, A. and Pochapsky, S. S. (1992). Optimized recording of heteronuclear multi-dimensional NMR spectra using pulsed field gradients. *J. Magn. Reson.*, **99**, 638–643

Bloch, F., Hansen, W. W. and Packard, M. (1946). Nuclear induction. *Phys. Rev.*, **69**, 127

Boucher, W., Laue, E. D., Campbell-Burk, S. and Domaille, P. J. (1992). Four-dimensional heteronuclear triple resonance NMR methods for the assignment of backbone nuclei in proteins. *J. Am. Chem. Soc.*, **114**, 2262–2264

Bystrov, V. F. (1976). Spin–spin coupling and the conformational states of peptide systems. *Progr. Nucl. Magn. Reson. Spectrosc.*, **10**, 41–81

Clore, G. M. and Gronenborn, A. M. (1989). Determination of three-dimensional structures of proteins and nucleic acids in solution by nuclear magnetic resonance spectroscopy. *CRC Crit. Rev. Biochem. Mol. Biol.*, **24**, 479–564

Clore, G. M. and Gronenborn, A. M. (1991). Applications of three- and four-dimensional NMR spectroscopy to protein structure determination. *Progr. NMR Spectrosc.*, **23**, 43–92

Clore, G. M., Kay, L. E., Bax, A. and Gronenborn, A. M. (1991). Four-dimensional ^{13}C/^{13}C-edited nuclear Overhauser enhancement spectroscopy of a protein in solution: Application to interleukin-1β. *Biochemistry*, **30**, 12–18

Clubb, R. T., Thanabal, V. and Wagner, G. (1992). A constant-time three-dimensional triple resonance pulse scheme to correlate intraresidue [1]H[N], ^{15}N, and ^{13}C chemical shifts in [15]N–[13]C labeled proteins. *J. Magn. Reson.*, **97**, 213–217

Davis, A. L., Boelens, R. and Kaptein, R. (1992). Rapid acquisition of three-dimensional triple-resonance experiments using pulsed field gradient techniques. *J. Biomol. NMR*, **2**, 395–400

Ernst, R. R. (1966). Sensitivity enhancement in magnetic resonance. *Adv. Magn. Reson.*, **2**, 1–137

Ernst, R. R., Bodenhausen, G. and Wokaun, A. (1987). *Principles of Nuclear Magnetic Resonance in One and Two Dimensions.* Clarendon Press, Oxford

Fesik, S. W. and Zuiderweg, E. R. P. (1988). Heteronuclear three-dimensional NMR spectroscopy: a strategy for the simplification of homonuclear two-dimensional NMR spectra. *J. Magn. Reson.*, **78**, 588–593

Griesinger, C., Sørensen, O. W. and Ernst, R. R. (1989). Three-dimensional Fourier spectroscopy. Application to high-resolution NMR. *J. Magn. Reson.*, **84**, 14–63

Grzesiek, S. and Bax, A. (1992a). Improved 3D triple resonance techniques applied to a 31 kDa protein. *J. Magn. Reson.*, **96**, 432–440

Grzesiek, S. and Bax, A. (1992b). Correlating backbone amide and sidechain resonances in larger proteins by multiple relayed triple resonance NMR. *J. Am. Chem. Soc.*, **114**, 6291–6293

Grzesiek, S. and Bax, A. (1992c). An efficient experiment for sequential backbone assignment of medium-sized isotopically enriched proteins. *J. Magn. Reson.*, **99**, 201–207

Grzesiek, S. and Bax, A. (1993). Amino acid type determination in the sequential assignment process of uniformly ^{13}C/^{15}N-enriched proteins. *J. Biomol. NMR*, **3**, 185–204

Grzesiek, S., Döbeli, H., Gentz, R., Garotta, G., Labhardt, A. M. and Bax, A. (1992). ^{1}H, ^{13}C, and ^{15}N NMR backbone assignments and secondary structure of human interferon-γ. *Biochemistry*, **31**, 8180–8190

Jeener, J. (1971). Unpublished lecture, Ampere Summer School, Basko Polje, Yugoslavia

Ikura, M., Kay, L. E. and Bax, A. (1990). A novel approach for sequential assignment of ^{1}H, ^{13}C and ^{15}N spectra of larger proteins: Heteronuclear triple resonance NMR spectroscopy. Application to calmodulin. *Biochemistry*, **29**, 4659–4667

Kaptein, R., Boelens, R., Scheek, R. and van Gunsteren, W. F. (1988). Protein structures from NMR. *Biochemistry*, **27**, 5389–5395

Kay, L. E., Bull, T. E., Nicholson, L. K., Griesinger, C., Schwalbe, H., Bax, A. and Torchia, D. A. (1992). Measurement of heteronuclear transverse relaxation times in AX_3 spin systems via polarization transfer techniques. *J. Magn. Reson.*, **100**, 538–558

Kay, L. E., Clore, G. M., Bax, A. and Gronenborn, A. M. (1990). Four-dimensional heteronuclear triple resonance NMR spectroscopy of interleukin-1β in solution. *Science*, **249**, 411–414

Kay, L. E., Torchia, D. A. and Bax, A. (1989). Backbone dynamics of proteins as studied by ^{15}N inverse detected heteronuclear NMR spectroscopy. Application to staphylococcal nuclease. *Biochemistry*, **28**, 8972–8979

LeMaster, D. M. and Richards, F. M. (1988). NMR sequential assignment of *Escherichia coli* thioredoxin utilizing random fractional deuteration. *Biochemistry*, **27**, 142–150

Lipari, G. and Szabo, A. (1982). Model-free approach to the interpretation of nuclear magnetic resonance relaxation in macromolecules. *J. Am. Chem. Soc.*, **104**, 4546–4558

Marion, D., Driscoll, P. C., Kay, L. E., Wingfield, P. T., Bax, A., Gronenborn, A. M. and Clore, G. M. (1989a). Overcoming the overlap problem in the assignment of ^{1}H NMR spectra of larger proteins using three-dimensional homonuclear Hartmann–Hahn and nuclear Overhauser ^{1}H–^{15}N heteronuclear multiple quantum coherence spectroscopy. *Biochemistry*, **29**, 6150–6156

Marion, D., Kay, L. E., Sparks, S. W., Torchia, D. A. and Bax, A. (1989b). Three-dimensional heteronuclear NMR of ^{15}N labelled proteins. *J. Am. Chem. Soc.*, **111**, 1515–1517

Markley, J. L. (1989). Two-dimensional nuclear magnetic resonance spectroscopy of proteins: An overview. *Methods Enzymol.*, **176**, 12–64

Oh, B. H., Westler, W. M., Derba, P. and Markley, J. L. (1988). Protein carbon-13 systems by a single two-dimensional nuclear magnetic resonance experiment. *Science*, **240**, 908–911

Ösapay, K. and Case, D. A. (1991). A new analysis of proton chemical shifts in proteins. *J. Am. Chem. Soc.*, **113**, 9436–9444

Oschkinat, H., Griesinger, C., Kraulis, P. J., Sørensen, O. W., Ernst, R. R., Gronenborn, A. M. and Clore, G. M. (1988). Three-dimensional NMR spectroscopy of a protein in solution. *Nature*, **332**, 374–376

Palmer, A. G. III, Cavanagh, J., Byrd, R. A. and Rance, M. (1992a). Sensitivity improvement in three-dimensional heteronuclear correlation NMR spectroscopy. *J. Magn. Reson.*, **96**, 416–424

Palmer, A. G. III, Fairbrother, W. J., Cavanagh, J., Wright, P. E. and Rance, M. (1992b). Improved resolution in three-dimensional constant-time triple resonance NMR spectroscopy of proteins. *J. Biomol. NMR*, **2**, 103–108

Purcell, E. M., Torrey, H. C. and Pound, R. V. (1946). Resonance absorption by nuclear magnetic moments in a solid. *Phys. Rev.*, **69**, 37–38

Sørensen, O. W. (1990). Aspects and prospects of multidimensional time-domain spectroscopy. *J. Magn. Reson.*, **89**, 210–216

Spera, S. and Bax, A. (1991). An empirical correlation between protein backbone conformation and Cα and Cβ chemical shifts. *J. Am. Chem. Soc.*, **113**, 5490–5492

Vuister, G. W., Boelens, R. and Kaptein, R. (1988). Non-selective three-dimensional NMR spectroscopy. The 3D NOE-HOHAHA experiment. *J. Magn. Reson.*, **80**, 176–185

Vuister, G. W., Boelens, R., Kaptein, R., Hurd, R. E., John, B. and van Zijl, P. C. M. (1991). Gradient-enhanced HMQC and HSQC spectroscopy. Applications to ^{15}N-labeled Mnt repressor. *J. Am. Chem. Soc.*, **113**, 9688–9690

Vuister, G. W., Delaglio, F. and Bax, A. (1992). An empirical correlation between $^{1}J_{C\alpha H\alpha}$ and protein backbone conformation. *J. Am. Chem. Soc.*, **114**, 9674–9675

Wagner, G., Thanabal, V., Stockman, B. J., Peng, J. W., Nirmala, N. R., Hyberts, S. G., Goldberg, M. S., Detlefson, D. J., Clubb, R. T. and Adler, M. (1992). NMR studies of structure and dynamics of isotope enriched proteins. *Biopolymers*, **32**, 381–390

Wishart, D. S., Sykes, B.D. and Richards, F.M. (1991). Relationship between nuclear magnetic resonance chemical shift and protein secondary structure. *J. Mol. Biol.*, **222**, 311–333

Wüthrich, K. (1986). *NMR of Proteins and Nucleic Acids*. Wiley, New York

Wüthrich, K. (1989). The development of nuclear magnetic resonance spectroscopy as a technique for protein structure determination. *Acc. Chem. Res.*, **22**, 36–44

Zuiderweg, E.R.P., Petros, A.M., Fesik, S.W. and Olejniczak, E.T. (1991). Four-dimensional [^{13}C, ^{1}H, ^{13}C, ^{1}H] HMQC-NOE-HMQC NMR spectroscopy: resolving tertiary NOE distance constraints in the spectra of larger proteins. *J. Am. Chem. Soc.*, **113**, 370–372

3

Determination of High-resolution NMR Structures of Proteins

David A. Case and Peter E. Wright

1 Introduction

As a result of continuing development over the past decade of increasingly sophisticated high-field NMR spectrometers and novel multidimensional experiments, NMR has emerged as a powerful method for determination of the three-dimensional structures of small proteins in solution. The first protein structures were reported in 1985 (Kaptein *et al.*, 1985; Williamson *et al.*, 1985) and were, not surprisingly, of relatively low resolution. Since then, intensive efforts have been directed towards development of methods for obtaining refined, high-resolution structures. In this chapter we describe our methods for structure determination and refinement, using the proteins plastocyanin, myoglobin and a zinc finger as examples. Some of our computation methods, such as distance geometry and restrained molecular dynamics, are by now relatively standard. However, refinements based on NOE intensities and chemical shifts are not yet commonplace, and we therefore describe our efforts in these directions in some detail.

NMR is unique as the only method for determination of the three-dimensional structures of proteins in solution. It is fully complementary to X-ray crystallography, but also provides direct and site-specific information on protein dynamics and provides a direct link to chemical properties related to function, such as the pK_as of active site groups (Dyson *et al.*, 1991). Refinement of structures to high resolution is clearly of great importance if maximal insight is to be obtained into the relationships between three-dimensional structure and biological function. Thus, identification of subtle structural differences between wild-type and mutant proteins or conformational changes that accompany ligand binding, protonation, and so on, requires determination of high-resolution structures. NMR provides a different view of protein structure from that provided by X-ray crystallography and in some instances, discussed at the end of this chapter, there are significant differences between the solution and crystal structures. Proper comparison of X-ray and NMR structures clearly requires that both be determined to high resolution.

2 Assignment Procedures and Data Processing

Determination of high-resolution structures requires that ^1H resonances be assigned as completely as possible. The procedure involves both specific assignment of the amino acid spin systems and the assignment of a high proportion of the NOESY cross-peaks. For proteins of molecular weight up to about 12 kD, homonuclear 2D NMR methods usually suffice. For example, we were able to obtain complete proton assignments for reduced French bean plastocyanin (99 amino acids: Chazin and Wright, 1988) and for the oxidized and reduced forms of *E. coli* thioredoxin (108 amino acids: Dyson *et al.*, 1989) using only homonuclear 2D NMR experiments. These included phase-sensitive NOESY, relayed COSY, TOCSY, double and triple quantum-filtered COSY, and two- and three-quantum experiments. Assignments were made using an integrated strategy (Chazin *et al.*, 1988b) in which magnetization is relayed from the backbone amide protons onto the side-chain, and from the side-chain terminus towards the backbone, to define overlapping spin subsystems that are combined to provide complete and reliable amino acid spin system assignments.

We use an iterative procedure to assign cross-peaks in ^1H 2D NOESY spectra (Dyson *et al.*, 1990). Initial assignments are based on chemical shifts and, where possible, ambiguities arising from resonance overlap are resolved on the basis of cross-peak shape and multiplet structure. These unambiguously assigned peaks are used to derive distance constraints that drive initial structure calculations. Additional NOESY peaks are assigned after each cycle of structure calculation on the basis of interproton distances in the current round of structures: additional cross-peaks are used to augment the constraint list only if a single assignment can be clearly distinguished from all other possible assignments. In this way a high proportion of the NOESY cross-peaks can be assigned to maximize the distance constraint density. Reliable use of such a strategy requires complete chemical shift assignments for all protons.

For determination of high-resolution structures, it is important to supplement the NOE-derived distance constraints with ϕ and χ_1 dihedral angle constraints derived from $^3J_{HN\alpha}$ and $^3J_{\alpha\beta}$ coupling constants measured from COSY- and ECOSY-type spectra. The availability of stereospecific assignments for β-methylene protons and valine methyl groups contributes significantly to the quality of the NMR structures by eliminating the need for pseudoatom corrections (Wüthrich *et al.*, 1983) which degrade the distance constraints.

For larger proteins, uniform ^{15}N and ^{13}C labelling and multidimensional heteronuclear NMR methods provide a powerful approach to assigning resonances (Clore and Gronenborn, 1991a). We have applied these methods to obtain complete ^{1}H, ^{13}C and ^{15}N assignments for the backbone and all aliphatic side-chains of the 162 amino acid enzyme IIA^{glc} domain from *B. subtilis* (Fairbrother *et al.*, 1991, 1992b). Assignments were derived from a minimal set of 3D spectra, comprising ^{1}H–^{15}N NOESY-HMQC and TOCSY-HMQC (Driscoll *et al.*, 1990; Fesik and Zuiderweg, 1990), HCCH-COSY and HCCH-TOCSY (Bax *et al.*, 1990a,b), triple resonance HCA(CO)N (Kay *et al.*, 1990; Powers *et al.*, 1991; Palmer *et al.*, 1992) and ^{13}C-edited NOESY spectra. Despite the relatively well resolved proton spectra and narrow resonances of the *B. subtilis* IIA^{glc} domain, extensive assignments would have been difficult, if not impossible, using only conventional homonuclear 2D NMR experiments. Approximate interproton distance constraints for structure determination were obtained by integration of cross-peak volumes in 3D ^{1}H–^{15}N NOESY-HMQC and ^{1}H–^{13}C NOESY-HSQC spectra (Fairbrother *et al.*, 1992a). Ambiguities in the 3D ^{1}H–^{13}C NOESY spectrum were resolved by use of 4D ^{1}H–^{13}C NOESY-HSQC data. The availability of uniformly ^{15}N-labelled protein greatly facilitates determination of φ and $χ_1$ dihedral angle constraints and stereospecific assignments of β-methylene protons. We have found the methods of Neri *et al.* (1990) and Chary *et al.* (1991) to be especially useful for proteins as large as enzyme IIA^{glc} (162 residues).

Multidimensional heteronuclear NMR methods should allow structure determination for proteins that are of molecular weight greater than 25 kD and can be labelled uniformly with both ^{15}N and ^{13}C. For smaller proteins, for which structures could be determined, using only 2D ^{1}H NMR methods, application of heteronuclear 3D NMR techniques allows assignment of a higher proportion of NOE cross-peaks and allows measurement of heteronuclear coupling constants to define side-chain rotamer preferences. Many aspects of heteronuclear methods are covered in other chapters in this volume.

3   Methods for Structure Determination and Refinement

The general procedure for determination of the three-dimensional structure of proteins is by now fairly well known: distance constraints based on nuclear Overhauser effect intensities and angular constraints based on coupling constants are used to first deduce and then refine a three-dimensional structure consistent both with general stereochemical constraints derived from the covalent structure and with the particular experimental data for the protein in question. These methods have received a number of recent reviews (Braun, 1987; Altman and Jardetzky, 1989; Clore and Gronenborn, 1989, 1991a; Wüthrich, 1989; Gippert *et al.*, 1990; Brünger and Karplus, 1991; Havel, 1991), and we do not intend to provide a comprehensive overview here. Rather, we outline some directions that may be useful in extending the precision and accuracy of protein NMR structures.

It is useful to divide protein structure determination into two parts. In the first part, an initial set of structures is generated that approximately satisfy the covalent and experimental constraints; this is followed by a second step in which these are refined against the experimental data in various ways. A similar division exists in protein crystallography, where an initial stage involves the generation of an atomic model that can be used to interpret the electron density, followed by refinement cycles that consider statistical features of the level of agreement between calculated and observed scattering intensities. Although this dividing line can sometimes be blurred in NMR refinements, it is a useful concept, because the algorithmic requirements are quite different in the two steps. In the initial phase, one is concerned with probing wide regions of conformational space and with generating ensembles of conformers with reasonably random distributions, given the constraints. The second step is aimed not only at 'improving' the initial models, but also at generating measures of the quality of the fit to experimental data and of the uncertainty of the structural models at various points in the molecule.

Distance Geometry Methods

The most popular method for carrying out the initial search step is based on a metric matrix or 'distance geometry' approach (Havel *et al.*, 1983; Crippen and Havel, 1988). Although this method has been in use for many years, some recent developments that appear to improve its sampling characteristics are worth noting.

If we consider describing a macromolecule in terms of the distances between atoms, it is clear that there are many constraints that these distances must satisfy, since for N atoms there are $N(N-1)/2$ distances but only $3N$ coordinates. General considerations for the conditions required to 'embed' a set of interatomic distances into a realizable three-dimensional object forms the subject of distance geometry (Blumenthal, 1953). The basic approach starts from the *metric matrix* that contains the scalar products of the vectors x_i that give the positions of the atoms:

$$g_{ij} \equiv x_i \cdot x_j \tag{3.1}$$

These matrix elements can be expressed in terms of the distances d_{ij}, d_{i0} and d_{j0}:

$$g_{ij} = \tfrac{1}{2}(d_{i0}^2 + d_{j0}^2 - d_{ij}^2) \tag{3.2}$$

If the origin ('0') is chosen at the centroid of the atoms, then it can be shown that distances from this point can be computed from the interatomic distances alone (Havel *et al.*, 1983):

$$d_{i0}^2 = \frac{1}{N} \sum_{j=1}^{N} d_{ij}^2 - \frac{1}{N^2} \sum_{k>j=1}^{N} d_{jk}^2 \tag{3.3}$$

A fundamental theorem of distance geometry states that a set of distances can correspond to a three-dimensional object only if the metric matrix g is rank 3, i.e. if it has three positive and $N-3$ zero eigenvalues. This is not a trivial theorem, but it may be made plausible by thinking of the eigenanalysis as a principal component analysis: all of the distance properties of the molecule should be describable in terms of three 'components', which would be the x, y and z coordinates. If we denote the eigenvector matrix as **w** and the eigenvalues λ_k, then the metric matrix can be written in two ways:

$$g_{ij} = \sum_{k=1}^{3} x_{ik} x_{jk} = \sum_{k=1}^{3} w_{ik} w_{jk} \lambda_k \tag{3.4}$$

The first equality follows from the definition of the metric tensor (Equation 3.1); the upper limit of 3 in the second summation reflects the fact that a rank 3 matrix has only 3 non-zero eigenvalues. Equating these two provides an expression for the coordinates x_{ik} ($k = 1$–3) in terms of the eigenvalues and eigenvectors of the metric matrix:

$$x_{ik} = \lambda_k^{1/2} w_{ik} \tag{3.5}$$

This, then, is a prescription for determining coordinates from distances. If the distances are not exact, then, in general, the metric matrix will have more than three non-zero eigenvalues, but an approximate scheme can be made by using Equation (3.5) with the three largest eigenvalues. Since information is lost by discarding the remaining eigenvectors, the resulting distances will not agree with the input distances, but will approximate them in a certain optimal fashion. A further 'refinement' of these structures in three-dimensional space can then be used to improve agreement with the input distances.

In practice, even approximate distances are not known for most atom pairs; rather, one can set upper and lower bounds on acceptable distances, based on the covalent structure of the protein and on the observed NOE cross-peaks. Then particular instances can be generated by choosing (often randomly) distances between the upper and lower bounds, and embedding the resulting metric matrix. The most commonly used programs to accomplish this are DISGEO (Havel and Wüthrich, 1984), DSPACE (Nerdal *et al.*, 1989), DGEOM (Blaney *et al.*, 1990), X-PLOR (Brünger, 1992) and DG-II (Havel, 1991). In addition, G. P. Gippert and M. Christiansen have prepared *cdisgeo*, a C-language version of the first of these, which will be distributed by the Quantum Chemistry Program Exchange. This version includes the updated procedures and libraries described earlier (Gippert *et al.*, 1990), and takes advantage of vector hardware on machines having this capability.

Considerable attention has been paid recently to improving the performance of distance geometry by examining the ways in which the bounds are 'smoothed' and by which distances are selected between the bounds (Havel, 1990, 1991; Oshior *et al.*, 1991; Kuszewski *et al.*, 1992). The use of triangle bound inequalities to improve consistency among the bounds has been used for many years; tetrangle inequalities represent another step in this direction, but available implementations are still computationally very expensive. We have been exploring the use of systematic search procedures in torsion angle space as a way of generating additional relations among bounds. This procedure is an extension of methods used to establish stereospecific assignments (Güntert *et al.*, 1989; Nilges *et al.*, 1990), generating a grid-sampling of all allowable conformers of a small portion (e.g. di- or tripeptide) of the protein, and from these the best upper and lower bounds for each atom pair within the fragment. By carrying out overlapping searches along the entire sequence, a bounds matrix can be generated that is often considerably tighter than that produced by conventional methods, since it includes important local correlations among the bounds. The triangle inequality can then be applied to obtain longer-range smoothing. Our current procedure (developed by G. P. Gippert) is called *xtags* (extended torsion-angle grid search), and searches four torsions at a time, doing the bookkeeping necessary to carry out the overlapping calculations and to incorporate the results into an overall bounds matrix. More efficient systematic search procedures are being developed (Dammkoehler *et al.*, 1989), which, along with faster computers, should allow the extension of this procedure to much larger fragments. It may also prove useful in conjunction with 'metrization' protocols (Havel, 1990, 1991; Kuszewski *et al.*, 1992), which interleave bound smoothing with specific distance selection to improve the sampling characteristics of the metric matrix technique. We have generally found metric matrix techniques to be a useful tool for generating initial structures, at least if the density of constraints is not too low, and view them as an integral part of our overall protocol for structure determination.

Molecular Dynamics Refinements

Refinements in three-dimensional space typically start from distance geometry structures, and attempt both to preserve the covalent geometry of the protein and to satisfy the constraints derived from the NMR measurements. The relative weighting of these two components is an important feature of the function to be optimized, which is almost always a sum of terms representing the energetics of the molecule *per se* and the experimental constraints. The molecular mechanics energy terms E_{mm} may be represented by simplified functions that maintain bond distances and angles and prevent non-bonded overlaps (Nilges *et al.*, 1988) or by use of more realistic molecular mechanics expressions (Weiner *et al.*, 1986). The penalty functions arising from NMR observations commonly have the following form:

$$E_{NOE} = \tfrac{1}{2} K_d (d - d^u)^2 \text{ for } d > d^u$$

$$= \tfrac{1}{2} K_d (d - d^l)^2 \text{ for } d < d^l$$

$$= 0 \text{ otherwise} \tag{3.6}$$

where d^u and d^l are the upper and lower bounds on a particular distance. Although standard procedures (such as conjugate gradients) may be used to minimize the sum of E_{MM} and E_{NOE}, these typically get trapped in nearby local minima, and it has become common practice to use molecular dynamics or Monte Carlo schemes to obtain more robust optimization (Brünger and Karplus, 1991). We have discussed our experience with annealing protocols elsewhere (Gippert *et al.*, 1990), and just repeat one point here. The cooling or annealing phase is crucial to the success of the procedure, and should be carried out as slowly as is practical. We use a modified dynamics algorithm that removes kinetic energy in an exponential fashion (Berendsen *et al.*, 1984), controlled by a time constant τ. (Because of partitioning between kinetic and potential energies, the actual time constant for the temperature decay is approximately 2τ.) Most of our calculations to date have used values in the range 0.5–2 ps, and the larger values nearly always give better results. It is often useful to repeat the cycle of heating and cooling, starting from the end-point of an initial annealing, but calculating more than two or three cycles does not appear to be productive. Our current 'default' protocol uses two cycles of simulated annealing of 14 ps each, about 70% of which is devoted to the cooling portion.

There is a natural question about the relative weights of the molecular mechanics and constraint portions of the optimization function. It is clear that some compromise between these two is required, and sample calculations with various weights bear out the expected behaviour in limiting situations. If the experimental constraints are too weak, then significant deviations from the observed spectrum will be obtained; in addition, further weakening of the NOE-based distance restraints will exacerbate this problem, with only a slight improvement in molecular mechanics energy. Analogous behaviour occurs when the experimental restraints are very strong: poor geometries are obtained, and further strengthening of the restraints increases these errors while providing little improvement in the agreement of calculated and observed spectral intensities. This behaviour is illustrated schematically in Figure 3.1. As proposed in the figure, there is a reasonable intermediate region that avoids the less desirable behaviour at the edges of the plots. Exploratory calculations on plastocyanin and calbindin indicate that values of K_d near 30 kcal/mol-Å^2 satisfy this criterion.

Refinement by NOE Intensities

Two-dimensional ^{1}H NOE spectra contain valuable information about molecular structures and dynamics, and a number of efforts are under way to use this information in a more quantitative manner. The 2-D NOE cross-peak intensity I_{ij} for a mixing time τ_m is given by (Macura and Ernst, 1980; Keepers and James, 1984)

$$I_{ij} = \exp(-\boldsymbol{R}\tau_m)_{ij} \tag{3.7}$$

where $\boldsymbol{R}$ is the dipolar longitudinal relaxation matrix, which depends on the set of interproton distances and the motional characteristics of the

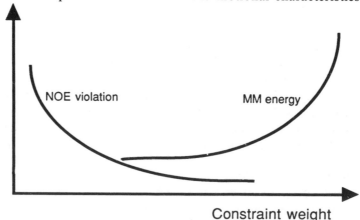

Constraint weight

Figure 3.1 Schematic dependence of the NOE violation energy and the molecular mechanics energy on the relative weights given to penalty functions

protein (Macura and Ernst, 1980; Lane, 1988). If the cross-peak intensities can be measured accurately for a sufficiently small mixing time, we have as a linear approximation to Equation (3.7)

$$I_{ij} = - R_{ij}\tau_{\mathrm{m}} \tag{3.8}$$

and the matrix elements can be directly related to the intensities. It is possible to calibrate the rate matrix elements against known distances, such as that between two methylene protons, between two ring protons in phenylalanine or tyrosine, or between interresidue distances that are approximately fixed in regions of regular secondary structure.

However, practical considerations can severely compromise the simplicity of the above method. The intensities at small mixing times may be strongly influenced by noise, especially for weak peaks corresponding to large (> 3.5 Å) distances. This can make it difficult to determine the range of validity of the linear approximation, Equation (3.8). Peaks whose primary intensities arise from indirect (multiple-spin) effects may have very short linear regimes (< 20 ms) that may be missed entirely. A number of authors have considered the likely effect of such errors on estimates of interproton distances (Clore and Gronenborn, 1985; Olejniczak *et al.*, 1986; Madrid and Jardetzky, 1988; Thomas *et al.*, 1991; Kominos *et al.*, 1992). A general conclusion is that longer distances may be significantly underestimated, often by more than 1 Å. While such uncertainties can be offset by assigning distance bounds deliberately higher than estimates of the true distances, this can entail a significant loss of information.

Alternative methods avoid the linear approximation and deal directly with Equation (3.7), using penalty functions based on the difference between calculated and observed intensities:

$$P = \sum_{\mathrm{peaks},\tau_{\mathrm{m}}} W_{ij} f(I_{ij}^{\mathrm{exp.}} - I_{ij}^{\mathrm{calc.}}) \tag{3.9}$$

The function f will have a minimum when the calculated and experimental estimates agree; its functional form will be discussed below. The gradient of P with respect to proton coordinates provides a contribution to the force in a molecular dynamics annealing scheme that is used to improve the structures. It can be generated by diagonalizing the rate matrix as $R = L\lambda L^{\mathrm{T}}$, where L is the unitary eigenvalue matrix, and λ the diagonal eigenvalue matrix. The I_{ij} can be written as $L \exp(-\lambda\tau_{\mathrm{m}})L^{\mathrm{T}}$, and the exponential of the diagonal matrix λ is easily determined. The gradient of I_{ij} with respect to nuclear coordinates is then given by (Yip and Case, 1989):

$$\nabla_\mu I_{ij} = \sum_{pqlm} L_{ip} L_{pq}^{\mathrm{T}} \nabla_\mu R_{ql} L_{lm} L_{mj}^{\mathrm{T}} \left[\frac{\exp(-\lambda_p \tau_{\mathrm{m}}) - \exp(-\lambda_{\mathrm{m}} \tau_{\mathrm{m}})}{\lambda_p - \lambda_{\mathrm{m}}} \right] \tag{3.10}$$

where λ_i is the *i*th eigenvalue and μ is any proton coordinate. $\nabla_\mu R_{ql}$ can easily be computed, since the elements of **R** are simple functions of distances.

This calculation of I_{ij} and its gradient is currently incorporated into several codes, including AMBER (Pearlman *et al.*, 1991), X-PLOR (Nilges *et al.*, 1991) and DIANA (Mertz *et al.*, 1991), and approximate or numerical derivatives have also been used (Baleja *et al.*, 1990b; Bonvin *et al.*, 1991; Stawarz *et al.*, 1992). In its exact form, this computation (especially for the gradient) can be quite expensive, and a variety of approaches have been explored to accelerate the computations. We are currently using a combination of two schemes. First, a set of overlapping 'sub-molecules' is defined, such that each cross-peak is influenced only by the magnetization dynamics of nearby spins, rather than by the whole molecule. For plastocyanin, a protein of about 100 amino acids, we define about 150 such subspaces, each with about 50–80 protons; this generally covers all interactions less than 6 Å, and is roughly equivalent to a sparse matrix implementation of the original problem (Mertz *et al.*, 1991). Second, we have found that a perturbation expansion in the off-diagonal elements of the rate matrix (Yip, 1989) can yield significant speed-ups. Development of the expansion through third order in the off-diagonal elements is accurate up to about 200 ms mixing time for a protein with a rotational tumbling time comparable to that of plastocyanin. Since the gradient is the expensive part of the calculation, we actually use the exact expression for the intensities (and, hence, for the penalty function), and use the perturbation results for the gradient. At 200 ms the mean error in the derivatives is about 5%, and in our experience, simulated annealing protocols are tolerant to such errors. The speed-up gained by using approximate derivatives is of nearly an order of magnitude; details will be presented elsewhere.

Refinement schemes based on Equation (3.9) require accurate measurements of cross-peak intensities (Weiss *et al.*, 1992), and can also be only as accurate as the models used for the relaxation processes. The next few paragraphs consider some of the issues involved in predicting cross-relaxation rates for proteins.

Overall Rotation

All quantitative approaches depend on the ability to estimate the cross-relaxation rate matrix appropriate for a given conformation. This is not completely straightforward, since (even in the absence of conformational heterogeneity) these rates depend not only upon distances between spins, but also on the rate of overall molecular tumbling and on properties of internal motion (Grant *et al.*, 1991). At short mixing times, cross-relaxation rates for macromolecules are nearly linearly related to the (isotropic) tumbling time. Since there is, in addition, an unknown scaling factor between theoretical and experimental intensities, it can often be difficult to determine the tumbling behaviour by fits to NOESY intensities, and other sources of information (such as measurements of fluorescence anisotropy decay, heteronuclear spin relaxation, dynamic light scattering or hydrodynamic calculations) are often needed. Considerable complication can also be introduced if the overall rotational motion is not isotropic, since then the orientation of the internuclear vector with respect to the rotational axes can have an important effect on cross-relaxation rates (Duben and Hutton, 1990; Withka *et al.*, 1991). Our approach to date, in the plastocyanin and zinc-finger examples discussed below, has been to use an isotropic model derived from fluorescence anisotropy data, using an assumed dependence of the rotational correlation time on η/T to correct for viscosity and temperature differences between fluorescence and NMR experiments, including (where appropriate) the increased viscosity of D_2O relative to H_2O. All current analyses assume that overall rotation can be separated from internal motion, discussed next.

Effects of Internal Motion

Several groups have investigated the effects of internal motions on proton–proton cross-relaxation in proteins. It has been recognized for some time that fluctuations in the angle between the internuclear vector and the external field tend to reduce the rate, whereas fluctuations in the distance between the spins generally increase cross-relaxation, relative to a rigid model with average distances (Olejniczak *et al.*, 1984; LeMaster *et al.*, 1988). Several recent solvated molecular dynamics simulations have confirmed this general picture and demonstrated that, for a large majority of proton–proton pairs, cross-relaxation rates predicted from the dynamics simulation are within about 30% of those that would be expected for a rigid model with average distances (Brüschweiler *et al.*, 1992; Palmer and Case, 1992; Post, 1992). For example, Figure 3.2 shows the distribution of 'correction' factors that relate rigid and non-rigid models found in a simulation of a zinc-finger peptide. We considered the 234 spin pairs actually used in an NMR structure determination of this peptide, eliminating methyl groups and intra-residue cross-peaks. The distribution is peaked near unity with a reasonably narrow width, suggesting that almost all distances determined by fitting cross-relaxation rates would suffer less than 10% error by neglect of internal motion. A small fraction of the distances (about 5% in this study) would have larger errors, and the impact these errors would have on a refined structure is not clear.

It is also useful to describe internal motions in terms of simplified models, such as restricted diffusion or jump models (Woessner, 1965; Wittebort and Szabo, 1978; Tropp, 1980; London, 1989; Olejniczak, 1989; Koning *et al.*, 1990). Jump models can be especially useful in describing methyl and aromatic group motions, where the various conformers are symmetrical and thus have equal populations. In the case where the rate of interconversion of conformations is much faster than, or much slower than, overall molecular tumbling, the resulting spectral densities become independent of the rate constants, so that no adjustable parameters are required to include these effects in a computational scheme (Yip and Case, 1991). For example, Figure 3.3 illustrates the expected effects of methyl rotation on cross-relaxation to a proton at various distances, estimated in the limit in which such rotation is fast relative to molecular tumbling, an approximation which should generally be fairly accurate. For distances in the 2.5–4 Å range that are typical

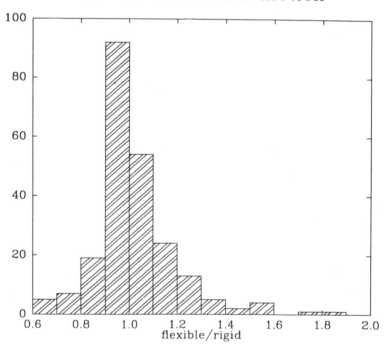

Figure 3.2 Overall effects of internal motion on proton–proton dipolar relaxation rates, computed from a molecular dynamics simulation of a zinc-finger peptide in water. The dynamics of 234 proton pairs used in the structure determination were examined to determine the ratio between the cross-relaxation rate predicted from the dynamics and that expected for a rigid protein with the same interproton distances. The histogram shows the number of ratios in various ranges. See Palmer and Case (1992) for details

in NMR structure determinations, the effect of methyl rotation is to reduce the cross-relaxation rate by 25–50%, relative to the value expected for a rigid methyl group.

It should be noted that the rate of methyl rotation may not always be sufficiently fast to be in the extreme narrowing limit. Simple models can be used for slower methyl rotations (Woessner, 1965; Kalk and Berendsen, 1976), but the importance of such effects is unclear, since methyl correlation times in proteins are not well characterized. Relatively 'slow' correlation times can be seen in solids (Torchia, 1984) and further study of this problem is needed.

Other Relaxation Effects

We do not have room to discuss here other potential relaxation pathways, or the effects of cross-correlation on dipolar relaxation (Bull, 1987; Grant *et al.*, 1991), although such effects are often expected to be small for proton relaxation in macromolecules. However, fits to ex-

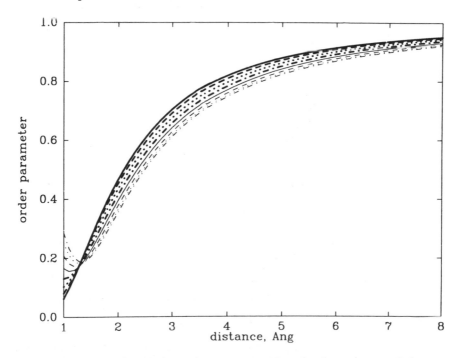

Figure 3.3 Effect of methyl rotation on cross-relaxation between a methyl group and a single proton, situated at various distances from the centre of the methyl group. The order parameter is defined as the ratio between the cross-relaxation rates expected for a methyl group whose internal rotation is in the fast motion limit and that for a rigid methyl group. The various curves are for different angles of approach; see Yip and Case (1991) for details

perimental intensities often appear to require a 'leakage' term, i.e. a general additional magnetization decay of unspecified origin, with relaxation times of the order of a few seconds. Such effects, if real, can become quantitatively important at mixing times of more than a few hundred milliseconds, and further investigations will be required if secure comparisons to experiment are to be made for such mixing times. Finally, we should mention again that (unrecognized) conformational disorder is an additional potential problem that needs to be addressed in a robust refinement algorithm.

Form of the Penalty Function

Even when spin diffusion is accounted for, and useful relations between structures and cross-relaxation rates are obtained, there remains the question of which statistic(s) relating calculated and observed intensities should be optimized in the refinement process. One could consider using a statistic akin to the R factor of X-ray crystallography, but this tends to give too little weight to weak cross-peaks that are important in defining the structure. Various proposals have been put forth for improved statistics (Borgias *et al.*, 1990; Gonzalez *et al.*, 1991). In our work to date, we have used a scheme like that suggested by Mertz *et al.* (1991), in which volume-based penalty functions are combined with conservative distance restraints, the latter serving to prevent large distance violations that might otherwise occur with little degradation of the R factor. As with protein crystallography, the development of generally accepted procedures for refinement will depend to a large extent on empirical tests of various schemes. Initial applications of these sorts of procedures (including the plastocyanin results outlined below) suggest that this general approach will yield useful measures of the extent to which the intensity data are modelled by particular structures (Baleja *et al.*, 1990a,c; Bonvin *et al.*, 1991; Mertz *et al.*, 1991; Nilges *et al.*, 1991). In this context, it is also worth noting that alternative methods exist for quantitative comparisons with NOESY intensities, based mainly on iterative refinements of the rate matrix itself (Boelens *et al.*, 1988; Borgias *et al.*, 1990; Koehl and Lefevre, 1990). Much work remains to be done to determine the best approach to refinement and the benefits of going beyond the present level of structure determinations using distance bounds derived by empirical calibration.

Refinements Based on Chemical Shifts

In principle, chemical shifts depend upon conformation in ways that might provide useful information, particularly if approximate structures are already known by methods outlined above. As with refinements based on NOE intensities, the principal question to be decided is whether the expressions used to compute the shifts for a particular structure are sufficiently accurate and reliable to be used as input in a refinement procedure. For chemical shift calculations, the situation is complicated by the fact that several physical interactions contribute to the observed values, not all of which are completely understood. Nevertheless, substantial progress has been made recently in the development and parametrization of useful relations between proton chemical shifts and structure.

Considerable attention has been paid to connections to secondary structure, particularly for the Hα shift. It has been recognized for some time that β sheet regions show downfield shifts for this resonance, while helical regions show upfield shifts (Dalgarno *et al.*, 1983; Szilágyi and Jardetzky, 1989; Pastore and Saudek, 1990; Williamson, 1990; Wishart *et al.*, 1991). The most extensive survey is that of Wishart *et al.* (1991), who considered shifts in over 70 proteins with assigned secondary structures. The mean Hα positions in helices and sheets differs by nearly 0.8 ppm and there is remarkably little overlap between the two distributions. These relations, and others seen for amide protons and ^{13}C and ^{15}N shifts (Kuntz *et al.*, 1991; Spera and Bax, 1991; Wishart *et al.*, 1991), are in many cases clear enough to drive secondary structure assignments (Wishart *et al.*, 1992), but do not identify the physical interactions responsible for the correlations seen.

It has been known for some time (see Haigh and Mallion, 1980, for a review) that protons near to aromatic rings exhibit shifts that can be understood from simple quantum mechanical formulas that include empirical parameters fitted to experiment. Similar interactions can also arise from amide or carbohydrate groups where the resonance interaction between valence bond structures leads to significant anisotropies in group magnetic susceptibilities (McConnell, 1957; Pauling, 1979). Another potential interaction involves local electric fields, which can polarize the electrons in the C–H bond and lead to increased shielding or deshielding of the protons (Buckingham, 1960). Recently two groups have shown that calculations that model these effects can explain a large portion of the chemical shift dispersion seen in globular proteins (Osapay and Case, 1991; Williamson and Asakura, 1991; Williamson *et al.*, 1992). Osapay and Case carried out an empirical analysis of proton chemical shifts from 17 proteins whose X-ray crystal structures had been determined, and showed that a significant improvement over ring current theories can be made by including the effects of the magnetic anisotropy of the peptide group and estimates of backbone electrostatic contributions. For 5678 protons bonded to carbon, they found a linear correlation coefficient of 0.88 between calculated and observed secondary shifts, with a root mean square error of 0.23 ppm; for side-chain protons in non-heme proteins, the rms error was 0.18 ppm and for methyl groups it was 0.13 ppm. Some conclusions from this analysis are summarized below.

Ring Current Calculations

The general form of empirical theories is $\sigma_{rc} = iBG(r)$, where r is the vector from the observed proton to the aromatic ring, $G(r)$ is a geometric factor and i and B are constants (Haigh and Mallion, 1980; Harris, 1986). In the Haigh–Mallion theory, which is one of the simplest to implement, the geometric factor is

$$G(r) = \sum_{ij} s_{ij} \left\{ \frac{1}{r_i^3} + \frac{1}{r_j^3} \right\} \qquad (3.11)$$

Here r_i and r_j are the distances from ring atoms i and j to the proton and s_{ij} is the area of the triangle formed by atoms i and j and the proton projected onto the plane of the aromatic ring. The sum is over the bonds in the ring. This formula (and those below) are easy to differentiate with respect to nuclear positions, and can form the basis of a refinement scheme using penalty functions analogous to those in Equation (3.9); these have been incorporated into the AMBER molecular modelling package (Pearlman *et al.*, 1991).

Peptide Group Contributions

It has been recognized for some time that the magnetic anisotropy of the peptide group is likely to contribute significantly to chemical shifts in proteins. When the observed proton and the 'source' of the magnetic anisotropy are far apart, McConnell (1957) has shown that the contribution to the local shielding tensor depends upon the magnetic anisotropy of the distant group:

$$\sigma_m = (3L_0 R^3)^{-1} \sum_{i = x, y, z} \chi_{ii}(1 - 3 \cos^2 \theta_i) \qquad (3.12)$$

Here L_0 is the Avogadro constant, R is the distance from the proton to the distant group, χ_{ii} is a component of the magnetic susceptibility tensor and θ_i is the angle between the i axis and the radius vector R. Since there is no direct method to measure the magnetic anisotropy of a peptide group within a protein, all estimates of these effects are to some extent empirical. Flygare's group has measured susceptibilities for a large number of molecules in the gas phase (Flygare, 1974), and has developed procedures to analyse these effects in terms of localized contributions (Schmalz *et al.*, 1973). Data for formamide (Tigelaar and Flygare, 1972) suggest that the peptide group is nearly axially symmetric about the out-of-plane axis, with an anisotropy $\Delta\chi$ of $-5.1 \pm 0.6 \times 10^{-6}$ erg/(G^2-mol). This value is in good agreement with an empirical theoretical approach developed by Pauling, which yields a value of -5.4×10^{-6} erg/(G^2-mol) (Pauling, 1979).

Electrostatic Contributions

A significant contribution to chemical shifts can also arise from distant polar groups, which can polarize the C–H bond and thereby increase or decrease the local shielding by electrons. The most significant term is expected to be proportional to the projection of the local electric field onto the C–H bond vector:

$$\sigma_{el} = A E(\text{C–H}) \tag{3.13}$$

Buckingham (1960) suggested that an appropriate value for A would be -2×10^{-12} e.s.u.$^{-1}$, but this magnitude clearly depends upon the way in which local electric fields are estimated. Most estimates to date have been based on Coulomb's law, using partial charge models from molecular mechanics force fields, but this approach ignores (potentially) important solvent effects. Calculations based on macroscopic dielectric models may lead to an improved description. It should be noted that electrostatic and solvation contributions are likely to be relatively more important for heavier nuclei (such as ^{13}C or ^{15}N), where the overall chemical shift dispersion is greater (Oldfield *et al.*, 1991).

4 Protein Structures

To illustrate applications of the above ideas, we discuss three examples of recent work on proteins. Plastocyanin and the zinc-finger peptides illustrate some of the interesting points that arise when relatively large numbers of constraints are available, and myoglobin illustrates some of the possibilities of using chemical shifts in a refinement procedure. These examples necessarily give only one view of the process of protein structure determination but, when considered alongside other work described in this volume, help illustrate many of the questions faced by workers in this field.

Plastocyanin

Solution structures of reduced (Cu I) French bean plastocyanin, a metalloprotein containing 99 amino acids, have been determined by use of a total of 1120 NOE-derived distance constraints and 103 dihedral angle constraints derived from coupling constants (Moore, *et al.*, 1991).

Stereospecific assignments were made for 26 β-methylene groups and the methyls of 11 valines. Pure absorption NOESY spectra were recorded for 12 mixing times in the range 10–400 ms in D_2O and for 14 mixing times in the range 10–300 ms in H_2O. The availability of these extensive build-up data allowed a tighter classification of backbone–backbone distances than is customary, and we used upper bounds of 2.4, 2.7, 3.0, 3.5 and 4.3 Å. Distance constraints were obtained from cross-peak volumes by calibration using known distances in regular secondary structure (Billeter *et al.*, 1982; Williamson *et al.*, 1985). A motional-averaging correction of 0.2 Å for backbone–backbone NOEs and 1.0 Å for NOEs involving side-chains was added; the molecular dynamics calculations described above suggest that these are conservative estimates. Upper bound distances were 3.0, 3.5, 4.0 and 5.0 Å for constraints involving side-chain protons.

No direct information on the coordination of the copper atom is available from the current NMR experiments. Distance geometry calculations carried out with omission of the metal were of high enough quality to identify the copper ligands (Cys 84, Met 92, His 37 and His 87) (Moore *et al.*, 1991). Additional distance geometry calculations with a combination of distance and chirality constraints that allow coordination of the copper to either Nε or Nδ of the two histidines showed clearly that both histidines bind through the Nδ position: the NOE and torsion angle constraints had defined the folded structure to sufficient resolution to allow identification of many side-chain positions and the specific atoms responsible for copper binding (Figure 3.4). This coordination was included in the molecular dynamics refinement step, using (loose) bond and angle constraints based on X-ray structures of model copper–amino acid complexes and, for the long bond to Met 92, the X-ray structure of the homologous poplar plastocyanin (Guss *et al.*, 1986).

Superimposed structures of French bean plastocyanin are shown in Figure 3.5. The conformation of the polypeptide backbone is well defined, with an average rms deviation from the mean of 0.45 Å for backbone heavy atoms (N, Cα, C and O); the corresponding value for all heavy atoms is 1.08 Å (Moore *et al.*, 1991). The maximum upper bound distance violation is 0.26 Å in the refined structures, compared with a maximum violation of more than 1 Å in the family of unrefined distance geometry structures. Deviations of bond lengths and angles from 'ideal' geometries are small, in accord with values obtained for well-refined crystallographic structures.

The NMR structures of French bean plastocyanin are of higher precision than those of an algal plastocyanin (from *Scenedesmus obliquus*) determined earlier (Moore *et al.*, 1988). The earlier structures showed an rms deviation of the backbone from the mean structure of 1.2 Å. The principal differences between the structure determinations lie in the larger density of constraints for the French bean structure (*c.* 12 constraints per residue as opposed to 5.9 for the algal protein), the use of tighter distance constraints for backbone protons and the availability

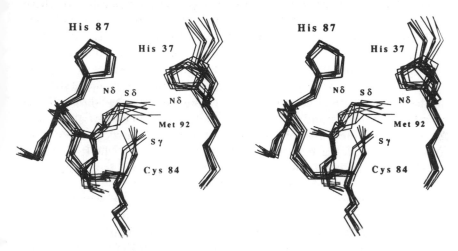

Figure 3.4 Stereoview of the copper ligands in the best 10 distance geometry structures for plastocyanin. No Cu coordination constraints were used in the calculations. The structures were superimposed for best fit of the backbone heavy atoms and side-chain β and γ heavy atoms of residues 37, 84, 87 and 92. From Moore *et al.* (1991): used with permission

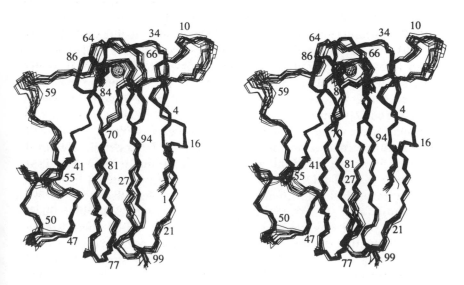

Figure 3.5 Stereoview showing the best-fit superposition of the backbone atoms of 16 MD-refined structures of plastocyanin. The position of the copper atom is indicated by a circle. From Moore *et al.* (1991): used with permission

of stereospecific assignments for many Hβ protons. Recognizing that simple statistics such as number of constraints or rms deviation tell only a part of the story, it nevertheless appears that improvements in the number and quality of the constraints plays a crucial role in determination of structures to a level close to that available from crystallographic analyses. We discuss below some other measures of the quality of the structures.

The precision of the French bean plastocyanin structures, as measured by the rms deviation among the structures in the family, is highest for the beta-strands that constitute the framework of the molecule and lowest for the turns and loops. There is an inverse correlation between the spread of the backbone conformations and the density of NOE constraints, but large backbone rms deviations in the NMR structures correlate poorly with backbone temperature factors from the X-ray structure of the homologous poplar plastocyanin.

The conformations of many side-chains, both in the hydrophobic core and on the surface, are well defined in these structures. Inclusion of stereospecific assignments and χ_1 constraints helps greatly to define the side-chain conformations (Figure 3.6). The χ_1 angles for most side-chains are close to the classical staggered rotamer angles. However, in our experience, refinement to a single χ_1 rotamer should not be interpreted as proof of a fixed side-chain conformation unless $^3J_{\alpha\beta}$ coupling constants preclude the possibility of conformational averaging. There are several examples in French bean plastocyanin, where a single χ_1 rotamer is highly preferred in the refined structures but where $^3J_{\alpha\beta}$ coupling constants indicate some conformational averaging (Moore *et al.*, 1991). In many cases this averaging may involve no more than limited fluctuations about a single minimum. However, for six surface residues with charged or polar side-chains, our refinement procedure produced single χ_1 rotamers, even though no input χ_1 constraints were imposed, and $^3J_{\alpha\beta}$ coupling constants provide clear evidence of extensive conformational averaging. This behaviour may arise from errors in the potential functions used in the final refinement, from deficiencies in the sampling characteristics of the refinement algorithm, or from the implicit assumption that all NOE cross-peaks should be interpreted in terms of a single molecular conformation. Indeed, as the structures of both plastocyanin and the *Xfin* zinc finger have been refined to higher resolution, side-chains have been recognized for which there are mutually incompatible restraints. For Ile 55 in plastocyanin, for example, no single structure could be found that satisfies simultaneously all the NOE constraints and the χ_1 constraint. The data strongly suggest averaging between rotamers with χ_1 near +60° and 180°. At this stage, we have dealt with these problems simply by relaxation of the mutually incompatible NOE constraints. However, a more satisfactory approach

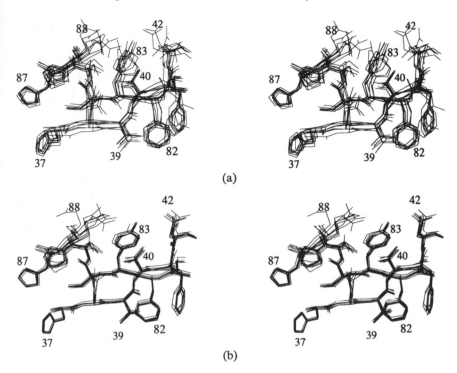

(a)

(b)

Figure 3.6 Stereoviews illustrating the quality of structures obtained without (a) and with (b) the inclusion of χ_1 constraints and stereospecific assignments for plastocyanin. The heavy line in (b) is from the poplar plastocyanin X-ray structure (Guss *et al.*, 1986)

might be to use time-averaged restraints (Torda *et al.*, 1989) or other methods to provide better sampling of the side-chain conformational ensemble.

The NMR structures of reduced French bean plastocyanin are similar to the X-ray structure of poplar plastocyanin (Figure 3.7), with a mean pairwise rms difference of 0.76 Å for backbone heavy atoms between the X-ray structure and 16 NMR structures (Moore *et al.*, 1991). Side-chain conformations are also generally similar (see Figure 3.6 for examples.) There is a good correlation between the mean spread of the side-chain heavy atoms in the solution structures and the side-chain temperature factors in the X-ray structure of the poplar protein: side-chains that have higher than average temperature factors tend to have higher than average spread among the NMR structures. For many of the disordered side-chains, the observation of averaged $^3J_{\alpha\beta}$ coupling constants provides direct evidence for conformational averaging.

One of the advantages of the use of a realistic force field (such as the AMBER one that we use) in restrained molecular dynamics refinement

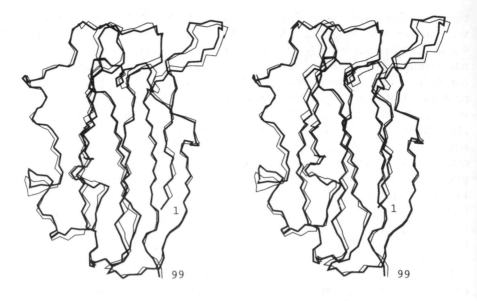

Figure 3.7 Backbone trace for the best NMR-derived structure (bold) and the poplar plastocyanin X-ray structure (Guss *et al.*, 1986, light)

appears to be its ability to model hydrogen bonding interactions. Although we often use hydrogen bond constraints deduced from amide exchange data in generating initial structures by distance geometry methods, all such constraints are omitted in the MD refinement to allow sampling of alternate hydrogen bonding conformations. For plastocyanin, 60 hydrogen bonds were found in 60% or more of the dynamics-refined structures, and 53 of these can be identified with slowly exchanging NH or OH protons. There is excellent agreement between the hydrogen bonding patterns found in the NMR structures and those identified in the crystal structure of poplar plastocyanin.

'Retrospective' Analysis

As an aid to understanding the performance of NMR refinement methods, we have considered various families of structures for plastocyanin which were generated in our laboratory over a period of 3 years with increasing numbers of NMR-derived constraints. We can compare predictions from these structures with the observed chemical shifts and with cross-peak intensities from NOESY spectra. The experimental data (which will be described in more detail elsewhere: Lepre, Yip, Case and Wright, manuscript in preparation) consisted of 7522 peak intensities, corresponding to about 700 spin pairs measured at different mixing times, as described above. The calculation of NOESY peak volumes was carried out, assuming an overall rotational correlation time of 4.2 ns (determined from fluorescence anisotropy decay,) and an external 'leakage' rate of $0.5\,\mathrm{s}^{-1}$ (estimated from the decay of a small number of resolved diagonal peaks, and assumed to be the same for all protons). Only one adjustable parameter was introduced: the overall scaling between experimental and theoretical NOE intensities. We evaluated the extent to which calculated intensities in various families of structures matched the experimental data, and, for our best set of conventionally refined structures, carried out 500 steps of conjugate gradient minimization using the penalty functions of Equation (3.9). For the function f we used the absolute difference between calculated and observed intensities, and W_{ij} (in kcal/mol) was taken to be numerically equal to the inverse of the normalized experimental intensity, with a maximum value of 40 kcal/mol and a minimum value of 5 kcal/mol.

Table 3.1 presents some features of this analysis. We analysed six sets of plastocyanin structures developed over a 3 year period. Three sets of constraints were considered, each set being used to generate initial structures using the DISGEO program, and refined structures following dynamical simulated annealing protocols discussed above. Details of the constraints and results are given elsewhere (Moore *et al.*, 1991), but an outline of the data is as follows: the earliest structures (labelled '88' in Table 3.1) were derived from those NOESY cross-peaks that could be identified unambiguously (i.e. with no potential overlaps), and no torsional information on side-chain dihedral angles. This data set had about 6–7 constraints per residue. A second set ('89') added torsional constraints for about half the χ_1 side-chain angles (determined from ECOSY experiments), plus some additional constraints that could be identified clearly by analysis of the initial structures. The final set ('90') arose from careful re-examination of spectra at several temperatures and at many mixing times, and consisted of about 12 constraints per residue. Our goal was to look back at the structures we had obtained, to see whether adding additional distance and torsion constraints during the refinement process had indeed led to structures whose predicted spectra were in better agreement with experiment. Several statistical tests were carried out (see Table 3.1), all of which pointed to an affirmative answer to this question. Several points are worth noting.

(1) For a given level of constraint information (i.e. for the 1988, 1989 or 1990 data sets) progressing from DG to SA reduces the overall structural spread in the family of structures, and also improves the quantitative agreement with the NOESY volumes.

Table 3.1 Progress of plastocyanin structure refinements

Year	Type	No. of constraints NOE	ϕ	χ_1	Pairwise rmsd	Mean AMBER energy	Mean NOE violation energy
88	DG	606	48	0	1.41		
89	DG	926	47	44	1.16		
90	DG	1110	59	44	1.00		
88	SA	667	47	0	1.30	−975.9	53.1
89	SA	939	47	44	0.79	−1117.7	28.3
90	SA	1120	59	44	0.66	−1204.9	7.6

Year	Type	NOESY volume analysis Pearson r	rms error	R-factor	Chemical shifts Pearson r	rms error
88	DG	0.430	0.0347	0.806	0.327	0.452
89	DG	0.578	0.0281	0.626	0.512	0.379
90	DG	0.584	0.0286	0.628	0.455	0.392
88	SA	0.523	0.0283	0.652	0.504	0.388
89	SA	0.727	0.0211	0.496	0.663	0.327
90	SA	0.733	0.0206	0.476	0.728	0.297
91	RM	0.876	0.0150	0.277	0.756	0.283

Notes

Type of refinement is 'DG' for distance geometry using *DISGEO*; 'SA' for molecular dynamics–simulated annealing using distance and angle restraints within *AMBER*; 'RM' for 400 steps of iterative relaxation matrix refinement starting from the 1990 SA structures, using 7522 NOESY volume restraints.

Pairwise rmsd gives the mean distance between pairs of structures within a family; here 'distance' is measured as the root mean square deviation of the backbone heavy atoms ($C\alpha$, C, N and O).

NOESY volume analysis gives statistics (linear correlation coefficient, root mean square error and *R*-factor) for interresidue cross-peaks. Experimental values were scaled to an 'absolute' scale, i.e. so that the intensity of a diagonal peak at zero mixing time is unity. The *R*-factor is a statistic akin to that used in crystallography:

$$R \equiv \frac{\sum\limits_{n} |I^0 - I^e|}{\sum\limits_{n} I^e}$$

Chemical shifts give statistics for 226 side-chain protons bonded to carbon atoms. The structures were not refined against any chemical shift data.

(2) Adding stereospecific assignments and χ_1 constraints (going from the 1988 to the 1989 data set) significantly tightens the structures and improves the agreement with the NOESY volumes.

(3) As the number of constraints increased, the average AMBER energy of the resulting structures moved to lower values; furthermore, the penalty energy associated with the NOE-derived distance constraints also became smaller, even though there were more terms in the penalty function. Figure 3.8 shows that the relation between the constraint

energy and the molecular mechanics energy was approximately linear over a wide range of structures. This suggests that we do not have a 'competition' between the molecular mechanics force field and the NMR constraints (which might tend to raise the molecular mechanics energy as the number of constraints became greater), but rather a 'co-operation', in which the NMR constraints help guide molecular folding into conformations that achieve favourable molecular mechanics interactions and minimal constraint violations.

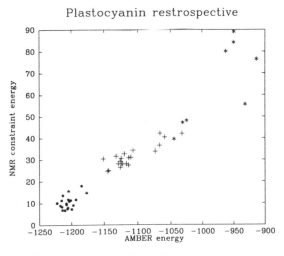

Figure 3.8 Constraint energy plotted against molecular mechanics energy for plastocyanin MD-refined structures. Dots, crosses and asterisks show results from the '90', '89' and '88' structure families, respectively (see Table 3.1)

(4) Comments (1) and (2) apply to estimates of chemical shifts as well as to NOESY volumes: when the structures are improved, the computed shifts much more closely resemble those seen experimentally. Note that none of the constraint sets included any penalty functions based on chemical shift information; structure refinement on the basis of chemical shifts is in progress.

(5) Only relatively small adjustments in the structures (of < 0.2 Å) are required for relaxation matrix refinements to improve the agreement with NOESY volumes significantly. Larger structural adjustments may be found when more volume data are available for analysis, or when the refinements are carried further.

Xfin Zinc Finger

The *Xfin-31* zinc finger represents another example of a structure that has been refined to high resolution in our laboratory. The earliest structures (Lee *et al.*, 1989, 1990) were determined by use of only 6.4 distance and dihedral angle constraints per residue. The structure has since been refined to higher resolution by addition of NOE constraints, χ_1 constraints and stereospecific Hβ assignments. Increasing the number of constraints from 6.4 to 16 per residue results in a significant tightening of the structures (Figure 3.9), with the average rms deviation from the mean structure decreasing from 0.62 Å to 0.35 Å for all backbone heavy atoms of residues 1–23. Residues 24 and 25 at the C terminus are disordered in the NMR structures. Detailed measurements of ^{13}C relaxation times and NOE enhancements confirm that these C-terminal residues are highly flexible in solution, having Cα–H order parameters S^2 less than 0.5. By contrast, the moderately flexible loop from residues Phe 10 to Lys 13 has an average S^2 of 0.77, and the remainder of the backbone Cα–H vectors show an average order parameter of 0.89 (Palmer *et al.*, 1991). Refinement of the *Xfin-31* zinc finger structures to higher resolution is important in allowing meaningful comparisons to be made between wild-type and mutant peptides (Lee, Morthishire-Smith and Wright, manuscript in preparation.) The NMR structures of *Xfin-31* are closely similar to the structures determined by X-ray crystallography for the three zinc fingers of *zif268* complexed with DNA (Pavletich and Pabo, 1991). Indeed, solution of the X-ray structure was facilitated by fitting the NMR structure into the electron density. As with plastocyanin (Figure 3.8), the zinc finger refinements show a strong correlation between the AMBER energy and the NMR-derived constraint energy in the final refined structures.

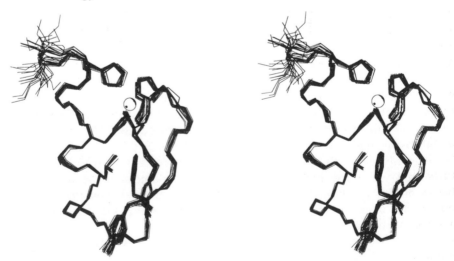

Figure 3.9 Stereoview of 40 refined structures of the *xfin-31* zinc-finger peptide. The backbone is shown, together with the zinc atom (sphere in upper centre) and the side-chains of Tyr 1, Cys 3 and 6, Phe 10, Leu 16, and His 19 and 23

Myoglobin

Myoglobin was the first protein for which an X-ray crystal structure was determined, and has served for many years as a testing ground for structure–function relations. Although it is fairly large by NMR standards, ring current shifts arising from the heme prosthetic group contribute to good chemical shift dispersion, and a portion of the spectrum was assigned from proton methods alone (Dalvit and Wright, 1987). The backbone assignments have subsequently been completed, using ^{15}N-labelled material and heteronuclear 3D NMR experiments (Theriault *et al.*, 1993). A substantial fraction of the side-chain resonances have also been assigned. Since the ring current contributions are so large, and since they are dominated by a single group whose shifts have been empirically calibrated (Cross and Wright, 1985), myoglobin offers an excellent opportunity to apply chemical shift restraints to protein structure refinement. Our preliminary calculations suggest that this will be a successful procedure. The current constraint set consists of 1384 distance upper bounds based on NOESY cross-peaks, plus penalty functions based on 409 chemical shifts. Although it is clear that there are not major differences between the solution and crystal structures, some local structure variations are evident. For example, when the X-ray structure of MbCO (Kuriyan *et al.*, 1986) is compared with the solution NMR data, we find 23 distance violations greater than 0.4 Å, with 3 greater than 2 Å. The root mean square error between calculated and observed chemical shifts is 0.28 ppm, with 30 predictions in error by more than 0.5 ppm. After one cycle with our 'default' annealing protocol (described above), the largest residual distance violation was 0.35 Å, and the rms error in chemical shifts was reduced to 0.14 ppm, with only 6 shifts more than 0.5 ppm in error.

Our current data set has about 9 NOE-derived constraints per residue, which suggests that it should be possible to generate reasonable structures without reference to the crystal coordinates. This is indeed the case: a family of 22 structures determined from distance geometry and dynamical simulated annealing showed an average rms deviation from the mean structure of 0.78 Å for the backbone atoms of the helices; the turns were less well-defined, so that the comparable figure for all backbone heavy atoms was 1.08 Å. The helices of these structures were on average 1.07 Å from the MbCO crystal structure. Figure 3.10 shows a subset of these structures, along with the X-ray structure for comparison. It is clear that, at the present level of information, several of the loops connecting helices are poorly defined. Nevertheless, in parallel calculations in which the chemical shift restraints were removed, the resulting backbones were further from the X-ray structure and there were significant levels of disagreement between calculated and observed

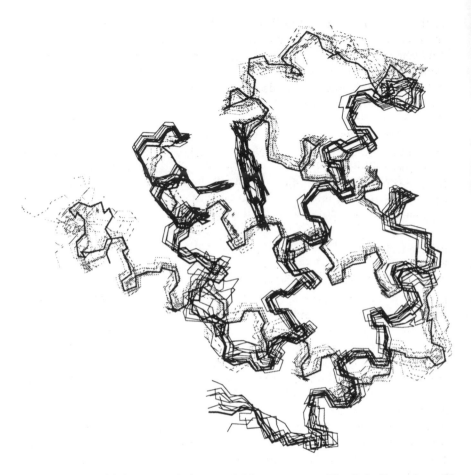

Figure 3.10 Initial sperm-whale myoglobin structures. The light lines show 12 DG/SA structures, using distance and chemical shift constraints. For comparison, the heavy line shows the X-ray structure of MbCO (Kuriyan *et al.*, 1986). The heme group and proximal histidine are in the upper centre of the picture

shifts. Further, many side-chain conformations in the vicinity of the heme could be established with some confidence from these refinements. While we do not yet have 'high-resolution' solution structures for myoglobin, the combined use of NOE, coupling constant and chemical shift restraints appears to be a promising approach.

5 Comparison of Solution and Crystal Structures

It is instructive to compare the structures determined for proteins in solution with the X-ray structures of the same or homologous proteins in the crystalline state. We discussed the case of plastocyanin above; at the present stage of refinement, no significant differences can be detected between the solution and crystal structures, and this seems to be true even in regions where there are intermolecular contacts within the crystal lattice. The agreement extends not only to the backbone conformation, but also to almost all details of the hydrogen bonding pattern and to the great majority of side-chain torsional angles. More careful refinements (against improved data sets, or making more quantitative use of cross-peak intensities) may alter this conclusion, but it is clear that the differences seen will almost certainly be minor. Broadly similar results have been obtained in comparisons of crystal and solution structures of tendamistat (Billeter *et al.*, 1989) and interleukin-1β (Clore and Gronenborn, 1991b).

In contrast, for some proteins there are clearly more significant differences between solution and crystalline behaviour. We mention briefly three examples from work in our laboratory. The first is thioredoxin from *E. coli*, where NMR structures for the reduced form of the protein have been determined; the average backbone RMS deviation from the mean structure is 0.56 Å for residues 3–108 in the NMR structures (Dyson *et al.*, 1990). The overall conformation is similar to that determined for the oxidized protein by X-ray crystallography (Holmgren *et al.*, 1975; Katti *et al.*, 1990). However, significant local conformational differences are observed that can be attributed to intermolecular contacts in the crystal. It has so far proved possible to crystallize thioredoxin only in the presence of Cu^{2+} ions that cross-link two molecules by coordination at Ser 1, Asp 2 and Asp 10. (This requirement for Cu^{2+} precludes crystallization of the reduced protein containing a dithiol pair at the active site.) There are two independent protein molecules in the unit cell that are very similar in overall conformation but differ in local details, and the rms difference between the two is 0.94 Å (Katti *et al.*, 1990). In some regions the deviation between the two independent structures in the crystal is greater than 3 Å, and is significantly larger than the spread of structures within the NMR family. The side-chain of His 6 is disordered in the crystal structure, and adopts two alternate conformations with χ_1 near $-60°$ and $180°$. Only a single conformation with χ_1 near $-60°$ is evident in solution for both oxidized and reduced thioredoxin. The NMR experiments also suggest an explanation for the requirement for Cu^{2+} ions for crystallization. In solution, residues 1–3 are clearly disordered. This is indicated both by the scatter observed in the NMR structures for the reduced protein (Dyson *et al.*, 1990) and by order parameters determined from ^{15}N relaxation measurements for both oxidation states (Stone *et al.*, 1993). This N-terminal region is ordered in the crystal, with backbone temperature factors close to or only slightly above the average (Katti *et al.*, 1990); copper binding at residues 1 and 2 presumably stabilizes this region in an ordered β-strand structure in the crystal through a cross-link to a neighbouring molecule. For thioredoxin, we conclude that both crystal packing and binding of the Cu^{2+} ions required for crystallization cause significant local perturbations in both backbone and side-chain structure.

In the case of the anaphylatoxin C3a, even more dramatic differences are seen between the solution (Chazin *et al.*, 1988a; Kalnik *et al.*, 1991) and crystal structures (Huber *et al.*, 1980). The overall conformation of the disulphide-bridged core is similar in the two environments. In the crystal a long C-terminal helix extends from Glu 47 to Ser 71; in solution this helix is frayed beyond Arg 65. The five amino acids essential for biological activity, Leu–Gly–Leu–Ala–Arg, are at the C terminus of the protein and adopt a turn-like structure in the crystalline state. In solution the last ten residues at the C terminus are dynamically disordered and there is no evidence for formation of the turn-like structure observed in the crystal. Other significant differences between the crystal and solution structures are observed at the N terminus. Residues 1–12, for which no electron density was observed in the X-ray diffraction experiments, are partly structured in solution. Residues 5–15 are helical, although the turn of helix between residues 5 and 8 appears to form only transiently. The N terminus appears to be quite flexible in solution, with dynamic fraying of the helix and with fluctuation of the helix relative to the core of the protein. Long-mixing-time NOESY experiments indicate that a transient structure is formed in which the N-terminal helix packs against the core of the protein (Chazin *et al.*, 1988a). In the crystal two C3a molecules associate closely to form a dimer with the C-terminal helices packed in an antiparallel arrangement (Huber *et al.*, 1980). It is presumably these packing interactions that stabilize the helix in the crystalline state. The packing of the molecules in the crystal would also prevent docking of an N-terminal helix onto the core of the protein.

Finally, we compare briefly the solution and crystal structures of the *B. subtilis* enzyme IIA[glc] domain. A low-resolution structure based on 1669 NOE constraints has been reported (Fairbrother *et al.*, 1992a), and refinement is in progress. It is worth noting that constraints involving backbone protons only, 815 of which were obtained in the course of making sequential assignments with ^{15}N-labelled protein, were insufficient to fold the protein correctly by distance geometry methods; additional side-chain–side-chain constraints obtained by use of ^{13}C-labelled protein were essential for calculation of a correctly folded structure. The overall fold seen in the solution structure is similar to that determined independently for *B. subtilis* IIA[glc] by X-ray crystallography (Liao *et al.*, 1991). However, a significant difference occurs at the N terminus, which is ordered in the X-ray structures and is highly disordered in solution, with order parameters (S^2) less that 0.6 for the backbone [^{15}N]–H vectors of residues 1–12 (Stone *et al.*, 1992). These differences appear not to be associated with crystal packing (Liao *et al.*, 1991).

6 Concluding Remarks

In this chapter we have described our approaches to determination of high-resolution solution structures of proteins from NMR data. It is now clear that, in favourable cases, structures can be determined at a resolution comparable to 1.5–2.0 Å resolution crystal structures. It remains to be seen just how much the quality of NMR structures can be further improved by the implementation of iterative relaxation matrix refinement procedures and by refinement against chemical shifts. At the very least, iterative NOE refinement protocols are desirable as aids in eliminating errors or overinterpretations in distance constraints. However, many problems remain to be solved, particularly with regard to conformational heterogeneity and variable order parameters arising from internal motion. Considerable work will be required before these methods can be applied routinely to NMR structure determination.

Several criteria can and should be applied to evaluate the quality of NMR structures. In the absence of a universally accepted protocol for calculating a figure of merit based on agreement between experimental NOE intensities and those calculated from the structures, these criteria are necessarily subjective. The most common criterion of 'quality' is the average root mean square deviation of a family of structures from their mean. Although this is a useful parameter, it is highly subjective and is dependent on the density and tightness assumed for the constraints as well as on the sampling procedures used. A small rmsd is sometimes found in regions with large constraint violations, perhaps as a result of the use of unrealistic upper distance bounds or due to the presence of conformational averaging, leading to mutually exclusive NOE or dihedral angle violations; an example from plastocyanin was discussed above. Thus, the rmsd by itself is inadequate as a criterion of quality of NMR structures and must be considered together with the number and size of residual constraint violations. A low potential energy for the resulting structure is also important, although it is sometimes difficult to know in advance what target should be met (see Figure 3.8). We have found it important to examine carefully constraint violations that occur in many structures in a family, even if they are small (0.1–0.2 Å), since these are often indicators of local problems in the structures. Examination of Ramachandran ϕ, ψ plots and deviations from idealized covalent geometries are also useful tests, as indeed they are for crystal structures (Morris *et al.*, 1992). Comparisons of calculated and observed chemical shifts may also prove to be of some use in evaluating structural quality. The increasing tendency to compute and report *R*-factor-like statistics should lead to improved ways of comparing structural resolution. As in crystallography, no single '*R*-factor' statistic is likely to suffice, but our experience agrees with that of others (Gonzalez *et al.*, 1991), that many reasonable statistics move in parallel, at least at current levels of refinement. Further progress in developing methods to assess the accuracy and precision of NMR structures will rest upon careful data collection procedures, increased exploitation of heteronuclear multidimensional NMR experiments, theoretical analyses of the connection between structure, dynamics and cross-relaxation rates, and the development of improved computational tools to tie these together.

Acknowledgements

Reviews like this would not be possible without the expert assistance of our collaborators whose work we cite. This work was supported by NIH grants GM45811 and HL40453 (to D.A.C.) and GM36643 and DK34909 (to P.E.W.).

References

Altman, R. B. and Jardetzky, O. (1989). Heuristic refinement method for determination of solution structure of proteins from nuclear magnetic resonance data. *Meth. Enzymol.*, **177**, 218–246

Baleja, J. D., Germann, R. W., Sande, J. H. van de and Sykes, B. D. (1990a). Solution conformation of purine–pyrimidine DNA octamers using nuclear magnetic resonance, restrained molecular dynamics and NOE-based refinement. *J. Mol. Biol.*, **215**, 411–428

Baleja, J. D., Moult, J. and Sykes, B. D. (1990b). Distance measurement and structure refinement with NOE data. *J. Magn. Reson.*, **87**, 375–389

Baleja, J. D., Pon, R. T. and Sykes, B. D. (1990c). The solution structure of phage λ half-operator DNA using NMR, restrained molecular dynamics and NOE-based refinement. *Biochemistry*, **29**, 4828–4839

Bax, A., Clore, G. M., Driscoll, P. C., Gronenborn, A. M., Ikura, M. and Kay, L. E. (1990a). Practical aspects of proton–carbon–carbon–proton three-dimensional correlation spectroscopy of ^{13}C-labeled proteins. *J. Magn. Reson.*, **87**, 620–627

Bax, A., Clore, G. M. and Gronenborn, A. M. (1990b). ^{1}H–^{1}H correlation via isotopic mixing of ^{13}C magnetization, a new three-dimensional approach for assigning ^{1}H and ^{13}C spectra of ^{13}C enriched proteins. *J. Magn. Reson.*, **88**, 425–431

Berendsen, H. J. C., Postma, J. P. M., Gunsteren, W. F. van, DiNola, A. and Haak, J. R. (1984). Molecular dynamics with coupling to an external bath. *J. Chem. Phys.*, **81**, 3684–3690

Billeter, M., Braun, W. and Wüthrich, K. (1982). Sequential resonance assignments in protein ^{1}H nuclear magnetic resonance spectra, computation of sterically allowed proton–proton distances, and statistical analysis of proton–proton distances in single crystal protein conformations. *J. Mol. Biol.*, **155**, 321–346

Billeter, M., Kline, A. D., Braun, W., Huber, R. and Wüthrich, K. (1989). Comparison of the high-resolution structures of the α-amylase inhibitor tendamistat determined by nuclear magnetic resonance in solution and by X-ray diffraction in single crystals. *J. Mol. Biol.*, **206**, 677–687

Blaney, J. M., Crippen, G. M., Dearing, A. and Dixon, J. S. (1990). *DGEOM—Distance Geometry*. Quantum Chemistry Program Exchange, Bloomington, Indiana

Blumenthal, L. (1953). *Theory and Applications of Distance Geometry*. Cambridge University Press, Cambridge

Boelens, R., Koning, T. M. G. and Kaptein, R. (1988). Determination of biomolecular structures from proton–proton NOE's using a relaxation matrix approach. *J. Mol. Struct.*, **173**, 299–311

Bonvin, A. M. J. J., Boelens, R. and Kaptein, R. (1991). Direct NOE refinement of biomolecular structures using 2D NMR data. *J. Biomol. NMR*, **1**, 305–309

Borgias, B. A., Gochin, M., Kerwood, D. J. and James, T. L. (1990). Relaxation matrix analysis of 2D nmr data. *Prog. NMR Spectrosc.*, **22**, 83–100

Braun, W. (1987). Distance geometry and related methods for protein structure determination from NMR data. *Quart. Rev. Biophys.*, **19**, 115–157

Brünger, A. T. (1992). *X-PLOR Manual, Version 3.0. A System for Crystallography and NMR*. Yale University

Brünger, A. T. and Karplus, M. (1991). Molecular dynamics simulations with experimental restraints. *Acc. Chem. Res.*, **24**, 54–61

Brüschweiler, R., Roux, B., Blackledge, M., Griesinger, C., Karplus, M. and Ernst, R. R. (1992). Influence of rapid intramolecular motion on NMR cross-relaxation rates. A molecular dynamics study of antamanide in solution. *J. Am. Chem. Soc.*, **114**, 2289–2302

Buckingham, A. D. (1960). Chemical shifts in the nuclear magnetic resonance spectra of molecules containing polar groups. *Can. J. Chem.*, **38**, 300–307

Bull, T. E. (1987). Cross-correlation and 2D NOE spectra. *J. Magn. Reson.*, **72**, 397–413

Chary, K. V. R., Otting, G. and Wüthrich, K. (1991). Measurement of small heteronuclear ^{1}H–^{15}N coupling constants in ^{15}N-labelled proteins by 3D $H_N NH_{AB}$–COSY. *J. Magn. Reson.*, **93**, 218–224

Chazin, W. J., Hugli, T. E. and Wright, P. E. (1988a). ^{1}H studies of human C3a anaphylatoxin in solution: Sequential resonance assignments, secondary structure, and global fold. *Biochemistry*, **27**, 9139–9148

Chazin, W. J., Rance, M. and Wright, P. E. (1988b). Complete assignment of the ^{1}H nuclear magnetic resonance spectrum of French bean plastocyanin. Application of an integrated approach to spin system identification in proteins. *J. Mol. Biol.*, **202**, 603–622

Chazin, W. J. and Wright, P. E. (1988). Complete assignment of the ^{1}H nuclear magnetic resonance spectrum of French bean plastocyanin. Sequential resonance assignments, secondary structure and global fold. *J. Mol. Biol.*, **202**, 623–636

Clore, G. M. and Gronenborn, A. M. (1985). Assessment of errors involved in determination of interproton distance ratios and distances by means of one- and two-dimensional NOE measurements. *J. Magn. Reson.*, **61**, 158–164

Clore, G. M. and Gronenborn, A. M. (1989). Determination of the three-dimensional structures of proteins and nucleic acids in solution by nuclear magnetic resonance spectroscopy. *CRC Crit. Rev. Biochem. Mol. Biol.*, **24**, 479–564

Clore, G. M. and Gronenborn, A. M. (1991a). Applications of three- and four-dimensional heteronuclear NMR spectroscopy to protein structure determination. *Prog. NMR Spectrosc.*, **23**, 43–92

Clore, G. M. and Gronenborn, A. M. (1991b). Comparison of the solution nuclear magnetic resonance and X-ray crystal structures of human recombinant interleukin-1β. *J. Mol. Biol.*, **221**, 47–53

Crippen, G. M. and Havel, T. F. (1988). *Distance Geometry and Molecular Conformation.* Research Studies Press, Taunton, UK

Cross, K. J. and Wright, P. E. (1985). Calibration of ring-current models for the heme ring. *J. Magn. Reson.*, **64**, 220–231

Dalgarno, D. C., Levine, B. A. and Williams, R. J. P. (1983). Structural information from NMR secondary chemical shifts of peptide α-CH protons in proteins. *Biosci. Rep.*, **3**, 443–452

Dalvit, C. and Wright, P. E. (1987). Assignment of resonances in the ^{1}H nuclear magnetic resonance spectrum of the carbon monoxide complex of sperm whale myoglobin by phase-sensitive two-dimensional techniques. *J. Mol. Biol.*, **194**, 313–327

Dammkoehler, R. A., Karasek, S. F., Shands, E. F. B. and Marshall, G. R. (1989). Constrained search of conformational hyperspace. *J. Comput.-aided Mol. Des.*, **3**, 3–21

Driscoll, P. C., Clore, G. M., Marion, D., Wingfield, P. T. and Gronenborn, A. M. (1990). Complete resonance assignment for the polypeptide backbone of interleukin-1β using three-dimensional heteronuclear NMR spectroscopy. *Biochemistry*, **29**, 3542–3556

Duben, A. J. and Hutton, W. C. (1990). Spatial aspects of homonuclear, proton NMR cross-relaxation. I. The effects of molecular shape and internal motion. *J. Am. Chem. Soc.*, **112**, 5917–5924

Dyson, H. J., Gippert, G. P., Case, D. A., Holmgren, A. and Wright, P. E. (1990). Three-dimensional solution structure of the reduced form of *Escherichia coli* thioredoxin determined by nuclear magnetic resonance spectroscopy. *Biochemistry*, **29**, 4129–4136

Dyson, H. J., Holmgren, A. and Wright, P. E. (1989). Assignment of the proton NMR spectrum of reduced and oxidized thioredoxin: Sequence-specific assignments, secondary structure and global fold. *Biochemistry*, **28**, 7074–7087

Dyson, H. J., Tennant, L. L. and Holmgren, A. (1991). Proton-transfer effects in the active-site region of *Escherichia coli* thioredoxin using two-dimensional ^{1}H NMR. *Biochemistry*, **30**, 4262–4268

Fairbrother, W. J., Cavanagh, J., Dyson, H. J., Palmer, A. G., III, Sutrina, S. L., Reizer, J., Saier, M. H., Jr. and Wright, P. E. (1991). Polypeptide backbone resonance assignments and secondary structure of *Bacillus subtitlis* enzyme III[glc] determined by two-dimensional and three-dimensional heteronuclear NMR spectroscopy. *Biochemistry*, **30**, 6896–6907

Fairbrother, W. J., Gippert, G. P., Reizer, J., Saier, M. H., Jr. and Wright, P. E. (1992a). Low resolution structure of the *Bacillus subtilis* glucose permease IIA domain derived from heteronuclear three-dimensional NMR spectroscopy. *FEBS Lett.*, **296**, 148–152

Fairbrother, W. J., Palmer, A. G., III, Rance, M., Reizer, J., Saier, M. H., Jr. and Wright, P. E. (1992b). Assignment of the aliphatic ^1H and ^{13}C resonances of the *Bacillus subtitlis* glucose permease IIA domain using double- and triple-resonance heteronuclear three-dimensional NMR spectroscopy. *Biochemistry*, **31**, 4413–4425

Fesik, S. W. and Zuiderweg, E. R. P. (1990). Heteronuclear three-dimensional NMR spectroscopy of isotopically labelled biological macromolecules. *Quart. Rev. Biophys.*, **23**, 97–131

Flygare, W. H. (1974). Magnetic interactions in molecules and an analysis of molecular electronic charge distribution from magnetic parameters. *Chem. Rev.*, **74**, 653–687

Gippert, G. P., Yip, P. F., Wright, P. E. and Case, D. A. (1990). Computational methods for determining protein structures from NMR data. *Biochem. Pharmacol.*, **40**, 15–22

Gonzalez, C., Rullmann, J. A. C., Bonvin, A. M. J. J., Boelens, R. and Kaptein, R. (1991). Toward an NMR R factor. *J. Magn. Reson.*, **91**, 659–664

Grant, D. M., Mayne, C. L., Liu, F. and Xiang, T.-X. (1991). Spin–lattice relaxation of coupled nuclear spins with application to molecular motion in liquids. *Chem. Rev.*, **91**, 1591–1624

Güntert, P., Braun, W., Billeter, M. and Wüthrich, K. (1989). Automated stereospecific ^1H NMR assignments and their impact on the precision of protein structure determinations in solution. *J. Am. Chem. Soc.*, **111**, 3997–4004

Guss, J. M., Harrowell, P. R., Murata, M., Norris, V. A. and Freeman, H. C. (1986). Crystal structure analysis of reduced (CuI) poplar plastocyanin at six pH values. *J. Mol. Biol.*, **192**, 361–387

Haigh, C. W. and Mallion, R. B. (1980). Ring current theories in nuclear magnetic resonance. *Prog. NMR Spectrosc.*, **13**, 303–344

Harris, R. K. (1986). *Nuclear Magnetic Resonance Spectroscopy, A Physicochemical View.* Longmans, Harlow, UK

Havel, T. F. (1990). The sampling properties of some distance geometry algorithms applied to unconstrained polypeptide chains. A study of 1830 independently computed conformations. *Biopolymers*, **29**, 1565–1585

Havel, T. F. (1991). An evaluation of computational strategies for use in the determination of protein structure from distance constraints obtained by nuclear magnetic resonance. *Prog. Biophys. Mol. Biol.*, **56**, 43–78

Havel, T. F., Kuntz, I. D. and Crippen, G. M. (1983). The theory and practice of distance geometry. *Bull. Math. Biol.*, **45**, 665–720

Havel, T. and Wüthrich, K. (1984). A distance geometry program for determining the structures of small proteins and other macromolecules from nuclear magnetic resonance measurements of intramolecular $^1H–^1H$ proximities in solution. *Bull. Math. Biol.*, **46**, 673–698

Holmgren, A., Söderberg, B.-O., Ecklund, H. and Bränden, C.-I. (1975). Three-dimensional structure of *Escherichia coli* thioredoxin-S_2 to 2.8 Å resolution. *Proc. Natl Acad. Sci. USA*, **72**, 2305–2309

Huber, R., Schloze, H. and Deisenhoffer, J. (1980). Crystal structure analysis and molecular model of human C3a anaphylatoxin. *Hoppe-Seyler's Z. Physiol. Chem.*, **361**, 1389–1399

Kalk, A. and Berendsen, H. J. C. (1976). Proton magnetic relaxation and spin diffusion in proteins. *J. Magn. Reson.*, **24**, 343–366

Kalnik, M. W., Chazin, W. J. and Wright, P. E. (1991). The three-dimensional solution structure of human anaphylatoxin C3a. In Villafranca, J. J. (Ed.), *Techniques in Protein Chemistry II.* Academic Press, San Diego, pp. 393–400

Kaptein, R., Zuiderweg, E. R. P., Scheek, R. M., Boelens, R. and Gunsteren, W. F. van (1985). A protein structure from nuclear magnetic resonance data. Lac repressor headpiece. *J. Mol. Biol.*, **182**, 179–182

Katti, S. K., LeMaster, D. M. and Eklund, H. (1990). Crystal structure of thioredoxin from *E. coli* at 1.68 Å resolution. *J. Mol. Biol.*, **212**, 167–184

Kay, L. E., Ikura, M., Tschudin, R. and Bax, A. (1990). Three-dimensional triple resonance NMR spectroscopy of isotopically-enriched proteins. *J. Magn. Reson.*, **89**, 496–514

Keepers, J. W. and James, T. L. (1984). A theoretical study of distance determinations from NMR. Two-dimensional nuclear Overhauser effect spectra. *J. Magn. Reson.*, **57**, 404–426

Koehl, P. and Lefevre, J.-F. (1990). The reconstruction of the relaxation matrix from an incomplete set of nuclear Overhauser effects. *J. Magn. Reson.*, **86**, 565–583

Kominos, D., Suri, A. K., Kitchen, D. B., Bassolino, D. and Levy, R. M. (1992). Simulating the effect of the two-spin approximation on the generation of protein structures from NOE data. *J. Magn. Reson.*, **97**, 398–410

Koning, T. M. G., Boelens, R. and Kaptein, R. (1990). Calculation of the nuclear Overhauser effect and the determination of proton–proton distances in the presence of internal motions. *J. Magn. Reson.*, **90**, 111–123

Kuntz, I. D., Kosen, P. A. and Craig, E. C. (1991). Amide chemical shifts in many helices in peptides and proteins are periodic. *J. Am. Chem. Soc.*, **113**, 1406–1408

Kuriyan, J., Wilz, S., Karplus, M. and Petsko, G. A. (1986). X-ray structure and refinement of carbon-monoxy Fe(II) myoglobin at 1.5 Å resolution. *J. Mol. Biol.*, **192**, 133–154

Kuszewski, J., Nilges, M. and Brünger, A. T. (1992). Sampling and efficiency of metric matrix distance geometry: A novel partial metrization algorithm. *J. Biomol. NMR*, **2**, 33–56

Lane, A. N. (1988). The influence of spin diffusion and internal motions on NOE intensities in proteins. *J. Magn. Reson.*, **78**, 425–439

Lee, M. S., Gippert, G. P., Soman, K. V., Case, D. A. and Wright, P. E. (1989). Three-dimensional solution structure of a single zinc finger DNA-binding domain. *Science*, **245**, 635–637

Lee, M. S., Gippert, G. P., Soman, K. V., Case, D. A. and Wright, P. E. (1990). Proton nuclear magnetic resonance assignments and solution structure of a synthetic zinc finger from *Xfin*. In Sarma, R. H. and Sarma, M. H. (Eds), *Structure and Methods*, Vol. 2: *DNA Protein Complexes and Proteins*. Adenine Press, Albany, NY, pp. 83–91

LeMaster, D. M., Kay, L. E., Brünger, A. T. and Prestegard, J. H. (1988). Protein dynamics and distance determination by NOE measurements. *FEBS Lett.*, **236**, 71–76

Liao, D. -I., Kapadia, G., Reddy, P., Saier, M. H., Jr., Reizer, J. and Herzberg, O. (1991). Structure of the IIA domain of the glucose permease of *Bacillus subtilis* at 2.2-Å resolution. *Biochemistry*, **30**, 9583–9594

London, R. E. (1989). Interpreting protein dynamics with nuclear magnetic resonance relaxation measurements. *Meth. Enzymol.*, **176**, 358–375

McConnell, H. M. (1957). Theory of nuclear magnetic shielding in molecules. I. Long-range dipolar shielding of protons. *J. Chem. Phys.*, **27**, 226–229

Macura, S. and Ernst, R. R. (1980). Elucidation of cross relaxation in liquids by two-dimensional N.M.R. spectroscopy. *Mol. Phys.*, **41**, 95–117

Madrid, M. and Jardetzky, O. (1988). Comparison of experimentally determined protein structures by solution of Bloch equations. *Biochem. Biophys. Acta*, **953**, 61–69

Mertz, J. E., Güntert, P., Wüthrich, K. and Braun, W. (1991). Complete relaxation matrix refinement of NMR structures of proteins using analytically calculated dihedral angle derivatives of NOE intensities. *J. Biomol. NMR*, 1, 257–269

Moore, J. M., Case, D. A., Chazin, W. J., Gippert, G. P., Powls, R. and Wright, P. E. (1988). Three-dimensional solution structure of plastocyanin from the green alga *Scenedesmus obliquus*. *Science*, 240, 314–317

Moore, J. M., Lepre, C., Gippert, G. P., Chazin, W. J., Case, D. A. and Wright, P. E. (1991). High-resolution solution structure of reduced French bean plastocyanin and comparison with the crystal structure of poplar plastocyanin. *J. Mol. Biol.*, 221, 533–555

Morris, A. L., MacArthur, M. W., Hutchinson, E. G. and Thornton, J. M. (1992). Stereochemical quality of protein structural coodinates. *Proteins Struct. Funct. Genet.*, 12, 345–364

Nerdal, W., Hare, D. R. and Reid, B. R. (1989). Solution structure of the *Eco*RI DNA sequence: Refinement of NMR-derived distance geometry structures by NOESY spectrum back calculations. *Biochemistry*, 28, 10008–10021

Neri, D., Otting, G. and Wüthrich, K. (1990). New nuclear magnetic resonance experiment for measurements of the vicinal coupling constants $^3J_{HN\alpha}$ in proteins. *J. Am. Chem. Soc.*, 112, 3663–3665

Nilges, M., Clore, G. M. and Gronenborn, A. M. (1988). Determination of three-dimensional structures of proteins from interproton distance data by hybrid distance geometry–dynamical simulated annealing calculations. *FEBS Lett.*, 239, 317–324

Nilges, M., Clore, G. M. and Gronenborn, A. M. (1990). ^{1}N–NMR stereospecific assignments by conformational data-base searches. *Biopolymers*, 29, 813–822

Nilges, M., Habazettl, J., Brünger, A. T. and Holak, T. A. (1991). Relaxation matrix refinement of the solution structure of squash trypsin inhibitor. *J. Mol. Biol.*, 219, 499–510

Oldfield, E., Guo, K., Augspurger, J. D. and Dykstra, C. E. (1991). A molecular model for the major conformational substrates in heme proteins. *J. Am. Chem. Soc.*, 113, 7537–7541

Olejniczak, E. T. (1989). Including methyl rotation in simulations of spin–lattice relaxation experiments. *J. Magn. Reson.*, 81, 392–394

Olejniczak, E. T., Dobson, C. M., Karplus, M. and Levy, R. M. (1984). Motional averaging of proton nuclear Overhauser effects in proteins. Predictions from a molecular dynamics simulation of lysozyme. *J. Am. Chem. Soc.*, 106, 1923–1930

Olejniczak, E. T., Gampe, R. T., Jr. and Fesik, S. W. (1986). Accounting for spin diffusion in the analysis of 2D NOE data. *J. Magn. Reson.*, 67, 28–41

Osapay, K. and Case, D. A. (1991). A new analysis of proton chemical shifts in proteins. *J. Am. Chem. Soc.*, 113, 9436–9444

Oshior, C. M., Thomason, J. and Kuntz, I. D. (1991). Effects of limited input distance constraints upon the distance geometry algorithm. *Biopolymers*, 31, 1049–1064

Palmer, A. G. and Case, D. A. (1992). Molecular dynamics analysis of NMR relaxation in a zinc-finger peptide. *J. Am. Chem. Soc.*, 114, 9059–9067

Palmer, A. G., III, Fairbrother, W. J., Cavanagh, J., Wright, P. E. and Rance, M. (1992). Improved resolution in three-dimensional constant-time triple-resonance NMR spectroscopy of proteins. *J. Biomol. NMR*, 2, 103–108

Palmer, A. G., Rance, M. and Wright, P. E. (1991). Intramolecular motions of a zinc finger DNA-binding domain from xfin characterized by proton-detected natural abundance ^{13}C heteronuclear NMR spectroscopy. *J. Am. Chem. Soc.*, **113**, 4371–4380

Pastore, A. and Saudek, V. (1990). The relationship between chemical shift and secondary structure in proteins. *J. Magn. Reson.*, **90**, 165–176

Pauling, L. (1979). Diamagnetic anisotropy of the peptide group. *Proc. Natl Acad. Sci. USA*, **76**, 2293–2294

Pavletich, N. P. and Pabo, C. O. (1991). Zinc finger-DNA recognition: Crystal structure of a Zif268–DNA complex at 2.1 Å. *Science*, **252**, 809–817

Pearlman, D. A., Case, D. A., Caldwell, J. C., Seibel, G. L., Singh, U. C., Weiner, P. and Kollman, P. A. (1991). *AMBER 4.0*. University of California, San Francisco

Post, C. B. (1992). Internal motional averaging and three-dimensional structure determination by NMR. *J. Mol. Biol.*, **224**, 1087–1101

Powers, R., Gronenborn, A. M., Clore, G. M. and Bax, A. (1991). Three dimensional triple resonance NMR of ^{13}C/^{15}N-enriched proteins using constant-time evolution. *J. Magn. Reson.*, **94**, 209–213

Schmalz, T. G., Norris, C. L. and Flygare, W. H. (1973). Localized magnetic susceptibility anisotropies. *J. Am. Chem. Soc.*, **95**, 7961–7967

Spera, S. and Bax, A. (1991). Empirical correlation between protein backbone conformation and Cα and Cβ ^{13}C nuclear magnetic resonance chemical shifts. *J. Am. Chem. Soc.*, **113**, 5490–5492

Stawarz, B., Genest, M. and Genest, D. (1992). A new constraint potential for the structure refinement of biomolecules in solution using experimental nuclear Overhauser effect intensity. *Biopolymers*, **32**, 633–642

Stone, M. J., Chandrasekhar, I., Holmgren, A., Wright, P. E. and Dyson, H. J. (1993). Comparison of backbone and tryptophan side-chain dynamics of reduced and oxidized *Escherichia coli* thioredoxin using ^{15}N NMR relaxation measurements. *Biochemistry*, **32**, 426–435

Stone, M. J., Fairbrother, W. J., Palmer, A. G., III, Reizer, J., Saier, M. H., Jr. and Wright, P. E. (1992). Backbone dynamics of the *Bacillus subtilis* glucose permease IIA domain determined from ^{15}N NMR relaxation measurements. *Biochemistry*, **31**, 4394–4406

Szilágyi, L. and Jardetzky, O. (1989). α-proton chemical shifts and secondary structure in proteins. *J. Magn. Reson.*, **83**, 441–449

Theriault, Y., Pochapsky, T. C., Chiu, M., Sligar, S. G. and Wright, P. E. (1993). Resonance assignments for the carbon monoxide complex of sperm whale myoglobin determined by two and three dimensional NMR spectroscopy (submitted)

Thomas, P. D., Basus, V. J. and James, T. L. (1991). Protein solution structure determination using distances from two-dimensional nuclear Overhauser effect experiments: Effect of approximations on the accuracy of derived structures. *Proc. Natl Acad. Sci. USA*, **88**, 1237–1241

Tigelaar, H. L. and Flygare, W. H. (1972). Molecular Zeeman effect in formamide and the α-proton chemical shift in poly(L-alanine). *J. Am. Chem. Soc.*, **94**, 343–346

Torchia, D. A. (1984). Solid state NMR studies of protein internal dynamics. *Ann. Rev. Biophys. Bioeng*, **13**, 125–144

Torda, A. E., Scheek, R. M. and VanGunsteren, W. F. (1989). Time-dependent distance restraints in molecular dynamics simulations. *Chem. Phys. Lett.*, **157**, 289–294

Tropp, J. (1980). Dipolar relaxation and nuclear Overhauser effects in nonrigid molecules: The effect of fluctuating internuclear distances. *J. Chem. Phys.*, **72**, 6035–6043

Weiner, S. J., Kollman, P. A., Nguyen, D. T. and Case, D. A. (1986). An all-atom force field for simulations of proteins and nucleic acids. *J. Computat. Chem.*, **7**, 230–252

Weiss, G. H., Kiefer, J. E. and Ferritti, J. A. (1992). Accuracy and precision in the estimation of internuclear distances for structure determinations. *J. Magn. Reson.*, **97**, 227–234

Williamson, M. P. (1990). Secondary-structure dependent chemical shifts in proteins. *Biopolymers*, **29**, 1423–1431

Williamson, M. P. and Asakura, T. (1991). Calculation of chemical shifts of protons on alpha carbons in proteins. *J. Magn. Reson.*, **94**, 557–562

Williamson, M. P., Asakura, T., Nakamura, E. and Demura, M. (1992). A method for the calculation of protein α-CH chemical shifts. *J. Biomol. NMR*, **2**, 83–98

Williamson, M. P., Havel, T. F. and Wüthrich, K. (1985). Solution conformation of proteinase inhibitor IA for bull seminal plasma by ^{1}H nuclear magnetic resonance and distance geometry. *J. Mol. Biol.*, **182**, 295–315

Wishart, D. S., Sykes, B. D. and Richards, F. M. (1991). Relationship between nuclear magnetic resonance chemical shift and protein secondary structure. *J. Mol. Biol.*, **222**, 311–333

Wishart, D. S., Sykes, B. D. and Richards, F. M. (1992). The chemical shift index: A fast and simple method for the assignment of protein secondary structure through NMR spectroscopy. *Biochemistry*, **31**, 1647–1651

Withka, J., Swaminathan, S., Beveridge, D. L. and Bolton, P. H. (1991). Time dependence of nuclear Overhauser effects of duplex DNA from molecular dynamics trajectories. *J. Am. Chem. Soc.*, **113**, 5041–5049

Wittebort, R. J. and Szabo, A. (1978). Theory of NMR relaxation in macromolecules: Restricted diffusion and jump models for multiple internal rotations in amino acid side chains. *J. Chem. Phys.*, **69**, 1722–1736

Woessner, D. E. (1965). Nuclear magnetic dipole–dipole relaxation in molecules with internal motion. *J. Chem. Phys.*, **42**, 1855–1859

Wüthrich, K. (1989). Protein structure determination in solution by nuclear magnetic resonance spectroscopy. *Science*, **243**, 45–50

Wüthrich, K., Billeter, M. and Braun, W. (1983). Pseudostructures for the 20 common amino acids for use in studies of protein conformations by measurements of intramolecular proton–proton distance constraints with nuclear magnetic resonance. *J. Mol. Biol.*, **169**, 949–961

Yip, P. F. (1989). Calculating NOESY intensities by perturbation expansion. *Chem. Phys. Lett.*, **161**, 50–54

Yip, P. and Case, D. A. (1989). A new method for refinement of macromolecular structures based on nuclear Overhauser effect spectra. *J. Magn. Reson.*, **83**, 643–648

Yip, P. and Case, D. A. (1991). Incorporation of internal motion in NMR refinements based on NOESY spectra. In Hoch, J. (Ed.), *Computational Aspects of the Study of Biological Macromolecules by NMR Spectroscopy*. Plenum Press, New York, pp. 317–330

4

Multidimensional NMR Studies of Immunosuppressant/Immunophilin Complexes

Stephen W. Fesik and Placido Neri

1 Introduction

Biology and Chemistry of the Immunosuppressants and Immunophilins

Cyclosporin A (CsA) and FK506 are two chemically unrelated immuno-suppressants (Figure 4.1) that act in a similar way by blocking specific signal transduction pathways at an early stage of T-cell activation. They inhibit the transcription of genes that produce interleukins (IL-2, IL-3, IL-4), granulocyte-macrophage colony stimulating factor, and inter-feron-γ (Bierer *et al.*, 1990a; Dumont *et al.*, 1990a) by inhibiting specific transcriptional regulators such as nuclear factor of activated T cells (NF-AT) (Emmel *et al.*, 1989). Although CsA and FK506 have similar biological activities, they bind to different cellular receptors, termed immunophilins. CsA binds to cyclophilin (17.8 kD, 165 aa) (Handschu-macher *et al.*, 1984), and FK506 binds to FKBP (11.8 kD, 107 aa) (Harding *et al.*, 1989; Siekierka *et al.*, 1989). Both cyclophilin and FKBP are peptidyl–prolyl *cis-trans* isomerases (PPIase) that are inhibited by CsA (Fischer *et al.*, 1989; Takahashi *et al.*, 1989) or FK506 (Harding *et al.*, 1989; Siekierka *et al.*, 1989), respectively. In an initial attempt to explain the immunosuppressant activity of CsA and FK506, it was hypothesized that PPIase inhibition was the cause of immunosuppression by inhibiting the folding of proteins involved in the early stages of T-cell activation. However, several pieces of evidence now suggest that the inhibition of PPIase activity is insufficient to cause an immunosup-pressive effect (Bierer *et al.*, 1990b). Instead, it has been proposed that it is the CsA/CyP and immunosuppressant/FKBP complexes that

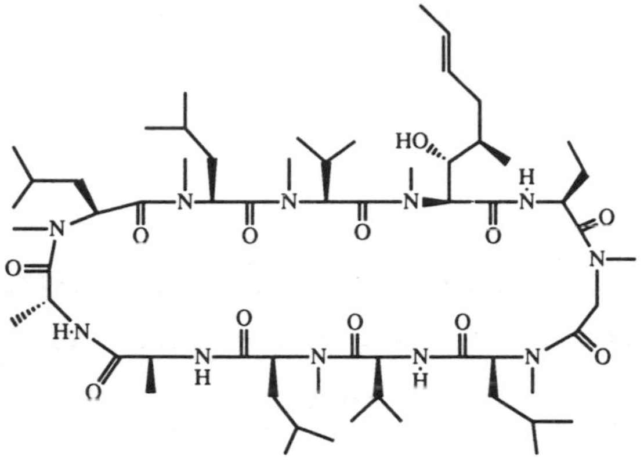

Cyclosporin A

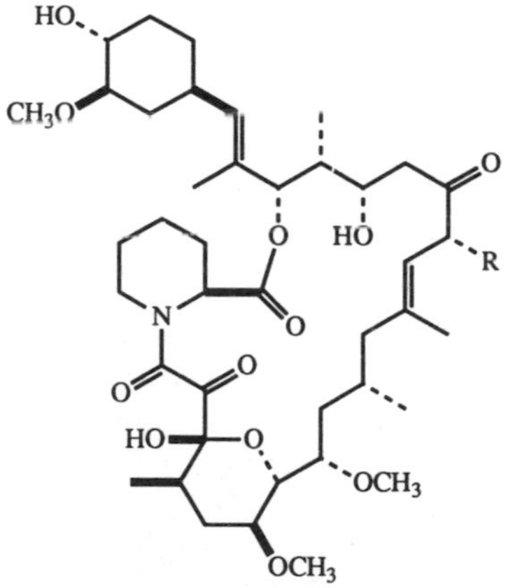

FK506 R=$CH_2CH{=}CH_2$
Ascomycin R=CH_2CH_3

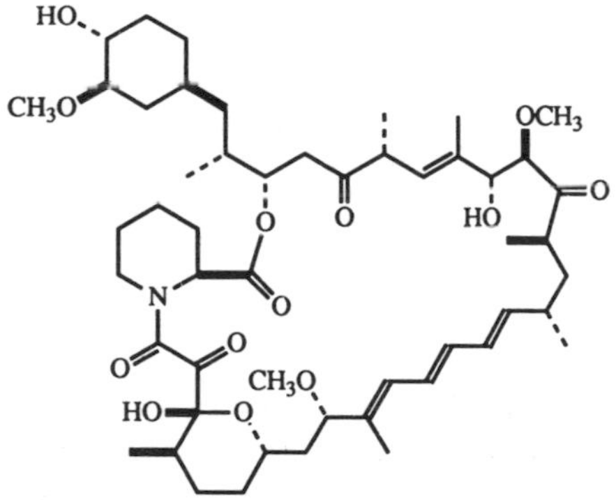

Rapamycin

Figure 4.1 Structures of immunosuppressants

are the biological effectors involved in immunosuppression by binding to proteins involved in T-cell activation (Schreiber, 1991). Indeed, it was recently discovered (Liu *et al.*, 1991) that the CsA/CyP and FK506/FKBP complexes bind to and inhibit the calcium- and calmodulin-dependent phosphatase, calcineurin; whereas CsA, CyP, FK506 or FKBP alone do not bind to this protein. Further evidence to support the relevance of calcineurin inhibition in immunosuppression is the demonstration (Liu *et al.*, 1992) of a strong correlation between the immunosuppressive activity, calcineurin inhibition and inhibition of the transcriptional activation of NF-AT by CsA and FK506 analogues whose immunosuppressive activities did not correlate with their binding affinity for the immunophilins. Figure 4.2 schematically depicts the currently accepted pathway for CsA- and FK506-mediated immunosuppression.

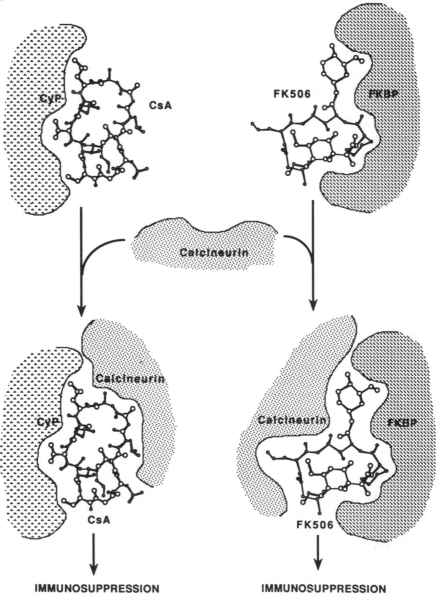

Figure 4.2   Currently accepted pathway for CsA- and FK506-mediated immunosuppression

In contrast to CsA and FK506, rapamycin, another immunosuppressant which is structurally similar to FK506 (Figure 4.1) and binds to FKBP, does not inhibit IL-2 production and expression. Rapamycin acts at a later stage in T-cell activation by inhibiting the response to IL-2 (Bierer *et al.*, 1990b; Dumont *et al.*, 1990b). These differences in the immunosuppressive activities displayed by rapamycin as compared with CsA and FK506 are consistent with the inability of the FKBP/rapamycin complex or rapamycin alone to bind to calcineurin, and suggest that a different protein may bind to the FKBP/rapamycin complex. Alternatively, the different biological activities displayed by FK506 and rapamycin may be explained by their selectivity for binding to different FKBPs. Recently a 25 kDa FKBP was isolated that binds to rapamycin two orders of magnitude better than FK506 (Galat *et al.*, 1992). However, additional work is required to prove whether or not the differences in binding affinities have any functional significance.

Rationale for the Structural Studies

Although cyclosporin A is currently the therapeutic agent of choice for the prevention of graft rejection following organ and bone marrow transplantation (Kahan, 1989), CsA's toxic side-effects, which include nephrotoxicity, hepatotoxicity and CNS disturbances, limit its clinical utility and make cyclosporin unsuitable for the treatment of autoimmune diseases. FK506 is more potent than CsA (Thomson, 1990); however, it also suffers from toxicity problems (Thomson, 1990).

Structural information on immunosuppressant/immunophilin complexes could be a valuable aid in the design of CsA or FK506 analogues with reduced toxicity and potentially greater clinical utility. In particular, it would be important to determine the conformation of these immunosuppressants when bound to their receptors (immunophilins) and to identify those portions of CsA and FK506 that are required for binding. This information may aid in the design of new molecules with different molecular frameworks that have the functional groups for binding held in their experimentally determined spatial orientation (Fesik, 1989, 1991). These new molecules could have better pharmacological profiles, less toxic side-effects (if not linked to the immunosuppressive effects) or improved catabolic resistance with higher bioavailability, or may be easier or less expensive to synthesize.

Complementary to the structural information on regions of the ligand that bind to the receptor is the identification of the solvent-exposed portions of the ligand. Typically, this information could be used in the design of analogues that are modified at these sites to improve their physical properties (e.g. water solubility), without affecting their binding affinity (Fesik, 1991). However, for the CsA/CyP and immunosuppressant/FKBP complexes, those portions of the ligands that are exposed to solvent may be very important for their function (Schreiber, 1991) by interacting with other proteins such as calcineurin (Liu *et al.*, 1991).

In addition to structural information on the ligand, it is also important to obtain information about the structure of the receptor when bound to the ligand by determining the complete three-dimensional structure of immunosuppressant/immunophilin complexes. Information on the structure of the binding pocket could be used in the *de novo* design of analogues that are very different from the parent molecules. This might be accomplished by characterizing the structural features of the binding site, using the GRID (Goodford, 1985) or DOCK (Kuntz *et al.*, 1982) algorithms, followed by a 3D database search using programs such as ALADDIN (Van Drie *et al.*, 1989) or CAVEAT (Bartlett *et al.*, 1989) for compounds that have the proper functional groups and geometric properties for binding to this site. The complete 3D structure may also be useful to locate additional binding sites that would not be revealed in structural studies focused on the ligand and may be used to explain differences in activity between structurally similar analogues.

General Strategy

In order for structural information to be of value in structure-based drug design, it must be obtained rapidly to impact the project at an early stage. Information that can be obtained rapidly by NMR is the bound conformation of the ligand and the identification of those portions of the ligand that interact with its receptor. In these studies the proton NMR signals of the ligand are selectively observed from the bulk of the protein resonances, which dramatically simplifies the analysis of the spectra and allows the structural information to be rapidly obtained. The selective observation of the ligand can be achieved by labelling one of the components of the drug/receptor complex. Several approaches have been proposed for studying drug/receptor complexes by NMR, using isotope labelling (Fesik, 1989, 1991). Figure 4.3 depicts four possible schemes of isotope labelling for studying drug/receptor complexes (Fesik, 1991). In one approach (Figure 4.3A), the ligand is labelled with ^{13}C or ^{15}N, and the proton NMR signals of the isotopically labelled ligand are selectively observed by isotope-edited 2D (Fesik *et al.*, 1988; Wider *et al.*, 1990) or heteronuclear 3D NMR techniques (Fesik and Zuiderweg, 1988; Marion *et al.*, 1989b). Alternatively, the ligand could be deuterated (Figure 4.3B) and structural information obtained from 2D NOE difference spectra recorded on complexes with protonated or deuterated ligands (Fesik and Zuiderweg, 1989). In those cases in which it is too difficult or too expensive to label the ligand, the receptor could be perdeuterated to allow for the selective observation

Figure 4.3 Possible labelling strategies used in NMR studies of molecular complexes

of the ligand (Figure 4.3C) (Seeholzer *et al.*, 1986). Although information cannot be obtained on those portions of the ligand that contact the receptor when this approach is used, the method has its advantages for determining the bound conformation of a series of analogues (Fesik, 1991). Finally, the receptor could be ^{13}C- and ^{15}N-labelled (Figure 4.3D), and the unlabelled ligand signals selectively observed by isotope-filtering techniques that eliminate the NMR signals of the protons attached to the ^{13}C- and ^{15}N-labelled nuclei of the protein (Wider *et al.*, 1990; Gemmecker *et al.*, 1992; Ikura and Bax, 1992).

In our initial investigations on the conformation of the immunosuppressants when bound to their binding proteins, we chose to study ^{13}C-labelled ligands complexed with unlabelled immunophilins. After assigning the NMR signals of the bound ligand, the bound conformation was determined from intramolecular proton–proton distance restraints derived from NOEs obtained in heteronuclear 3D NOE spectra. Intermolecular NOEs were used to identify those regions of the ligand in close proximity to the protein. This was followed by the identification of the solvent-exposed regions of the receptor-bound ligand by comparing the proton relaxation rates of the ligand in the presence and absence of a paramagnetic agent. More recently (Petros *et al.*, 1992b), we have studied

unlabelled ligands bound to uniformly ^{13}C- and ^{15}N-labelled immunophilins, using isotope-filtering techniques in which the labelled protein
signals are removed (Figure 4.3D). This approach has the advantage that
labelled ligands, which may be difficult or impossible to prepare, are not
required and information on the binding site can be obtained by use of
this approach, unlike with NMR studies employing deuterated protein.

Structural studies that are focused on the ligand have the advantage
that they can be carried out in a short period of time (a few weeks).
However, the information obtained from these studies is not complete.
High-resolution structures of the entire immunosuppressant/immuno-
philin complexes can provide additional, valuable structural information
for drug design. The procedure that we use in these studies involves:
(1) the assignment of both the ligand and protein NMR signals; (2)
extraction of the relevant parameters that define the three-dimensional
structure (NOEs and three-bond coupling constants); and (3) structure
calculations using NOEs-derived distance and J-coupling-derived dihedral angle restraints. Owing to the spectral overlap observed in the NMR
studies of the immunosuppressant/immunophilin complexes, ^{13}C- and
^{15}N-labelled immunophilins and heteronuclear 3D (Fesik and Zuiderweg,
1988; Marion *et al.*, 1989a) and 4D NMR methods (Kay *et al.*, 1990a;
Clore *et al.*, 1991; Zuiderweg *et al.*, 1991) were employed.

The utility of heteronuclear multidimensional NMR experiments for
resolving spectral overlap is illustrated in Figure 4.4 (Fesik, 1991). In a
2D NOE experiment, NOEs between overlapping protons Ha/Hb and
Hc/Hd cannot be unambiguously identified. By editing in f_1 using the
^{13}C chemical shifts in a 3D NOE experiment, the NOEs can be resolved
in different planes. For example, NOEs involving Ha can be distinguished from those involving Hb by their appearance on different planes
corresponding to the different frequencies of their attached carbons (Ca,
Cb). However, the NOE cross-peaks in f_2 are only defined by their proton
frequencies. Thus, the NOE between Ha and Hc or Hd appearing on plane
Ca cannot be uniquely defined. By further editing of the proton signals in
f_2 in a 4D NOE experiment, the NOE between Ha and Hc is unambiguously identified, since it appears on plane Cc and not Cd. In this way an
NOE cross-peak will be unambiguously identified by four frequencies:
those of the two interacting protons and their attached heteronuclei.

In this review we describe our NMR studies of the CsA/cyclophilin
and FKBP/ascomycin complexes. Although we have focused the discussion on *our* work, it is important to note that other groups have carried
out NMR and X-ray crystallographic studies of similar systems. For
example, X-ray crystal structures of uncomplexed cyclophilin (Ke *et al.*,
1991) and a cyclophilin/tetrapeptide complex (Kallen *et al.*, 1991; Kallen
and Walkinshaw, 1992) have been reported as well as X-ray structures
of FK506/FKBP (Van Duyne *et al.*, 1991a) and rapamycin/FKBP (Van

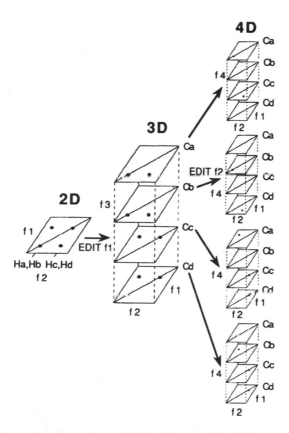

Figure 4.4 Schematic illustration of 2D, 3D and 4D NOE spectra containing NOEs Ha/Hc, Hb/Hd and Hc/Hd. From Fesik (1991), with permission from *Journal of Medicinal Chemistry*

Duyne *et al.*, 1991b). In addition to *our* NMR studies (Fesik *et al.*, 1991a, 1992; Neri *et al.*, 1991), the conformation of CsA when bound to cyclophilin has been determined independently by Weber *et al.* (1991), and a model of the CsA/CyP complex has also been proposed by Spitzfaden *et al.* (1992). Furthermore, two low-resolution structures of uncomplexed FKBP have been determined by NMR (Michnick *et al.*, 1991; Moore *et al.*, 1991).

2 Cyclosporin A/Cyclophilin Complex

Bound Conformation of Cyclosporin A

Early investigations on the conformation of cyclosporin A (CsA) were conducted using uncomplexed CsA dissolved in apolar solvents (Loosli *et al.*, 1985; Kessler *et al.*, 1990) or in the crystalline state (Loosli *et al.*, 1985). Under either of these experimental conditions, the conformation of uncomplexed CsA (Figure 4.5) was found to consist of an antiparallel β-pleated sheet from residues 11 to 7 and a type-II′ β turn at residues 2–5. The remaining residues (8–10) formed part of a loop with a *cis* amide bond between MeLeu[9] and MeLeu[10]. Four intramolecular

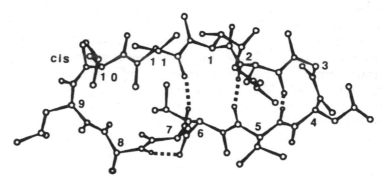

Figure 4.5 X-ray structure of uncomplexed cyclosporin A (Loosli *et al.*, 1985)

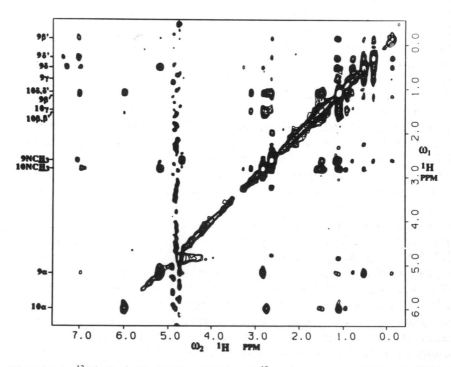

Figure 4.6 ^{13}C-edited 2D NOE spectrum of ^{13}C-labelled CsA bound to cyclophilin. Only residues 9 and 10 of CsA were uniformly ^{13}C-labelled. From Fesik *et al.* (1990b), with permission from *Science*

hydrogen bonds between the NH of residues 2, 5, 7 and 8 and the carbonyl of residues 5, 2, 11 and 6, respectively (Loosli *et al.*, 1985), were found to stabilize the three-dimensional structure of uncomplexed CsA (Figure 4.5).

The conformation of CsA determined in the absence of its receptor was an important discovery that was used to interpret structure–activity relationships for CsA analogues and to aid in the design of new molecules (Wenger, 1986; Quesniaux *et al.*, 1987; Rich *et al.*, 1989; Aebi *et al.*, 1990). However, even though the conformation of uncomplexed CsA appeared to be stable, since it was observed under a variety of experimental conditions, it was not necessarily the same as the bioactive conformation of CsA. Therefore, we embarked on NMR studies to determine the conformation of CsA when bound to its receptor, cyclophilin. Initial studies (Fesik *et al.*, 1990b) were conducted with a CsA/ CyP complex composed of unlabelled cyclophilin and cyclosporin A that was ^{13}C-labelled at the 9 and 10 positions. After assigning the ^{1}H and ^{13}C signals of the labelled MeLeu^9 and MeLeu^{10} CsA residues, information on the bound conformation of CsA was obtained from an analysis of isotope-edited 2D NOE spectra of the [^{13}C$_{9, 10}$]-CsA/CyP complex (Figure 4.6) in which NOEs were observed between protons attached to the ^{13}C-labelled nuclei of CsA (ω_1) and other nearby protons of CsA and cyclophilin (ω_2). The intense NOE observed between MeLeu^9(H$^\alpha$) and MeLeu^{10}(NMe) and the lack of an NOE between the α protons of the two labelled MeLeu residues (Figure 4.6) indicated that CsA must adopt a *trans* 9–10 amide bond when bound to CyP. These findings were in marked contrast to the 9,10 *cis* peptide bond observed in uncomplexed CsA and provided the first experimental evidence (Fesik *et al.*, 1990b) that the bound conformation of CsA was significantly different from that of uncomplexed CsA (Loosli *et al.*, 1985).

The complete characterization of the conformation of CsA when bound to cyclophilin was obtained independently by two research groups from an analysis of isotope-edited 2D (Weber *et al.*, 1991) or hetero-nuclear 3D NMR spectra (Fesik *et al.*, 1991a) of the CsA/CyP complex prepared with uniformly ^{13}C-labelled cyclosporin A. After assigning the CsA NMR signals from a variety of 2D NMR experiments (Fesik *et al.*, 1991a; Weber *et al.*, 1991), the NOE data were interpreted. Figure 4.7 depicts several ^{1}H, ^{1}H planes from a 3D HMQC-NOESY spectrum of [U-^{13}C]-CsA bound to unlabelled cyclophilin. In these spectra, NOEs were observed that originated from protons attached to the ^{13}C-labelled nuclei of CsA to other nearby protons of CsA and cyclophilin. By editing in a third dimension by the ^{13}C frequencies in the 3D NMR experiment, many NOEs were well resolved and easily assigned. From the integrated 3D NOE cross-peak volumes, proton–proton distances were obtained by use of the two-spin approximation and used as

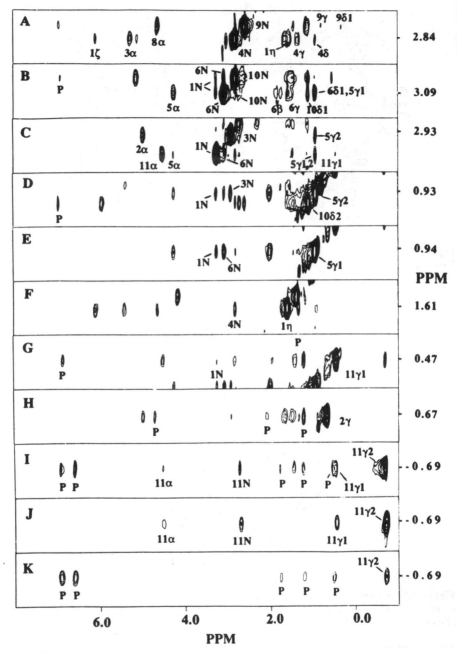

Figure 4.7 (A–I) Cross-sections (ω_2, vertical axis; ω_3, horizontal axis) from a 3D ^{13}C HMQC-NOESY data set of the [U-^{13}C]CsA/cyclophilin complex at the ^{13}C chemical shifts (ω_1): (A) 29.3, (B) 31.8, (C) 34.4, (D) 19.4, (E) 18.2, (F) 17.2, (*Cont. opposite page...*)

Figure 4.8 Stereoview of 20 NMR structures of CsA when bound to cyclophilin. From Fesik *et al.* (1991a), with permission from *Biochemistry*

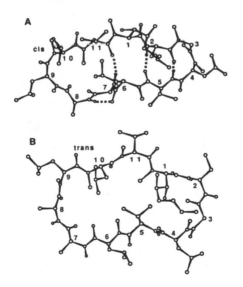

Figure 4.9 Comparison of the (A) X-ray structure of uncomplexed CsA (Loosli *et al.*, 1985) with the (B) NMR structure of CsA when bound to cyclophilin (Fesik *et al.*, 1991a)

(G) 18.8, (H) 9.8 and (I) 16.3 ppm. (J,K) Cross-sections (ω_2, ω_3) from the half-filtered 3D ^{13}C HMQC-NOESY data set of the [U-^{13}C]CsA bound to cyclophilin at the ^{13}C chemical shift (ω_1) of the γ^2 carbon of MeVal^{11} (16.3 ppm). From Fesik *et al.* (1991a), with permission from *Biochemistry*

restraints in the calculation of three-dimensional structures of CsA by a distance geometry (DG)/simulated annealing (SA) protocol (Nilges *et al.*, 1988). From 219 initial DG structures and subsequent SA refinement, a total of 95 were selected that best fitted the experimental NOE restraints. The average rms deviation obtained upon superimposition of these structures on the average structure for the backbone and heavy atoms was 0.37 Å (± 0.11) and 0.50 Å (± 0.14), respectively. In Figure 4.8, 20 structures with the lowest NOE energy contribution are shown which illustrate how well the NMR data were able to define the conformation of CsA when bound to cyclophilin. As expected on the basis of our initial studies (Fesik *et al.*, 1990b), the conformation of cyclosporin A when bound to its receptor (Fesik *et al.*, 1991a; Weber *et al.*, 1991) was very different from that of uncomplexed CsA (Loosli *et al.*, 1985). As shown in Figure 4.9, the four amide protons that participate in hydrogen bonds in uncomplexed CsA are not involved in intramolecular hydrogen bonds in the complex and point to the outside of the CsA ring. Several other differences in structure were observed, including a conformational change of the CsA side-chains. For example, the MeBmt[1] side-chain was found to be located on a different side of the macrocyclic ring in bound as opposed to free CsA (Fesik *et al.*, 1991a).

Portions of Cyclosporin A That Bind to Cyclophilin

In addition to intramolecular NOEs that define the bound conformation of CsA, intermolecular NOEs were also observed between CsA and CyP. These NOEs were used to identify those portions of CsA in contact with the protein. In principle, intra- and intermolecular NOEs could be distinguished from one another in a conventional 3D HMQC-NOESY spectrum, since the CsA/CyP NOEs would be expected to be observed in only one ^{13}C plane corresponding to the ^{13}C chemical shift of CsA (CyP is not labelled), whereas the CsA/CsA NOEs would be observed in two planes at both ^{13}C chemical shifts of CsA. However, the different types of NOEs may be difficult to distinguish in some cases, owing to spectral overlap. Another way to distinguish between intra- and intermolecular NOEs is through the use of a double half-filtered 2D NOESY (Wider *et al.*, 1990) or a half-filtered 3D HMQC-NOESY experiment (Fesik *et al.*, 1991a). By suitably combining two F_3-half-filtered 3D HMQC-NOESY data sets acquired with or without a 180° ^{13}C pulse in the half-filter, NOEs from the labelled ligand to other 'labelled' protons of the ligand or to the unlabelled protein can be selected, respectively. This is illustrated in Figure 4.7, which depicts NOEs involving the γ^2 protons of MeVal[11] of CsA that were observed in a conventional 3D HMQC-NOESY spectrum (Figure 4.7I) as compared with the half-filtered experiment that selects for the CsA/CsA (Figure 4.7J) or CsA/ CyP NOEs (Figure 4.7K) (Fesik *et al.*, 1991a).

From an analysis of the 3D NOE data, several CsA protons were found to be in close proximity to CyP (Fesik *et al.*, 1991a). These are indicated in Figure 4.10 by the filled and checkered carbon atoms. In general, those CsA residues in close proximity to the protein are those that are important for cyclophilin binding (Handschumacher *et al.*, 1984; Quesniaux *et al.*, 1987, 1988; Durette *et al.*, 1988) and immunosuppressive activity (Wenger, 1986; Aebi *et al.*, 1990). For example, modifications of CsA residues 1, 10 or 11 which are close to CyP yielded CsA analogues with a decreased binding affinity for CyP and reduced immunosuppressive activity (Wenger, 1986a,b; Durette *et al.*, 1988; Quesniaux *et al.*, 1988; Sigal *et al.*, 1991). On the other hand, CsA analogues with modifications at residues 4, 5, 6 or 8 that are further away from CyP were found to bind to cyclophilin with about the same affinity as native CsA and generally possessed comparable immunosuppressive activities. However, some of these CsA analogues (e.g. [MeAla^6]CsA) were less potent as immunosuppressants than expected on the basis of their ability to bind to cyclophilin (Sigal *et al.*, 1991). Although these results had been used to question the role of cyclophilin in CsA-mediated immunosuppression, the reduced immunosuppressive

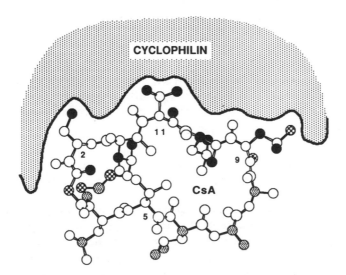

Figure 4.10 Ball-and-stick model of the NMR structure of CsA bound to cyclophilin. The protons involved in CsA/CyP NOEs are indicated by the filled and checkered carbon atoms. The protons attached to the jagged and checkered carbon atoms exhibited the largest change in relaxation rate upon the addition of HyTEMPO. From Fesik *et al.* (1991b), with permission from *Journal of the American Chemical Society*

activity displayed by these analogues can now be explained by a reduced affinity for calcineurin (Liu *et al.*, 1992).

Solvent-exposed Regions of CsA When Bound to Cyclophilin

In addition to identifying those portions of CsA in close proximity to CyP from the CsA/CyP NOE data, it was also important to identify those regions of CsA that were exposed to solvent. In most cases, this complementary information can be valuable for identifying those portions of the molecule that may be modified without affecting the binding affinity (Fesik, 1991). For CsA, however, these regions could be interacting with calcineurin (Liu *et al.*, 1991) and may therefore be important for the immunosuppressive activity.

In order to identify the solvent-exposed regions of CsA when bound to cyclophilin, a paramagnetic relaxation reagent was employed. The approach (Fesik *et al.*, 1991b) relies on the distance-dependent change in the proton longitudinal relaxation rates upon addition of 4-hydroxy-2,2,6,6-tetra-methylpiperidinyl-1-oxy (HyTEMPO). The solvent-exposed regions of CsA when bound to cyclophilin were expected to display a greater change in relaxation rate upon HyTEMPO addition. The longitudinal relaxation rates were determined by an exponential fit of the cross-peak intensities observed in a series of T_1-HMQC experiments recorded with different delays. In Figure 4.10 the CsA protons with the greatest change in T_1 upon the addition of HyTEMPO are indicated by the jagged and checkered carbon atoms (Fesik *et al.*, 1991b). Although most of these CsA protons were distant from CyP, as evidenced by the lack of intermolecular NOEs between these CsA protons and cyclophilin, a few of them that were strongly affected by HyTEMPO showed NOEs to CyP (those attached to the checkered carbon atoms) (Fesik *et al.*, 1991b).

3D Structure of CsA/CyP Complex

Although the NMR studies that were focused on the ligand rapidly provided important structural information on the bound conformation of CsA and the identification of the CsA residues that interact with the protein, very little information was obtained from these studies on the structure of the binding site which could also be a valuable aid in designing improved CsA analogues. In order to obtain the complete three-dimensional structure of the CsA/CyP complex, the assignment of the cyclophilin NMR signals was required. The CyP resonances of the backbone were assigned from a combination of heteronuclear 3D NMR experiments as previously described (Neri *et al.*, 1991). Two types of ^{15}N-resolved 3D NMR experiments were recorded in which

^{1}H–^{1}H NOESY or TOCSY spectra were edited in different planes by the ^{15}N chemical shifts of the amide nitrogens. By spreading the NOESY and TOCSY cross-peaks in several planes, spectral overlap was greatly reduced, facilitating the assignment process. As illustrated for residues M136–E140 in the ^{15}N-resolved 3D NOESY-HSQC spectrum shown in Figure 4.11, the α-helical regions of CyP were identified by matching the NH/NH cross-peaks in the NOESY-HSQC spectrum (Marion *et al.*, 1989a; Fesik and Zuiderweg, 1990), and, as shown for residues I156–D160, the adjacent CyP residues in the β-sheet regions were primarily identified from NH/H^α NOESY cross-peaks (Neri *et al.*, 1991). Ambiguities encountered in the analysis of the two ^{15}N-resolved 3D NMR experiments were alleviated by two triple resonance 3D NMR experiments applied to CsA/[U-^{15}N, ^{13}C]-CyP. In the 3D HNCA experiment (Figure 4.12, top), the amide ^{1}H(ω_3) and ^{15}N(ω_1) chemical shifts of the *i*th residue were correlated to the α-carbons (ω_2) of the *i* and *i* − 1 amino acids (Kay *et al.*, 1990c). In the complementary 3D HN(CO)CA experiment (Figure 4.12, bottom), the ^{1}H(ω_3) and ^{15}N(ω_1) chemical shifts of the amides (*i*) were correlated *exclusively* to the α-carbons (ω_2) of the *i* − 1 residues by magnetization transfer through the intervening carbonyl carbons (Bax and Ikura, 1991). Comparison of the HNCA and the HN(CO)CA experiments provided an unambiguous distinction between the C^α(*i*) and C^α(*i* − 1) cross-peaks.

From the four experiments described above, the H^N and ^{15}N frequencies of the amides were correlated to the H^α and C^α chemical shifts of the same (*i*) and the adjacent (*i* − 1) amino acids. However, in order to obtain the sequence-specific assignments, the spin systems had to be identified by amino acid type. This was accomplished from ^{1}H/^{15}N HSQC spectra of CsA/CyP complexes prepared with cyclophilins that were selectively ^{15}N-labelled by a specific type of amino acid (McIntosh *et al.*, 1990). Six different CsA/CyP samples were prepared, using cyclophilin selectively ^{15}N-labelled at Gly, Phe, Val, Ala, Lys and Ile. As shown in Figure 4.13A in the HSQC spectrum of CsA/[^{15}N-Phe, ^{13}C=O-Leu]-CyP, the amide proton and nitrogen chemical shifts of the Phe residues were unambiguously identified by use of this approach. In addition, since the carbonyl carbons of the Leu residues were also ^{13}C-labelled and only one Leu–Phe pair is present in cyclophilin (L24–F25), the ^{1}H and ^{15}N amide resonances of F25 were unambiguously assigned by the $^1J_{N, CO}$ coupling observed in the HSQC spectrum (see inset in Figure 4.13A). Another example of an HSQC spectrum of a selectively ^{15}N-labelled CyP is shown in Figure 4.13B. Although only ^{15}N-labelled Gly was added to the growth medium, additional peaks were observed in the spectrum that were attributed to cross-labelling with Cys and Ser (Neri *et al.*, 1991).

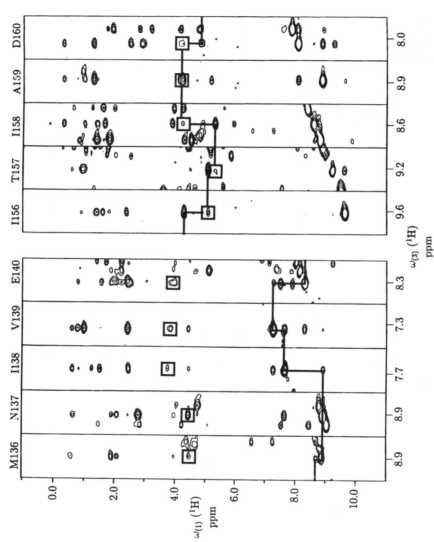

Figure 4.11 $^1H(\omega_1)$, $^1H(\omega_3)$ cross-sections corresponding to the ^{15}N frequencies (ω_2) of M136–E140 and I156–D160 of a 3D NOESY-HSQC spectrum of CsA/[U-^{15}N]-CyP. H^N/N^α cross-peaks that were observed in a 3D TOCSY-HSQC spectrum of CsA/[U-^{15}N]-CyP are indicated by boxes. From Neri *et al.* (1991), with permission from *FEBS Letters*

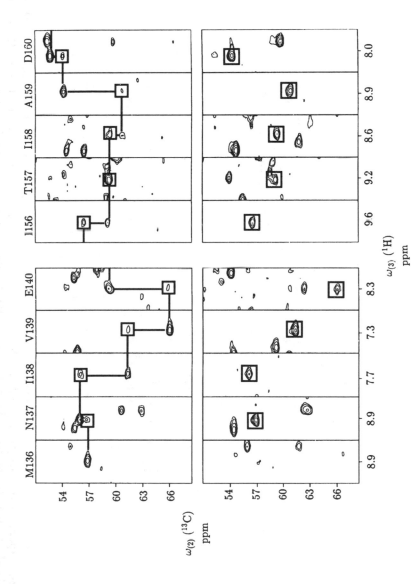

Figure 4.12 $C^\alpha(\omega_2)$, $H^N(\omega_3)$ cross-sections corresponding to the ^{15}N frequencies (ω_1) of M136-E140 and T156-D160 of a 3D HNCA (top) and 3D HN(CO)CA (bottom) spectrum of CsA/[U-^{15}N, ^{13}C]-CyP. $N(i)_\omega C^\alpha(i-1)/H^N(i)$ cross-peaks are indicated by boxes. From Neri *et al.* (1991), with permission from *FEBS Letters*

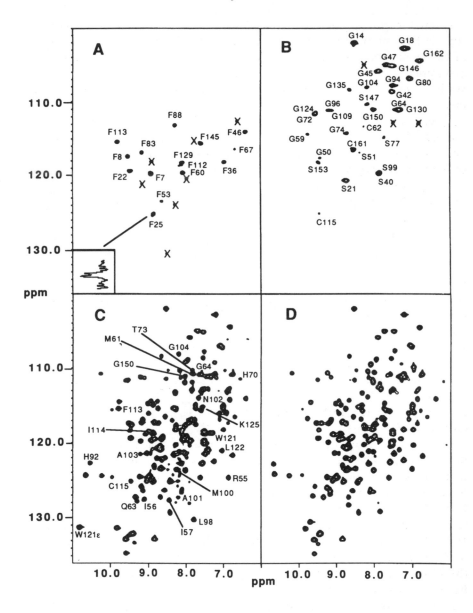

Figure 4.13 ^{1}H, ^{15}N HSQC spectra (^{1}H, horizontal axis; ^{15}N, vertical axis) of (A) [^{15}N-Phe, ^{13}C=O-Leu]-CyP/CsA, (B) [^{15}N-Gly, Ser, Cys]-CyP/CsA, (C) [U-^{15}N]-CyP/CsA and (D) [U-^{15}N]-CyP. The inset in (A) corresponds to a cross-section (vertical axis) of the amide ^{1}H and ^{15}N frequency of F25. The cross-peaks marked by an 'x' correspond to sample impurities. CyP amide resonances are labelled in (C) that do not match any of the cross-peaks in (D) within the range of ± 0.7 ppm for ^{15}N and/or ± 0.17 ppm for ^{1}H. From Neri *et al.* (1991), with permission from *FEBS Letters*

From the primary sequence of CyP, the identification of the amide signals by amino acid type from selective labelling and the correlation of the signals corresponding to the adjacent amino acids, the sequential assignments of the backbone nuclei were obtained (Neri *et al.*, 1991). With these assignments in hand, the regular secondary structural elements of cyclophilin were identified on the basis of multiple successive d_{NN}, $d_{\alpha N}$ and other characteristic NOEs (Wüthrich, 1986) observed in the 3D ^{15}N-resolved NOESY-HSQC spectra, $^3J_{H^N,H^\alpha}$ coupling constants, and amide exchange rates. The secondary structure of CyP when bound to CsA was found to be similar to that determined previously for uncomplexed CyP (Ke *et al.*, 1991; Wüthrich *et al.*, 1991), consisting of two α helices (residues 31–43 and 135–147), eight β strands forming an antiparallel β-sheet (residues 5–12, 15–24, 55–57, 61–64, 97–101, 112–116, 130–133 and 155–164), and a series of loops and turns.

Other important structural information that could be obtained at this stage in the 3D structure determination of the CsA/CyP complex was the characterization of the CsA binding site. On the basis of chemical shift differences between uncomplexed CyP (Figure 4.13D) and CyP bound to CsA (Figure 4.13C), the location of the CsA binding site was inferred (Kallen *et al.*, 1991; Neri *et al.*, 1991). These changes in chemical shift mainly occur in four regions of CyP composed of residues 55–74, 98–104, 112–116 and 120–126. Although these regions are far apart in the primary structure of CyP, the NOE data indicated that they were relatively close to one another in space (Figure 4.14). However, defining the location of the CsA binding site on the basis of the chemical shift data must be treated with caution, since the changes in CyP chemical shift upon binding CsA may arise from either a direct interaction between CsA and CyP or a conformational change in the protein distant from the CsA binding site. Furthermore, the chemical shift data cannot be used to determine the orientation of CsA in the CyP binding pocket. For a further characterization of the CsA binding site, CsA/CyP NOEs were assigned on the basis of the backbone assignments (Kallen *et al.*, 1991; Neri *et al.*, 1991) and a limited number of side-chain assignments of CyP. From an analysis of the NOE data, it was possible to orient the CsA molecule within the binding site, as schematically depicted in Figure 4.14 (Neri *et al.*, 1991).

In order to generate a model of the CsA/CyP complex, a starting structure of the complex was built by docking the bound NMR structure of CsA (Fesik *et al.*, 1991a) into the X-ray structure of free CyP (Ke *et al.*, 1991), using the CsA/CyP NOE data (Fesik *et al.*, 1992). The model of the complex was refined by restrained molecular dynamics in which CsA/CsA and CsA/CyP distance restraints were included in the calculations to define the bound conformation of CsA and orient CsA relative to CyP in the complex, respectively. Figure 4.15 depicts the

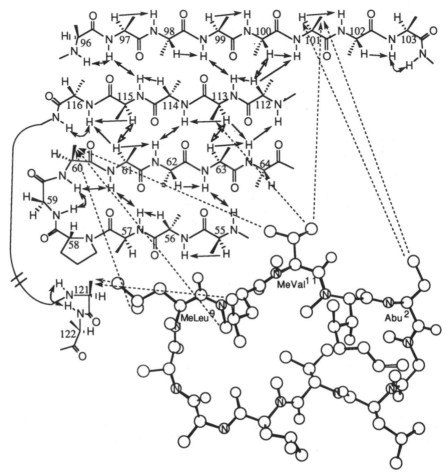

Figure 4.14 Secondary structure of the proposed CsA binding site as determined from NOE data. From Neri *et al.* (1991), with permission from *FEBS Letters*

model of the CsA/CyP complex generated from the combination of NMR and X-ray data. Consistent with the chemical shift data, the portions of cyclophilin that bind to CsA consist of four antiparallel β strands (H54–I57, F60–G64, G96–N102, Q111–T116) and residues W121–H126. It is interesting to note that the MeLeu⁹–MeLeu¹⁰– MeVal¹¹–MeBmt¹–Abu² residues of cyclosporin A bind to CyP in a similar region as a tetrapeptide substrate (*N*-acetyl-Ala–Ala–Pro–Ala-amidomethylcoumarin) for this enzyme, whose structure was determined by X-ray crystallography (Kallen *et al.*, 1991; Kallen and Walkinshaw, 1992). However, CsA is oriented in an N- to C-terminal orientation opposite to that of the tetrapeptide. From a comparison of a tetrapeptide/CyP X-ray structure (Figure 2 in Kallen *et al.*, 1991) and our model

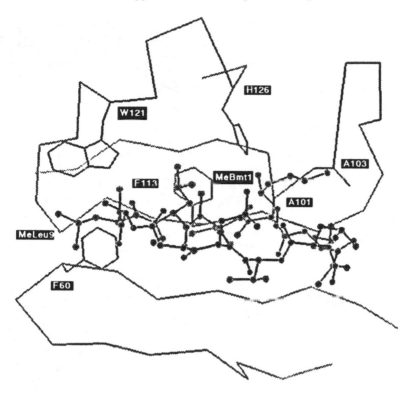

Figure 4.15 Portion of a model of the CsA/CyP complex generated from NOE data on the complex and the X-ray structure of free cyclophilin. From Fesik *et al.* (1992), with permission from *Journal of the American Chemical Society*

CsA/CyP structure, the MeBmt[1] carbinol group of CsA appeared to be located in approximately the same position as the carbonyl group of the Ala–Pro peptide bond of the tetrapeptide substrate that undergoes isomerization. These results suggested that the MeBmt[1] residue of CsA, which is required for immunosuppressive activity and CyP binding, could be mimicking an intermediate or transition state in the PPIase reaction (Fesik *et al.*, 1992). However, upon further analysis with a more refined X-ray structure of the tetrapeptide/CyP complex (Kallen *et al.*, 1991), it was found that these earlier conclusions may be incorrect. In the refined X-ray structure of the tetrapeptide/CyP complex (Kallen and Walkinshaw, 1992), the Ala–Pro peptide bond was found to be *cis* (it was *trans* in the unrefined structure), and the distance between the H126 ring and carbonyl carbon differed (Kallen *et al.*, 1991). From a comparison of the refined X-ray structure of the tetrapeptide/CyP complex and a model of the CsA/CyP complex generated by Wüthrich and co-workers (Spitzfaden *et al.*, 1992), it was concluded that the 11–1 peptide bond of CsA mimics the Ala–Pro peptide bond of the substrate and that CsA is not a transition state analogue (Spitzfaden *et al.*, 1992). However, the relevance of the X-ray structure of the tetrapeptide to the actual structures formed during the PPIase reaction remains unclear. In addition, only models of the CsA/CyP complex (Fesik *et al.*, 1992; Spitzfaden *et al.*, 1992) have been generated that are based on incomplete assignments. An experimentally determined three-dimensional structure of the CsA/CyP complex had not yet been reported at that time.

Attempts to generate a solution structure of the CsA/CyP complex without the X-ray data, using only proton–proton distance restraints derived from ^{15}N-resolved 3D NOESY experiments of CsA/[U-^{15}N]-CyP and NOEs involving CsA, failed. The 3D structures of the complex were poorly defined, especially in the loop regions. In order to obtain a well-defined NMR structure of the CsA/CyP complex by NMR, the CyP side-chain assignments and unambiguous identification of long-range NOEs involving side-chain protons were necessary. Owing to the broad linewidths of the NMR signals, ^{1}H, ^{1}H COSY and TOCSY experiments were largely unsuccessful at obtaining the side-chain assignments. However, by transferring the magnetization through the large one-bond ^{1}H–^{13}C and ^{13}C–^{13}C couplings (Bax *et al.*, 1990; Kay *et al.*, 1990b; Fesik *et al.*, 1990a), the scalar coupled ^{1}H and ^{13}C signals could be identified in 3D HCCH-TOCSY (Bax *et al.*, 1990) and 3D HCCH-COSY (Kay *et al.*, 1990b) experiments. This is illustrated in Figure 4.16 by the ^{1}H, ^{1}H cross-sections from a 3D HCCH-COSY spectrum of the CsA/[U-^{13}C, ^{15}N]-CyP complex extracted at the ^{13}C chemical shifts of the I156 spin system. By use of these methods, greater than 90% of the ^{1}H and ^{13}C NMR signals of the CyP side-chains were assigned. These assignments were necessary to interpret the heteronuclear 3D and 4D NOE

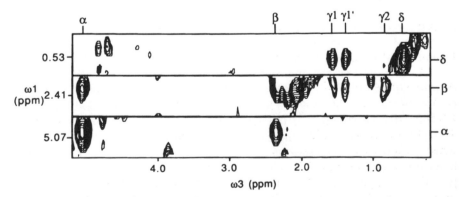

Figure 4.16 ^{1}H(ω_1), ^{1}H(ω_3) planes selected from the 3D HCCH-COSY spectrum of CsA/[U-^{15}N, ^{13}C]CyP at the ^{13}C chemical shifts (ω_2) corresponding to I156: (A) C$^\alpha$, 58.8; (B) C$^\beta$, 36.4; (C) C$^\delta$, 12.9 ppm.

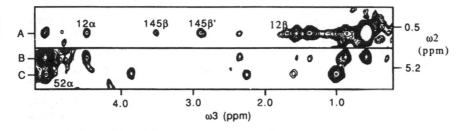

Figure 4.17 ^{1}H(ω_2), ^{1}H(ω_3) planes selected from the 3D HMQC-NOESY spectrum of CsA/[U-^{15}N, ^{13}C]CyP at the ^{13}C chemical shifts (ω_1) of 13 (top) and 59 (bottom) ppm, corresponding to (A) 156H$^\delta$, (B) 156H$^\alpha$, and (C) 157H$^\alpha$/6H$^\alpha$. Only the long-range NOEs are labelled in the figure

data of the CsA/CyP complex. In principle, 4D NOE experiments would be the method of choice for obtaining the proton–proton distance restraints used in defining the structure. However, owing to the low sensitivity of the 4D [^{13}C, ^{1}H, ^{13}C, ^{1}H] NOESY spectrum of CsA/[U-^{13}C, ^{15}N]-CyP, most of the NOE-derived distance restraints used in the structure calculations of the CsA/CyP complex were obtained primarily from 3D NOE spectra. Figure 4.17 depicts two ^{1}H, ^{1}H planes from the ^{13}C-resolved 3D NOESY data set that indicate some of the long-range NOEs helpful in defining the solution structure of the complex. From the distance restraints derived from the ^{13}C-resolved 3D NOE spectra, a high-resolution structure of the CsA/CyP complex was generated (Theriault *et al.*, 1993). The experimentally derived structure was found to be generally similar to previously published model structures (Fesik *et al.*, 1992; Spitzfaden *et al.*, 1992). However, some important differences have been noted (Theriault *et al.*, 1993).

3 Ascomycin/FKBP Complex

Bound Conformation of Ascomycin

Ascomycin (Arai *et al.*, 1962) is a potent immunosuppressant (Hatanka *et al.*, 1988) that binds tightly to FKBP. It is structurally similar to FK506, with an ethyl versus allyl substitution at the 21 position (Figure 4.1). In order to rapidly determine the conformation of ascomycin when bound to FKBP, ascomycin was biosynthetically ^{13}C-labelled, and heteronuclear 2D and 3D NMR experiments were performed on the [U-^{13}C]-ascomycin/FKBP complex. The approach used was similar to that used for determining the bound conformation of CsA. After assigning the bound ascomycin NMR signals from ^{1}H–^{13}C (HMQC) and ^{13}C–^{13}C (COSY) spectra (Petros *et al.*, 1991), proton–proton distance

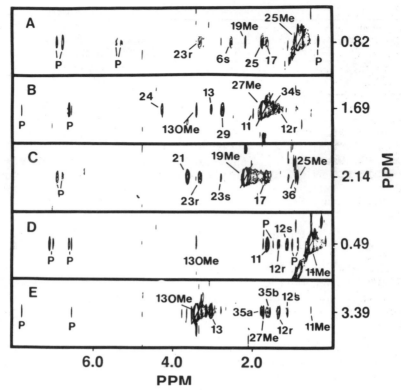

Figure 4.18 (A–E) Cross-sections (ω_2, vertical axis; ω_3 horizontal axis) from a 3D ^{13}C HMQC-NOESY spectrum of the [U-^{13}C]ascomycin/FKBP complex at the ^{13}C chemical shifts (ω_1): (A) 11.3, (B) 11.3, (C) 19.2, (D) 16.5 and (E) 56.8 ppm. From Petros *et al.* (1991), with permission from *Journal of Medicinal Chemistry*

restraints were obtained from an analysis of a ^{13}C-edited 3D NOE spectrum. Figure 4.18 depicts several planes from a 3D HMQC-NOESY spectrum of [U-^{13}C]-ascomycin/FKBP acquired with a mixing time of 40 ms. From this spectrum, 88 intramolecular NOEs were assigned and interpreted in terms of interproton distances, using the two-spin approximation (Petros *et al.*, 1991). With these restraints, 200 initial structures were generated by use of a metric matrix distance geometry program (DSPACE, Hare Research) and refined with a simulated annealing protocol (Nilges *et al.*, 1988) using XPLOR (Brünger, 1990). Figure 4.19 depicts the superimposition of 20 structures generated with the lowest NOE energy contribution. The structure of ascomycin when bound to FKBP was well defined by the NMR data, as evidenced by the low rmsd (0.45 ± 0.17) between the calculated structures.

As shown in Figure 4.20(A), the conformation of ascomycin when bound to FKBP was found to be different from the X-ray

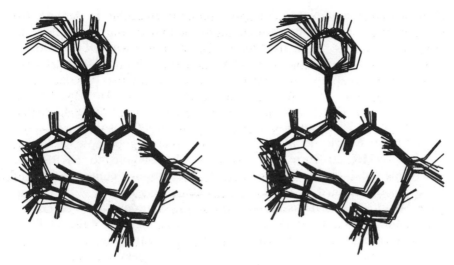

Figure 4.19 Stereoview of 20 NMR structures of ascomycin bound to FKBP. The superimposition of the heavy atoms of these structures on the calculated average gives an average rmsd of 0.45 ± 0.17 Å. From Petros *et al.* (1991), with permission from *Journal of Medicinal Chemistry*

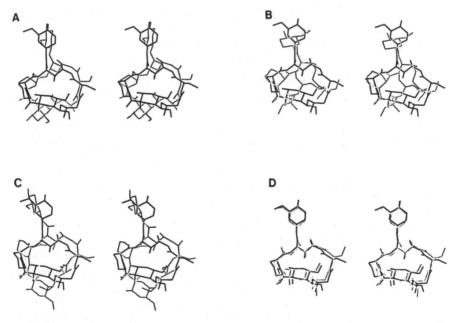

Figure 4.20 Stereoviews of the average NMR structure of bound ascomycin (bold) superimposed on (A) the X-ray structure of uncomplexed FK-506, (B) the NMR structure of the *cis* form of FK506 in solution, (C) the NMR structure of the *trans* form of FK506 in solution and (D) the X-ray structure of FK506 when bound to FKBP

structure of uncomplexed FK506 (Taga *et al.*, 1987). Among the differences was the *trans* vs. *cis* amide bond at the 7,8 position. In solution, both the *cis* (major) and *trans* (minor) forms of FK506 are present, and the conformation of both forms of free FK506 have been determined in solution by NMR (Karuso *et al.*, 1990; Mierke *et al.*, 1991). As shown in Figure 4.20, the bound conformation of ascomycin was found to be very different from either the *cis* (Figure 4.20B) or *trans* (Figure 4.20C) conformers of unbound FK506 in solution (Karuso *et al.*, 1990; Mierke *et al.*, 1991). It is interesting to note, however, that the NMR structure of ascomycin when bound to FKBP was nearly identical (rmsd = 0.56 Å) with the bound conformation of FK506 as determined by the X-ray crystallography (Figure 4.20D) (Van Duyne *et al.*, 1991a). Thus, irrespective of the method used to determine the three-dimensional structures, the bound conformations of the immunosuppressive ligands are similar but very different from the structure of the uncomplexed immunosuppressants. These results are analogous to those obtained for cyclosporin A (Fesik *et al.*, 1991a; Weber *et al.*, 1991) and highlight the importance of determining the bound conformation.

Portions of Ascomycin That Bind to FKBP

In addition to the intramolecular NOEs used in defining the bound conformation of ascomycin, intermolecular NOEs were observed between ascomycin and FKBP protons in the 3D HMQC-NOESY spectra (Petros *et al.*, 1991, 1992a). From an analysis of the intermolecular NOEs, those regions of ascomycin in close proximity to the protein were identified. These are indicated in Figure 4.21 by the filled and checkered carbon atoms. The portions of ascomycin in contact with FKBP include the piperidine ring (in agreement with the upfield chemical shifts caused by FKBP aromatic residues) (Wandless *et al.*, 1991), the pyranose ring as indicated by NOEs from 11-Me (Figure 4.18D), H-11, H-12, H-13 and 13-OMe (Figure 4.18E) to the protein, and part of the cyclohexyl ring as evidenced by NOEs between H-29, H-30, H-31 and 31-OMe of ascomycin and FKBP.

Solvent-exposed Regions of Ascomycin

The solvent-exposed regions of ascomycin when bound to FKBP were determined (Petros *et al.*, 1992c) by monitoring the change in proton relaxation rates upon the addition of HyTEMPO. As shown in Figure 4.21, the solvent-exposed regions of ascomycin included the ethyl group (35–36) and the atoms around it from position 15 to 24 as well as one side of the cyclohexyl ring (positions 32–34). From these results, it can

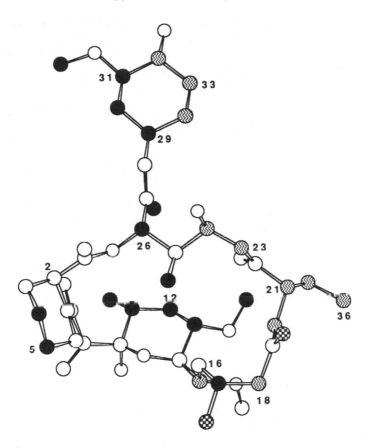

Figure 4.21 Ball-and-stick model of the NMR structure of ascomycin bound to FKBP in which the protons involved in ascomycin/FKBP NOEs are indicated by the filled and checkered carbon atoms. The protons attached to the jagged and checkered carbon atoms exhibited the largest change in relaxation rate upon the addition of HyTEMPO. From Petros *et al.* (1992a), with permission from *Journal of Biomolecular NMR*

be concluded that the region of ascomycin from position 15 to 24 is not directly involved in binding to FKBP (Petros *et al.*, 1992c) and could be modified without altering the binding affinity to FKBP. This is exemplified by FK506 and rapamycin, which are structurally very different in this region (Figure 4.1) but have similar FKBP binding affinity. Although not important for binding to FKBP, this portion of the ligand is very important for the fine tuning of the biological properties of these molecules (Schreiber, 1991), as suggested by the differences in immunosuppressive activity of FK506 and rapamycin. These differences have been attributed to the interaction of the FKBP/FK506 and FKBP/rapamycin complexes with different proteins. Indeed, the FKBP/FK506

complex binds to calcineurin, whereas the FKBP/rapamycin complex
does not (Liu *et al.*, 1991).

3D Structure of Ascomycin–FKBP Complex

The NMR studies involving ^{13}C-labelled ascomycin bound to unlabelled
FKBP were important for quickly providing useful structural information on the bound conformation of ascomycin and the identification of
those portions of the immunosuppressant that interact with its receptor.
Although more time-consuming, the complete structure determination
of the ascomycin/FKBP complex by NMR could produce additional
insights for designing improved immunosuppressants. In order to reduce
the time necessary for the analysis of the multidimensional NMR data
acquired on the ascomycin/FKBP complex, we used a computer program
featuring computer graphics and semiautomated assignment tools with
a database for the storage and retrieval of NMR data (Xu *et al.*, 1992a).
With this program, correlations observed in one 3D NMR spectrum
could rapidly be correlated to those from a different experiment using
the frequencies common to both spectra. The first step was to correlate
the H^N and ^{15}N chemical shifts of the amides that are the common
frequencies in the 3D experiments to the α protons and α carbons of
the same (*i*) and adjacent (*i* – 1) residues, applying the same approach
used in assigning the backbone resonances of the CsA/CyP complex. In
the next step, the ^{1}H and ^{13}C signals of the side-chains of the spin
systems were identified from correlations observed in 3D HCCH-TOCSY
(Bax *et al.*, 1990) and HCCH-COSY (Kay *et al.*, 1990b) experiments
using the H^α and C^α chemical shifts as starting points. From an analysis
of the ^{1}H and ^{13}C chemical shifts obtained from these experiments, the
amino acid spin systems were assigned by amino acid type and, together
with the identification of neighbouring spin systems, yielded the sequence-
specific ^{1}H, ^{13}C and ^{15}N assignments of the ascomycin/FKBP complex
(Xu *et al.*, 1992a). The stereospecific assignments of the 9 Val γ-methyl
groups and the 7 Leu δ-methyl groups of FKBP were obtained from an
analysis of a high-resolution heteronuclear single quantum correlation
(HSQC) spectrum of the FKBP/ascomycin complex composed of ^{13}C-
labelled FKBP prepared biosynthetically from a mixture of uniformly
^{13}C-labelled (20%) and unlabelled glucose by the method of Wüthrich
and co-workers (Neri *et al.*, 1989; Senn *et al.*, 1989).

The secondary structure of FKBP when bound to ascomycin was
determined on the basis of the NOE data, $^3J_{H^N,H^α}$ coupling constants,
amide exchange rates (Wüthrich, 1986), and C^α, C^β chemical shifts
(Spera and Bax, 1991). A summary of the data is given in Figure 4.22.
The NMR data indicate that FKBP is largely composed of β sheet. The
relative location of the β strands was determined from long-range NOEs

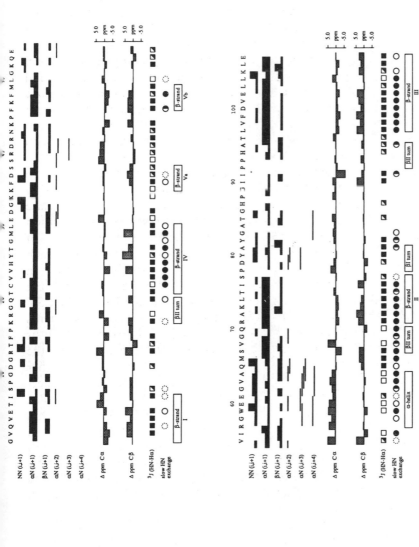

Figure 4.22 Summary of the NOEs, C$^\alpha$ and C$^\beta$ chemical shift data, $^3J_{H^N,H^\alpha}$ coupling constants (open squares, < 5.5 Hz; half-filled squares, 5.5–8.5 Hz, filled squares, > 8.5 Hz), amide exchange rates and secondary structure of FKBP when bound to ascomycin (Xu et al., 1992a)

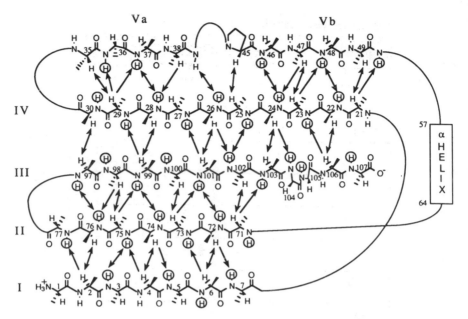

Figure 4.23 Interstrand NOEs (arrows) that define the relative location of the β strands in the β sheet. The circled amide protons exchange slowly. From Xu *et al.* (1992a)

between the individual β strands, as shown in Figure 4.23. The remainder of the protein consists of one α-helix (residues 57–64) and a series of loops and turns (Xu *et al.*, 1992a).

In order to determine a highly resolved 3D structure of the ascomycin/FKBP complex, we obtained proton–proton distance restraints from an analysis of 4D [^{13}C, ^{1}H, ^{15}N, ^{1}H] (Kay *et al.*, 1990a) and 4D [^{13}C, ^{1}H, ^{13}C, ^{1}H] NOESY spectra (Clore *et al.*, 1991; Zuiderweg *et al.*, 1991). Figure 4.24 depicts a series of ^{1}H(ω_2), ^{1}H(ω_4) planes from the 4D [^{13}C, ^{1}H, ^{13}C, ^{1}H] NOESY spectrum of [U-^{13}C, ^{15}N]-FKBP bound to ascomycin. Although some of the NOEs could be resolved in the heteronuclear 3D NOE spectrum (Figure 4.24, top), the increase in resolution obtained in the 4D experiment achieved by editing *both* protons involved in the NOE by the frequencies of *each* of their attached heteronuclei provided many more unambiguous proton–proton distance restraints (Meadows *et al.*, 1993).

In addition to the 1700 NOE-derived distance restraints, 123 dihedral angle restraints were included in the structure calculations. The ϕ angle restraints were derived from $^3J_{H^N,H^\alpha}$ coupling constants measured in a 2D HMQC-*J* spectrum (Kay and Bax, 1990), whereas the χ_1 restraints were obtained from $^3J_{H^\alpha,H^\beta}$ and $^3J_{N,H^\beta}$ coupling constants obtained from ECOSY-like spectra (Montelione *et al.*, 1989; Gemmecker and Fesik,

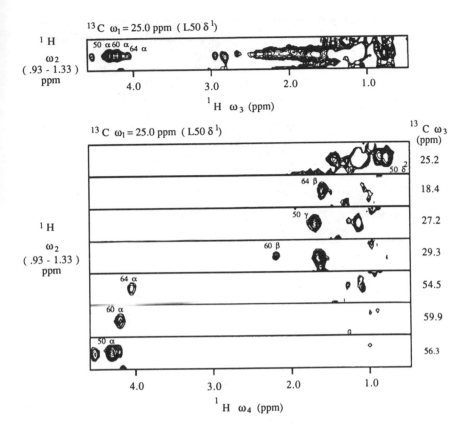

Figure 4.24 (top) ^{1}H(ω_2), ^{1}H(ω_3) planes selected from a 3D HMQC-NOESY spectrum of ascomycin/[U-^{15}N, ^{13}C]FKBP at 25.0 ppm. (bottom) ^{1}H(ω_2), ^{1}H(ω_4) planes selected from a 4D [^{13}C, ^{1}H, ^{13}C, ^{1}H] NOESY spectrum at the ^{13}C chemical shifts of ω_2 = 25.0 ppm and ω_3 = 25.2, 18.4, 27.2, 29.3, 54.5, 59.9 and 56.3 ppm

1991). The $^3J_{H^\alpha,H^\beta}$ coupling constants were measured in a ^{1}H-coupled 3D HCCH-TOCSY spectrum (Xu *et al.*, 1992b), as illustrated in Figure 4.25, and the $^3J_{N,H^\beta}$ coupling constants were measured in a ^{15}N-coupled 3D HSQC-TOCSY spectrum (Figure 4.26) (Montelione *et al.*, 1989). Hydrogen bond restraints (102) obtained from the amide exchange data were also included in the structure calculations. Finally, although not included as restraints in the structure calculations, the locations of the tightly bound water molecules were determined (Xu *et al.*, 1993) from an analysis of a ^{15}N-resolved 3D HMQC-NOESY spectrum (Clore *et al.*, 1990) of ascomycin/[U-^{15}N]-FKBP and ^{13}C-resolved 3D HMQC-NOESY spectra of ascomycin/[U-^{13}C, ^{15}N]-FKBP and [U-^{13}C]-ascomycin/FKBP.

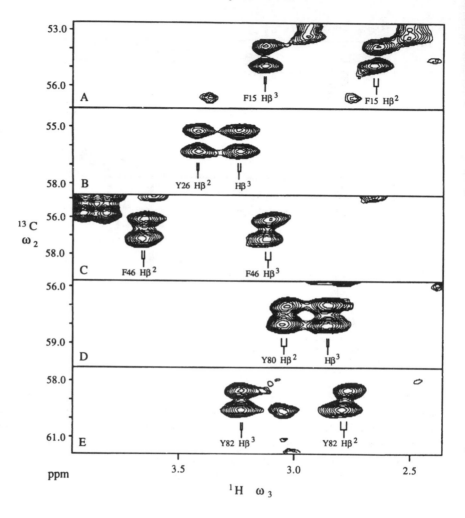

Figure 4.25 $^{13}C(\omega_2)$, $^1H(\omega_3)$ planes from the ^{1}H-coupled 3D HCCH-TOCSY spectrum extracted at (A) 5.27, (B) 6.72, (C) 4.98, (D) 4.49, (E) 4.28 ppm (ω_1). $^3J_{H^\alpha, H^\beta}$ coupling constants are measured from the frequency difference in ω_3 between the two ^{13}C-coupled α-proton peaks (ω_2). From Xu *et al.* (1992b)

Figure 4.27 depicts 20 preliminary structures of the ascomycin/FKBP complex generated from the NMR data. An rmsd of 0.6 Å for the backbone and 1.0 Å for all atoms was obtained for the structures superimposed on the average. In general, the structure of the complex was well defined by the NMR data, except for two solvent-accessible loops consisting of residues 10–14 and 49–55. For both of these regions, there is a lack of interresidue NOEs, which could explain the inability of the NMR data to define this part of the structure.

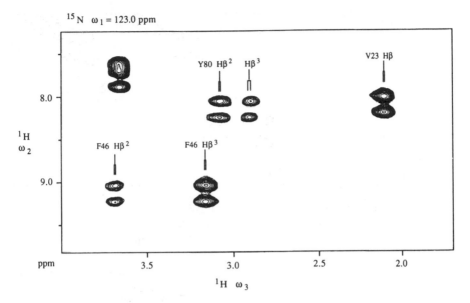

Figure 4.26 ^{1}H(ω_2), ^{1}H(ω_3) cross-section from the ^{1}H-coupled 3D HSQC-TOCSY spectrum extracted at 123.0 ppm (ω_1) that depicts H^N, H$^\beta$ cross-peaks. $^3J_{N,H^\beta}$ coupling constants are measured from the frequency difference in ω_3 between the two ^{15}N-coupled amide proton peaks (ω_2). From Xu *et al.* (1992b)

The structure of the ascomycin/FKBP complex determined in solution from heteronuclear 3D and 4D NMR data was much more well-defined than the NMR structures of free FKBP determined on the basis of 2D NMR experiments (Michnick *et al.*, 1991; Moore *et al.*, 1991). In particular, residues 35–44 and 78–96 are ill-defined by the NMR data in free FKBP, whereas, in the NMR structure of the ascomycin/FKBP complex, residues 37–39 form part of the 5th β strand, residues 40–45

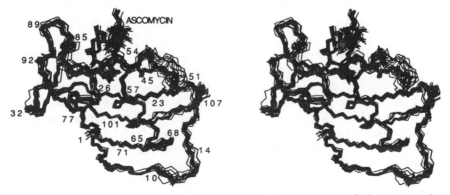

Figure 4.27 Stereoview of 20 preliminary NMR structures of the ascomycin/FKBP complex

adopt a well-defined yet irregular bulge, and the structure of the large loop between residues 78–96 is more well-defined.

The solution structure of the ascomycin/FKBP complex was found to be more similar to the previously reported X-ray structure of the FK506/ FKBP complex (Van Duyne *et al.*, 1991a). A detailed comparison of the NMR and X-ray structures has recently been reported (Meadows *et al.*, 1993).

4 Conclusions and Future Perspectives

In this report we have described our multidimensional NMR studies of the CsA/CyP and ascomycin/FKBP complexes. Our initial studies involved the use of ^{13}C-labelled immunosuppressants bound to unlabelled proteins. From this work, we were able to determine the bound conformation of the ligands and distinguish between those portions of the ligands in close proximity to the proteins from those that are exposed to solvent. In both cases the conformations of the immunosuppressants when bound to their receptors were found to be vastly different from the uncomplexed molecules. These results indicate that the three-dimensional structures of the ligands in the absence of their receptors may be unsuitable as templates for structure-based drug design and highlight the importance of determining the conformations of the ligands when bound to their target sites.

The methods used to simplify the NMR spectra of drug/receptor complexes through the use of isotopically labelled ligands are very powerful for rapidly providing valuable structural information. Another advantage is that pure protein is not required in these studies, since the labelled ligand will selectively bind to its receptor, and only the signals of the ligand and its vicinity will be detected.

The major disadvantage of these experiments is the requirement for ^{13}C-labelled ligands. Although it may be feasible to label one of the ligands used in the initial NMR experiments, it will generally be impractical to label all of the analogues for study by NMR. A more suitable approach would involve the study of unlabelled ligands bound to the uniformly ^{13}C- and ^{15}N-labelled receptor proteins. In these studies the NMR spectra can be simplified by eliminating the signals of the labelled protein (Wider *et al.*, 1990). Owing to recent advances in these isotope-editing techniques (Gemmecker *et al.*, 1992; Ikura and Bax, 1992), virtually all of the background protein signals can be removed, improving the utility of this method. Indeed, using this approach, we have been able to determine the conformation of a series of ascomycin analogues when bound to FKBP and identify those regions of the analogues that interact with FKBP in as little as 2 weeks (Petros *et al.*, submitted).

Although NMR studies that are focused on the structure of the ligand and its vicinity can be very useful, additional structural information important in drug design could be obtained from the complete 3D structure of the drug/receptor complex. Using recently developed heteronuclear 3D and 4D NMR methods, we were able to obtain high-resolution structures of both the CsA/CyP and ascomycin/FKBP complexes in solution. Currently, we are extending these structural studies to include new ascomycin analogues bound to FKBP and ascomycin bound to FKBP mutants in an effort to understand the structural features required for an active immunosuppressant. Since the assignments have already been obtained on the ascomycin/FKBP complex, these new structures can be determined in a short period of time, facilitating an iterative approach to the design of new analogues.

Acknowledgements

We thank R. Helfrich, R. Simmer, S. Pratt, T. Pilot-Matias, H. Smith, E. Gubbins, R. Edalji, J. Severin, T. Holzman, J. Hochlowski and M. Jackson for the preparation of cyclophilin, FKBP and ^{13}C-labelled CsA and ascomycin samples. We also acknowledge the members of the Abbott NMR group past and present who contributed to this work, including: R. Gampe, Jr., G. Gemmecker, H. Eaton, A. Petros, E. Olejniczak, R. Xu, T. Logan, Y. Theriault, L. Yu, R. Meadows and D. Nettesheim. This research was supported in part by the National Institute of General Medical Sciences (GM45351, awarded to S.W.F.).

References

Aebi, J. D., Deyo, D. T., Sun, C. Q., Guillaume, D., Dunlap, B. E. and Rich, D. H. (1990). Synthesis, conformation and immunosuppressive activities of three analogues of cyclosporin A modified in the 1-position. *J. Med. Chem.*, **33**, 999–1009
Arai, T., Koyama, Y., Suenaga, T. and Honda, H. (1962). Ascomycin, an antifungal antibiotic. *J. Antibiotics*, **15**, 231–232
Bartlett, P. A., Shea, G. T., Telfer, S. J. and Waterman, S. (1989). CAVEAT: A program to facilitate the structure-derived design of biologically active molecules. In Roberts, S. M. (Ed.), *Molecular Recognition: Chemical and Biological Problems*. Royal Society of Chemistry, London, Vol. 78, pp. 182–196
Bax, A., Clore, G. M. and Gronenborn, A. M. (1990). Proton–proton correlation via isotropic mixing of carbon-13 magnetization, a new three-dimensional approach for assigning proton and carbon-13 spectra of carbon-13-enriched proteins. *J. Magn. Reson.*, **88**, 425–431
Bax, A. and Ikura, M. (1991). An efficient 3D NMR technique for correlating the proton and nitrogen-15 backbone amide resonances with the α-carbon of

the preceding residue in uniformly nitrogen-15/carbon-13 enriched proteins. *J. Biomol. NMR*, **1**, 99–104

Bierer, B. E., Mattila, P. S., Standaert, R. F., Herzenberg, L. A., Burakoff, S. J., Crabtree, G. and Schreiber, S. L. (1990a). Two distinct signal transmission pathways in T lymphocytes are inhibited by complexes formed between an immunophilin and either FK506 or rapamycin. *Proc. Natl Acad. Sci. USA*, **87**, 9231–9235

Bierer, B. E., Somers, P. K., Wandless, T. J., Burakoff, S. J. and Schreiber, S. L. (1990b). Probing immunosuppressant action with a nonnatural immunophilin ligand. *Science*, **250**, 556–559

Brünger, A. T. (1990). *XPLOR Manual, Version 2.1*

Clore, G. M., Bax, A., Wingfield, P. T. and Gronenborn, A. M. (1990). Identification and localization of bound internal water in the solution structure of interleukin 1β by heteronuclear three-dimensional ^{1}H rotating-frame Overhauser ^{15}N–^{1}H multiple quantum coherence NMR spectroscopy. *Biochemistry*, **29**, 5671–5676

Clore, G. M., Kay, L. E., Bax, A. and Gronenborn, A. M. (1991). Four-dimensional ^{13}C/^{13}C-edited nuclear Overhauser enhancement spectroscopy of a protein in solution: Application to interleukin 1β. *Biochemistry*, **30**, 12–18

Dumont, F. J., Melino, M. R., Staruch, M. J., Koprak, S. L., Fischer, P. A. and Sigal, N. H. (1990a). The immunosuppressive macrolides FK-506 and rapamycin act as reciprocal antagonists in murine T cells. *J. Immunol.*, **144**, 1418–1424

Dumont, F. J., Staruch, M. J., Koprak, S. L., Melino, M. R. and Sigal, N. H. (1990b). Distinct mechanisms of suppression of murine T cell activation by the related macrolides FK-506 and rapamycin. *J. Immunol.*, **144**, 251–258

Durette, P. L., Boger, J., Dumont, F., Firestone, R., Frankshun, R. A., Koprak, S. L., Lin, C. S., Melino, M. R., Pessolano, A. A., Pisano, J., Schmidt, J. A., Sigal, N. H., Staruch, M. J. and Witzel, B. E. (1988). A study of the correlation between cyclophilin binding and *in vitro* immunosuppressive activity of cyclosporine A and analogues. *Transplant. Proc.*, **20**, 51–57

Emmel, E. A., Verweij, C. L., Durand, D. B., Higgins, K. M., Lacy, E. and Crabtree, G. R. (1989). Cyclosporin A specifically inhibits function of nuclear proteins involved in T cell activation. *Science*, **246**, 1617–1620

Fesik, S. W. (1989). Approaches to drug design using nuclear magnetic resonance spectroscopy. In Perun, T. J. and Propst, C. L. (Eds), *Computer-Aided Drug Design. Methods and Applications*. Marcel Dekker, New York, pp. 133–184

Fesik, S. W. (1991). NMR studies of molecular complexes as a tool in drug design. *J. Med. Chem.*, **34**, 2937–2945

Fesik, S. W., Eaton, H. L., Olejniczak, E. T., Zuiderweg, E. R. P., McIntosh, L. P. and Dahlquist, F. W. (1990a). 2D and 3D NMR spectroscopy employing ^{13}C–^{13}C magnetization transfer by isotropic mixing. Spin system identification in large proteins. *J. Am. Chem. Soc.*, **112**, 886–888

Fesik, S. W., Gampe, R. T. J., Eaton, H. L., Gemmecker, G., Olejniczak, E. T., Neri, P., Holzman, T. F., Egan, D. A., Edalji, R., Helfrich, R., Hochlowski, J. and Jackson, M. (1991a). NMR studies of [U-^{13}C]Cyclosporin A bound to cyclophilin. Bound conformation and portions of cyclosporin involved in binding. *Biochemistry*, **30**, 6574–6583

Fesik, S. W., Gampe, R. T., Jr., Holzman, T. F., Egan, D. A., Edalji, R., Luly, J. R., Simmer, R., Helfrich, R., Kishore, V. and Rich, D. H. (1990b). Isotope-edited NMR of cyclosporin A bound to cyclophilin: Evidence for a trans 9,10 amide bond. *Science*, **250**, 1406–1409

Fesik, S. W., Gemmecker, G., Olejniczak, E. T. and Petros, A. M. (1991b). Identification of solvent-exposed regions of enzyme-bound ligands by nuclear magnetic resonance. *J. Am. Chem. Soc.*, **113**, 7080–7081

Fesik, S. W., Luly, J. R., Erickson, J. W. and Abad, Z. C. (1988). Isotope-edited proton NMR study on the structure of a pepsin/inhibitor complex. *Biochemistry*, **27**, 8297–8301

Fesik, S. W., Neri, P., Meadows, R., Olejniczak, E. T. and Gemmecker, G. (1992). A model of the cyclophilin/cyclosporin A (CsA) complex from NMR and X-ray data suggests that CsA binds as a transition-state analogue. *J. Am. Chem. Soc.*, **114**, 3165–3166

Fesik, S. W. and Zuiderweg, E. R. P. (1988). Heteronuclear three dimensional NMR spectroscopy. A strategy for the simplification of homonuclear two dimensional NMR spectra. *J. Magn. Reson.*, **78**, 588–593

Fesik, S. W. and Zuiderweg, E. R. P. (1989). An approach for studying the active site of enzyme/inhibitor complexes using deuterated ligands and 2D NOE difference spectroscopy. *J. Am. Chem. Soc.*, **111**, 5013–5015

Fesik, S. W. and Zuiderweg, E. R. P. (1990). Heteronuclear three-dimensional NMR spectroscopy of isotopically labelled biological macromolecules. *Quart. Rev. Biophys.*, **23**, 97–131

Fischer, G., Wittmann-Liebold, B., Lang, K., Kiefhaber, T. and Schmid, F. X. (1989). Cyclophilin and peptidyl-prolyl *cis–trans* isomerase are probably identical proteins. *Nature*, **337**, 476–478

Galat, A., Lane, W. S., Standaert, R. F. and Schreiber, S. L. (1992). A rapamycin-selective 25-kDa immunophilin. *Biochemistry*, **31**, 2427–2434

Gemmecker, G. and Fesik, S. W. (1991). A method of measuring proton–proton coupling constants in carbon-13-labeled molecules. *J. Magn. Reson.*, **95**, 208–213

Gemmecker, G., Olejniczak, E. T. and Fesik, S. W. (1992). An improved method for selectively observing protons attached to ^{12}C in the presence of ^{1}H–^{13}C spin pairs. *J. Magn. Reson.*, **96**, 199–204

Goodford, P. J. (1985). A computation procedure for determining energetically favorable binding sites on biologically important macromolecules. *J. Med. Chem.*, **28**, 849–857

Handschumacher, R. E., Harding, M. W., Rice, J., Drugge, R. J. and Speicher, D. W. (1984). Cyclophilin: a specific cytosolic binding protein for cyclosporin A. *Science*, **226**, 544–547

Harding, M. W., Galat, A., Uehling, D. E. and Schreiber, S. L. (1989). A receptor for the immunosuppressant FK-506 is a *cis–trans* peptidyl-propyl isomerase. *Nature*, **341**, 758–760

Hatanka, H., Kino, T., Miyata, S., Inamura, N., Kuroka, A., Goto, T., Tanaka, H. and Okuhara, M. (1988). FR-900520 and FR-900523, novel immunosuppressants isolated from a Streptomyces. II. Fermentation, isolation and physicochemical and biological characteristics. *J. Antibiot.*, **41**, 1592–1601

Ikura, M. and Bax, A. (1992). Isotope-filtered 2D NMR of a protein–peptide complex: study of a skeletal muscle myosin light chain kinase fragment bound to calmodulin. *J. Am. Chem. Soc.*, **114**, 2433–2440

Kahan, B. B. (1989). Cyclosporine. *New Engl. J. Med.*, **321**, 1725–1738

Kallen, J., Spitzfaden, C., Zurini, M. G., Wider, G., Widmer, H., Wuthrich, K. and Walkinshaw, M. D. (1991). Structure of human cyclophilin and its binding site for cyclosporin A determined by X-ray crystallography and NMR spectroscopy. *Nature*, **353**, 276–279

Kallen, J. and Walkinshaw, M. D. (1992). The x-ray structure of a tetrapeptide bound to the active site of human cyclophilin A. *FEBS Lett.*, **300**, 286–290

Karuso, P., Kessler, H. and Mierke, D. F. (1990). Solution structure of FK-506 from nuclear magnetic resonance and molecular dynamics. *J. Am. Chem. Soc.*, **112**, 9434–9436

Kay, L. E. and Bax, A. (1990). New methods for the measurement of NH–C$_\alpha$H coupling constants in ^{15}N-labeled proteins. *J. Magn. Reson.*, **86**, 110–126

Kay, L. E., Clore, G. M., Bax, A. and Gronenborn, A. M. (1990a). Four-dimensional heteronuclear triple-resonance NMR spectroscopy of interleukin-1β in solution. *Science*, **249**, 411–414

Kay, L. E., Ikura, M. and Bax, A. (1990b). Proton–proton correlation via carbon–carbon couplings: A three-dimensional NMR approach for the assignment of aliphatic resonances in proteins labeled with carbon-13. *J. Am. Chem. Soc.*, **112**, 888–889

Kay, L. E., Ikura, M., Tschudin, R. and Bax, A. (1990c). Three-dimensional triple-resonance NMR spectroscopy of isotopically enriched proteins. *J. Magn. Reson.*, **89**, 496–514

Ke, H., Zydowsky, L. D., Liu, J. and Walsh, C. T. (1991). Crystal structure of recombinant human T-cell cyclophilin A at 2.5 Å resolution. *Proc. Natl Acad. Sci. USA*, **88**, 9483–9487

Kessler, H., Köck, M., Wein, T. and Geherke, M. (1990). Reinvestigation of the conformation of cyclosporin A in chloroform. *Helv. Chim. Acta*, **72**, 1818–1832

Kuntz, I. D., Blaney, J. M., Oatley, S. J., Langridge, R. and Ferrin, T. E. (1982). A geometric approach to macromolecule-ligand interactions. *J. Mol. Biol.*, **161**, 269–288

Liu, J., Albers, M. W., Wandless, T. J., Luan, S., Alberg, D. G., Belshaw, P. J., Cohen, P., MacKintosh, C., Klee, C. B. and Schreiber, S. L. (1992). Inhibition of T cell signaling by immunophilin–ligand complexes correlates with loss of calcineurin phosphatase activity. *Biochemistry*, **31**, 3896–3901

Liu, J., Farmer, J. D., Jr., Lane, W. S., Friedman, J., Weissman, I. and Schreiber, S. L. (1991). Calcineurin is a common target of cyclophilin–cyclosporin A and FKBP–FK-506 complexes. *Cell*, **66**, 807–815

Loosli, H. R., Kessler, H., Oschkinat, H., Weber, H. P., Petcher, T. J. and Widmer, T. J. (1985). The conformation of cyclosporin A in the crystal and in solution. *Helv. Chim. Acta*, **68**, 682–704

McIntosh, L. P., Wand, A. J., Lowry, D. F., Redfield, A. G. and Dahlquist, F. W. (1990). Assignment of the backbone ^{1}H and ^{15}N NMR resonances of bacteriophage T4 lysozyme. *Biochemistry*, **29**, 6341–6362

Marion, D., Driscoll, P. C., Kay, L. E., Wingfield, P. T., Bax, A., Gronenborn, A. M. and Clore, G. M. (1989a). Overcoming the overlap problem in the assignment of ^{1}H-NMR spectra of larger proteins by use of three-dimensional heteronuclear ^{1}H–^{15}N Hartmann–Hahn multiple quantum coherence and nuclear Overhauser-multiple quantum coherence spectroscopy. *Biochemistry*, **28**, 6150–6156

Marion, D., Kay, L. E., Sparks, S. W., Torchia, D. A. and Bax, A. (1989b). Three-dimensional heteronuclear NMR of ^{15}N-labeled proteins. *J. Am. Chem. Soc.*, **111**, 1515–1517

Meadows, R. P., Nettesheim, D. G., Xu, R. X., Olejniczak, E. T., Petros, A. M., Holzman, T. F., Severin, J., Gubbins, E., Smith, H. and Fesik, S. W. (1993). Three-dimensional structure of the FK506 binding protein/ascomycin complex in solution by heteronuclear three- and four-dimensional NMR. *Biochemistry*, **32**, 757–765

Michnick, S. W., Rosen, M. K., Wandless, T. J., Karplus, M. and Schreiber, S. L. (1991). Solution structure of FKBP, a rotamase enzyme and receptor for FK506 and rapamycin. *Science*, **252**, 836–839

Mierke, D. F., Schmieder, P., Karuso, P. and Kessler, H. (1991). Conformational analysis of the *cis*- and *trans*-isomers of FK-506 by NMR and molecular dynamics. *Helv. Chim. Acta*, **74**, 1027–1047

Montelione, G. T., Winkler, M. E., Rauenbuehler, P. and Wagner, G. (1989). Accurate measurements of long-range heteronuclear coupling constants from homonuclear 2D NMR spectra of isotope-enriched proteins. *J. Magn. Reson.*, **82**, 198–204

Moore, J. M., Peattie, D. A., Fitzgibbon, M. J. and Thomson, J. A. (1991). Solution structure of the major binding protein for the immunosuppressant FK506. *Nature*, **351**, 248–250

Neri, P., Meadows, R., Gemmecker, G., Olejniczak, E., Nettesheim, D., Logan, T., Simmer, R., Helfrich, R., Holzman, T., Severin, J. and Fesik, S. (1991). ^1H, ^{13}C and ^{15}N backbone assignments of cyclophilin when bound to cyclosporin A (CsA) and preliminary structural characterization of the CsA binding site. *FEBS Lett.*, **294**, 81–88

Neri, D., Szyperski, T., Otting, G., Senn, H. and Wüthrich, K. (1989). Stereospecific nuclear magnetic resonance assignments of the methyl groups of valine and leucine in the DNA-binding domain of the 434 repressor by biosynthetically directed fractional ^{13}C labeling. *Biochemistry*, **28**, 7510–7516

Nilges, M., Clore, G. M. and Gronenborn, A. M. (1988). Determination of three-dimensional structures of proteins from interproton distance data by hybrid distance geometry dynamical simulated annealing calculations. *FEBS Lett.*, **229**, 317–324

Petros, A. M., Gampe Jr., R. T., Gemmecker, G., Neri, P., Holzman, T. F., Edalji, R., Hochlowski, J., Jackson, M., McAlpine, J., Luly, J. R., Pilot-Matias, T., Pratt, S. and Fesik, S. W. (1991). NMR studies of an FK-506 analog, [U-^{13}C]ascomycin, bound to FKBP: Conformation and regions of ascomycin involved in binding. *J. Med. Chem.*, **34**, 2925–2928

Petros, A. M., Gemmecker, G., Neri, P., Olejniczak, E. T., Nettesheim, D., Xu, R. X., Gubbins, E. G., Smith, H. and Fesik, S. W. (1992a). NMR studies of an FK-506 analog, [U-^{13}C] ascomycin, bound to FK-506-binding protein. *J. Med. Chem.*, **35**, 2467–2473

Petros, A. M., Kawai, M., Luly, J. R. and Fesik, S. W. (1992b). Conformation of two non-immunosuppressive FK506 analogs when bound to FKBP by isotope-filtered NMR. *FEBS Lett.*, **308**, 309–314

Petros, A. M., Neri, P. and Fesik, S. W. (1992c). Identification of solvent-exposed regions of an FK-506 analog, ascomycin, bound to FKBP using a paramagnetic probe. *J. Biomol. NMR*, **2**, 11–18

Quesniaux, V. F. J., Schreier, M. H., Wenger, R. M., Hiestand, P. C., Harding, M. W. and Van Regenmortel, M. H. V. (1987). Cyclophilin binds to the region of cyclosporine involved in its immunosuppressive activity. *Eur. J. Immunol.*, **17**, 1359–1365

Quesniaux, V. F. J., Schreier, M. H., Wenger, R. M., Hiestand, P. C., Harding, M. W. and Van Regenmortel, M. H. V. (1988). Molecular characteristics of cyclophilin–cyclosporin interaction. *Transplantation*, **46**, 23S–27S

Rich, D. H., Sun, C.-Q., Guillaume, D., Dunlap, B., Evans, D. A. and Weber, A. (1989). Synthesis, biological activity, and conformational analysis of 2S,3R,4S-MeBmt1-cyclosporin, a novel 1-position epimer of cyclosporin A. *J. Med. Chem.*, **32**, 1982–1987

Schreiber, S. L. (1991). Chemistry and biology of the immunophilins and their immunosuppressive ligands. *Science*, **251**, 283–287

Seeholzer, S. H., Cohn, M., Putkey, J. A., Means, A. R. and Crespi, H. L. (1986). NMR studies of a complex of deuterated calmodulin with melittin. *Proc. Natl Acad. Sci. USA*, **83**, 3634–3638

Senn, H., Werner, B., Messerle, B., Weber, C., Traber, R. and Wüthrich, K. (1989). Stereospecific assignment of the methyl ^{1}H NMR lines of valine and leucine in polypeptides by nonrandom ^{13}C labelling. *FEBS Lett.*, **249**, 113–118

Siekierka, J. J., Hung, S. H. Y., Poe, M., Lin, C. S. and Sigal, N. H. (1989). A cytosolic binding protein for the immunosuppressant FK506 has peptidyl-prolyl isomerase activity but is distinct from cyclophilin. *Nature*, **341**, 755–757

Sigal, N. H., Dumont, F., Siekierka, J. J., Peterson, L., Rich, D. H., Dunlap, B. E., Staruch, M. J., Melino, M. R., Koprack, S. L., Williams, D., Witzel, B. and Pisano, J. M. (1991). Is cyclophilin involved in the immunosuppressive and nephrotoxic mechanism of action of cyclosporin A? *J. Exptl Med.*, **173**, 619–628

Spera, S. and Bax, A. (1991). Empirical correlation between protein backbone conformation and C^α and C^β ^{13}C nuclear magnetic resonance chemical shifts. *J. Am. Chem. Soc.*, **113**, 5490–5492

Spitzfaden, C., Weber, H.-P., Braun, W., Kallen, J., Wider, G., Widmer, H., Walkinshaw, M. D. and Wüthrich, K. (1992). Cyclosporin A–cyclophilin complex formation. A model based on x-ray and NMR data. *FEBS Lett.*, **300**, 291–300

Taga, T., Tanaka, H., Goto, T. and Tada, S. (1987). Structure of a new macrocylic antibiotic. *Acta Crystallog.*, **C43**, 751–753

Takahashi, N., Hayano, T. and Suzuki, M. (1989). Peptidyl-prolyl *cis–trans* isomerase is the cyclosporin A-binding protein cyclophilin. *Nature*, **337**, 473–475

Theriault, Y., Logan, T. M., Meadows, R., Yu, L., Olejniczak, E. T., Holzman, T., Simmer, R. L. and Fesik, S. W. (1993). Solution structure of the cyclosporin A/cyclophilin complex by NMR. *Nature*, **361**, 88–91

Thomson, A. W. (1990). FK-506: profile of an important new immunosuppressant. *Transplantation Rev.*, **4**, 1–13

Van Drie, J. H., Weininger, D. and Martin, Y. C. (1989). ALADDIN: An integrated tool for computer-assisted molecular design and pharmacophore recognition from geometric, steric, and substructure searching of three-dimensional molecular structures. *J. Comput.-aided Des.* **3**, 225–251

Van Duyne, G. D., Standaert, R. F., Karplus, P. A., Schreiber, S. L. and Clardy, J. (1991a). Atomic structure of FKBP–FK-506, an immunophilin–immuno-suppressant complex. *Science*, **252**, 839–842

Van Duyne, G. D., Standaert, R. F., Schreiber, S. L. and Clardy, J. (1991b). Atomic structure of the rapamycin human immunophilin FKBP-12 complex. *J. Am. Chem. Soc.*, **113**, 7433–7434

Wandless, T. J., Michnick, S. W., Rosen, M. K., Karplus, M. and Schreiber, S. L. (1991). FK506 and rapamycin binding to FKBP: common elements in immunophilin–ligand complexation. *J. Am. Chem. Soc.*, **113**, 2339–2341

Weber, C., Wider, G., von Freyberg, B., Traber, R., Braun, W., Widmer, H. and Wüthrich, K. (1991). The NMR structure of cyclosporin A bound to cyclophilin in aqueous solution. *Biochemistry*, **30**, 6563–6574

Wenger, R. M. (1986). Cyclosporine and analogues: Structural requirements for immunosuppressive activity. *Transplant. Proc.*, **18**, 213–218

Wider, G., Weber, C., Traber, R., Widmer, H. and Wüthrich, K. (1990). Use of a double-half-filter in two-dimensional ^{1}H nuclear magnetic resonance studies of receptor-bound cyclosporin. *J. Am. Chem. Soc.*, **112**, 9015–9016

Wüthrich, K. (1986). *NMR of Proteins and Nucleic Acids*. Wiley, New York

Wüthrich, K., Spitzfaden, C., Memmert, K., Widmer, H. and Wider, G. (1991). Protein secondary structure determination by NMR. Application with recombinant human cyclophilin. *FEBS Lett.*, **285**, 237–247

Xu, R. X., Meadows, R. and Fesik, S. W. (1993). Heteronuclear 3D NMR studies of water bound to an FK506 binding protein/immunosuppressant complex. *Biochemistry*, **32**, 2473–2480

Xu, R. X., Nettesheim, D., Olejniczak, E. T., Meadows, R., Gemmecker, G. and Fesik, S. W. (1992a). ^{1}H, ^{13}C and ^{15}C assignments and secondary structure of the FK506 binding protein when bound to ascomycin. *Biopolymers*, **33**, 535–550

Xu, R. X., Olejniczak, E. T. and Fesik, S. W. (1992b). Stereospecific assignments and χ_1 rotamers for FKBP when bound to ascomycin from $^3J_{H^\alpha, H^\beta}$ and $^3J_{N, H^\beta}$ coupling constants. *FEBS Lett.*, **305**, 137–143

Zuiderweg, E. R. P., Petros, A. M., Fesik, S. W. and Olejniczak, E. T. (1991). Four-dimensional [^{13}C, ^{1}H, ^{13}C, ^{1}H]HMQC-NOE-HMQC NMR spectroscopy: Resolving tertiary NOE distance constraints in the spectra of larger proteins. *J. Am. Chem. Soc.*, **113**, 370–372

5

NMR Studies of the Structure and Role of Modules Involved in Protein–Protein Interactions

Iain D. Campbell and Paul C. Driscoll

1 Introduction

Non-covalent molecular interactions in biological systems are ubiquitous and finely tuned. While enzyme substrate complexes are relatively well understood and considerable knowledge has been gained recently about the way proteins interact with DNA (Harrison, 1991), understanding of protein–protein interactions is relatively poor. Many complexes involving proteins are large and apparently intractable for structural studies. Examples of interest include: cell–cell interactions mediated by different adhesion molecules; a cytokine bound to a membrane receptor; a tyrosine kinase bound to a protein substrate; a blood clotting complex. In cases where there is information available about protein–protein interactions, e.g. in antibody/protein antigen complexes, the interacting protein surface patches seem to involve more than one peptide loop from different parts of the protein sequence (Chothia, 1991).

In view of this complexity, it is perhaps surprising that the study of protein–protein interactions is a particularly fertile area for research in the 1990s. The aspects that we wish to emphasize here are advances which have arisen from studies of protein 'modules'. Analysis of the rapidly expanding sequence database reveals consensus sequences that occur over and over again. In fact it has been proposed that there may be a relatively limited set of exons used in biological systems (e.g. Dorit and Gilbert, 1991). It is also established that many of these consensus sequences give rise to autonomously folding domains which often appear to have evolved by exon shuffling and duplication. Evidence for this comes from the fact that the phase at the intron–exon boundary is constant (Patthy, 1987). These exon-related domains will be called modules here, following Patthy (1985). It is now widely recognized that the specificity of many protein–protein interactions is determined by protein modules.

Another feature of modular protein–protein interactions is that one particular kind of protein module may have a variety of different functional surface patches. For example, epidermal growth factor-like (G) modules use different patches on the structure when the module is used to bind to a growth factor receptor and when it is used in an enzyme complex involved in blood clotting (Campbell and Baron, 1991; and see below). The notion of a relatively limited number of protein modules with multiple functions opens the way to a 'dissect and rebuild' strategy where information about important interactions in large complexes can be deduced from studies of individual modules, homology modelling and site-specific mutagenesis. Recent studies have been facilitated by the availability of new and improved structure determination methods such as high-resolution NMR and the ability to produce proteins, protein fragments and engineered variants in various cell expression systems. This approach is leading to a rapid and exciting increase in understanding of protein–protein interactions in numerous important areas of biology.

In this chapter we outline our strategy for investigating the structure and function of modular proteins and illustrate its application to the important biological fields of cell adhesion and blood clot homoeostasis.

2 The Study of Protein Modules

Some of the modules commonly found associated with protein–protein interactions are listed in Figure 5.1. The modules appear predominantly, but not exclusively, in extracellular proteins from multicellular organisms. Only a few years ago the structural information about these was very limited. This was partly because proteins made up from mosaics of different modules are glycosylated, membrane-bound and apparently flexible, thus making crystal formation difficult. In the last few years, however, many structures of protein fragments containing modules have appeared, often using material produced by expression techniques. These include recent X-ray structures of two domain, CD4 (I) (Ryu *et al.*, 1990; Wang *et al.*, 1990), kringle (K) modules (Soriano-Garcia *et al.*, 1989; de Vos *et al.*, 1992b); a γ-carboxyglutamate-containing module (Gla) (Soriano-Garcia *et al.*, 1992); a mannose-binding (Lec) module (Weis *et al.*, 1991); and cytokine receptor (CR) modules bound to a cytokine (de Vos *et al.*, 1992a). In addition, high-resolution NMR has been used to study various epidermal growth factors such as (G) modules

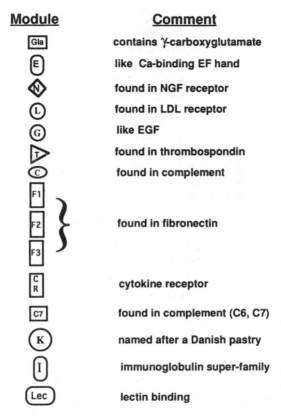

Module	Comment
Gla	contains γ-carboxyglutamate
E	like Ca-binding EF hand
N	found in NGF receptor
L	found in LDL receptor
G	like EGF
T	found in thrombospondin
C	found in complement
F1	
F2	found in fibronectin
F3	
C R	cytokine receptor
C7	found in complement (C6, C7)
K	named after a Danish pastry
I	immunoglobulin super-family
Lec	lectin binding

Figure 5.1 A list of protein modules identified by multiple-sequence alignment of mosaic proteins. Each module is related to a single exon in the gene. The intron/exon boundaries are usually phase 1. More details of these and other modules are given in the text and in Baron *et al.* (1991) and Patthy (1991). Further details about some of the modules shown can be obtained in the following references: L (Soutar and Knight, 1990); E (Engel, 1991); N (Mallett *et al.*, 1990); C7 (Reid and Day, 1989); T (Engel, 1991)

(Cooke *et al.*, 1987; Montelione *et al.*, 1987; Tappin *et al.*, 1989; Handford *et al.*, 1990, 1991; Selander *et al.*, 1990; Hommel *et al.*, 1991; Baron *et al.*, 1992b), complement control protein (C) modules (Norman *et al.*, 1991; Barlow *et al.*, 1992), K modules (Atkinson *et al.*, 1990; Byeon and Llinas, 1991), a fibronectin type I (F1) module (Baron *et al.*, 1990; Downing *et al.*, 1992), a fibronectin type II (F2) module (Constantine *et al.*, 1991), a fibronectin type III (F3) module (Baron *et al.*, 1992a) and an immunoglobulin superfamily (I) module (Driscoll *et al.*, 1991).

The strategy used in this laboratory to study modular proteins has been set out previously (Baron *et al.*, 1991). It is shown, in outline, in Figure 5.2. Numerous module families have been discovered and defined

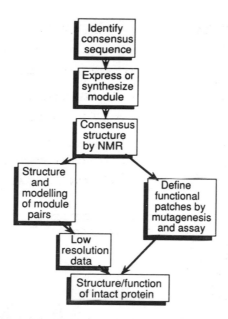

Figure 5.2 Outline of the strategy for studying modular proteins exercised in our laboratory (see also Baron *et al.*, 1991)

by consensus sequences (Williams and Barclay, 1988; Doolittle, 1989; Reid and Day, 1989; Baron *et al.*, 1991; Patthy, 1991). Single or multiple modules are chosen for study and produced for structural studies using a heterologous gene expression system in suitable cells such as yeast or *E. coli*. The expressed protein is purified and characterized and its structure is determined by NMR (see next section). The known structures of modules can be used to model other similar modules, and this, together with the structure of module pairs and low-resolution data, e.g. from electron microscopy, solution scattering or chemical cross-linking, can be used to build up a view of the structure of the intact protein.

Another important aspect of this work, however, is to relate the structure to the function, and here the 'dissect and rebuild' strategy is again a powerful approach. The functional role of individual modules and their site-specific variants can be assessed in an assay system and compared with results with the intact protein (e.g. Handford *et al.*, 1991). Both the structure and function of large biological complexes can thus be built up from information derived from studies of fragments and intact molecules. In the remainder of this review we shall briefly illustrate the methodology and strategy used in this laboratory to derive NMR structures of modular proteins. Four of the structures obtained in our laboratory are illustrated in Figure 5.3.

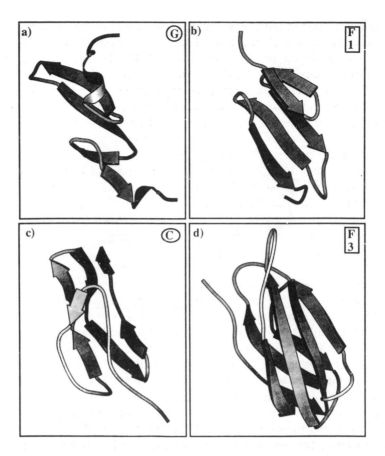

Figure 5.3 Schematic illustrating the backbone folds of the solution structures of four protein modules determined by NMR spectroscopy in our group: G module, human epidermal growth factor (Hommel *et al.*, 1992); F1 module from human tissue plasminogen activator (Downing *et al.*, 1992); tenth F3 module from human fibronectin (Baron *et al.*, 1992a; Main *et al.*, unpublished); and 16th C module from human factor H (Norman *et al.*, 1991). Regular β-sheet and α-helical secondary structure elements are indicated. The illustrations were composed using the program MOLSCRIPT (Kraulis, 1991)

3 NMR Methods

Our studies of the three-dimensional solution structures of protein modules are based on the methods of multidimensional NMR pioneered by Kurt Wüthrich (1986, 1989). There are now more than 100 structures of small proteins that have been determined by these techniques, and a more detailed discussion of the methods involved is given by Bax (1991), Clore and Gronenborn (1991a,b) and in Chapters 1–4. Here we give a brief summary of the particular techniques used in our laboratory; some of these have been influenced by personal experience (P.C.D.) of the techniques developed in the laboratory of Dr G. M. Clore and Dr A. M. Gronenborn at the National Institutes of Health.

All of our studies have been performed either on a home-built 500 MHz spectrometer equipped for inverse mode heteronuclear work and three-dimensional spectroscopy or on a Bruker AM series 600 MHz machine. Both spectrometers utilize 5 mm probeheads and a home-built temperature controller. 2D and 3D NMR data processing is routinely performed off-line, using the software package FELIX (Hare Research Inc.) on SUN Microsystems SPARC work stations.

The proteins that we study have all been produced in recombinant expression systems, using either yeast or bacterial cell culture. The yields of protein produced vary considerably but are typically in the region 1–5 mg pure protein per litre of culture. The NMR samples are usually 1–3 mM in concentration. After characterization of the protein spectrum and optimization of solution conditions (buffers, concentration, pH and temperature), using 1D and 2D spectroscopy in both H_2O and D_2O solution, sequence-specific resonance assignment is performed in the usual manner (Wüthrich, 1986). We usually combine 2D HOHAHA, NOESY and COSY spectra recorded in H_2O solution with 2D HOHA-HA, NOESY and DQF-COSY spectra recorded in D_2O solution to provide a minimum set of NMR data. The initial analysis identifies as many different amino acid spin systems as possible by recognition of through-bond connectivities. These are then assigned to specific amino acid spin systems by sequential NOEs in the H_2O NOESY spectrum. In our experience, there is no simple formula to the efficient completion of this stage of spectral assignment. The data have to be recorded under a variety of conditions and analysed as they are obtained, to derive an unambiguous self-consistent solution to the assignment problem. This strategy pays dividends during the later stages of the structure determination, when the resolution of closely overlapping long-range NOEs is very important.

A philosophy that we have found useful in our choice of NMR experiments, particularly in H_2O solution, is to avoid wherever possible presaturation of the solvent resonance, since this leads to bleaching of the $C\alpha H$ region of the spectrum, which is important for sequential assignment and characterization of secondary structure elements. The application of 'jump–return' (Plateau and Guéron, 1982) read pulses in H_2O NOESY and HOHAHA experiments is a convenient way of suppressing the solvent resonance while retaining spectral cross-peaks that connect $C\alpha H$ resonances near to the water peak (Driscoll *et al.*, 1989). Successful application of the jump–return pulse sequence requires the residual solvent resonance to be artificially removed from the FID by deconvolution in the time domain (Marion *et al.*, 1989). The spectra are processed to optimize the base-plane in the downfield region of the spectrum (aromatics and amide NH signals). The upfield aliphatic regions of jump–return spectra are difficult to process with optimal base-plane characteristics and this region is best analysed by other experiments in D_2O solution. The jump–return pulse sequence is not readily applicable in the COSY experiment, but we find that application of the SCUBA method (Brown *et al.*, 1987) to recover 'bleached' cross-peaks in H_2O solution works well.

$^3J_{NH}{}^\alpha$ spin–spin coupling constants can be determined from F_2 cross-sections of high-resolution COSY or DQF-COSY spectra of the protein in H_2O. We utilize a line shape simulation program to best-fit these cross-sections with theoretical line shapes that take into account the data processing used (Redfield and Dobson, 1990). For analysis of χ_1 side-chain torsion angles, $^3J_{\alpha\beta}$ spin–spin coupling constants are obtained from measurement of the cross-peak fine structure of PECOSY spectra recorded in D_2O solution (Müller, 1987; Marion and Bax, 1988).

It is valuable to isotopically label the protein with ^{15}N whenever possible. We do this by growing the cells in minimal medium supplemented with a ^{15}N ammonium salt as the sole nitrogen source. In two cases described in more detail below, the rat CD2 domain 1 protein and the tenth type III repeat of human fibronectin, 2D and 3D heteronuclear $^{15}N-{}^1H$ NMR spectroscopy was utilized to perform the sequential assignment. First the 2D $^{15}N-{}^1H$ heteronuclear single quantum correlation experiment is used to characterize the protein and count amino acid residues and side-chain NH_2 groups. (At this stage occasional mistakes in sequencing can be picked up; for example, in the case of the CD2 domain we found an asparagine residue instead of the predicted aspartic acid at one position.) Conditions of pH, temperature and concentration are then identified to optimize the appearance of the 2D heteronuclear spectrum before the acquisition of 3D $^{15}N-{}^1H$ heteronuclear NOESY-HMQC and HOHAHA-HMQC spectra.

Analysis of 3D spectra enhances significantly the speed and reliability of the sequential assignment process for proteins in the size range 10–20 kDa (Clore and Gronenborn, 1991b) through the increased resolution provided by the extra dimension. In keeping with the philosophy on the saturation of the solvent resonance, we find that the application of 'trim' pulses in these heteronuclear experments is an important addition to avoid saturation of CαH resonances (Messerle *et al.*, 1989). The ^{15}N-labelled proteins are also useful for a number of other experiments. Thus, the HMQC-J (Kay and Bax, 1990) experiment is a very good source of $^3J_{NH}{}^\alpha$ spin–spin coupling constants. The F_1 doublet line shapes obtained for each non-glycine amide NH of the protein can be fitted to a theoretical line shape using a non-linear simplex-based fitting program (Redfield and Dobson, 1990; Norwood *et al.*, 1992). ^{15}N-labelled proteins are also candidates for studies of backbone dynamics through the measurement of ^{15}NT_1, T_2 and heteronuclear ^{1}H–^{15}N NOE and time-resolved amide NH/solvent exchange kinetics (Kay *et al.*, 1989).

4 Structure Calculation

Our protein structures have been calculated on the basis of the NOE-derived distance restraints and spin–spin coupling constant derived dihedral angle restraints mainly using a dynamical simulated annealing approach with the program X-PLOR (Brünger *et al.*, 1987; Brünger, 1988). The protocol used is similar to YASAP (*Yet Another Simulated Annealing Protocol*), a program written by Michael Nilges and described in the X-PLOR 2.1 manual, which is in turn derived from the protocol and restraints potentials developed by Nilges *et al.* (1988). In outline, the method involves using high-temperature molecular dynamics to fold a protein from a conformation with randomized backbone torsion angles and extended side-chains under the influence of forces that maintain correct covalent geometry and chirality of the chemical structure, soft square well NOE and torsion angle potentials that drive the protein chain towards the correctly folded structure, and a soft quartic repulsive term that serves to represent the van der Waals (non-bonded) component of the target function. No electrostatic or explicit hydrogen-bonding terms are used during the calculations. Several picoseconds of dynamics at 1000 K are simulated during which the van der Waals force constant is kept low, allowing the protein to rearrange its conformation under the influence of the experimental restraint terms. The van der Waals force constant is then slowly raised, after which the system is slowly cooled to 300 K, followed by restrained minimization. Structures that have small violations of the NOE and constrained dihedral terms of the target function, and good covalent geometry and van der Waals contacts are selected. Structures that have successfully converged are those that exhibit no significantly large violations of any one distance or dihedral restraint and have low values of the van der Waals term. When a structure has been successfully refined, the convergence rate of the calculation increases above ~50%, and a large family of structures can be generated from different random starting conformations. The degree of definition of each part of the structure can be determined by best-fitting all the structures to one another. Regions of the protein structure that are well defined show only a small root mean square difference (rmsd), while those parts of the structure that have relatively few restraints will show a broader range of conformations. In many cases (but not necessarily all), these poorly defined regions of the structure may correspond to regions of the protein structure that are relatively mobile in solution. A single representative structure, useful for display purposes, can be generated by averaging the coordinates of the fitted structures, followed by restrained minimization. This structure is known as the restrained minimized average structure (Clore *et al.*, 1986).

The process of refinement of the NMR structure is one where, on the basis of structures computed with a limited set of experimentally derived restraints, the spectra can be re-examined to resolve overlapping or ambiguous NOEs that have not been included previously in the calculations (Clore and Gronenborn, 1991b). In the latter stages of the structure refinement we utilize systematic comparison of short-range distance restraints with a systematic database of all ϕ, ψ, χ_1 conformations (in 10° steps) of a Gly–X–Gly tripeptide segment using the program STEREOSEARCH (Nilges *et al.*, 1990). Implicit in this type of analysis is the stereospecific assignment of pro-chiral groups in the protein. This approach leads to high-resolution solution structures, a goal that we are currently pursuing for a number of the protein modules whose low-resolution structures are described in this chapter.

5 Cell Adhesion

Multicellular organisms contain specialized cells which interact with each other and with a complex network of secreted protein and carbohydrates that make up the extracellular matrix. A very schematic and simplified example is shown in Figure 5.4, where a T-lymphocyte cell is shown interacting with an antigen-presenting cell and with one molecule in the extracellular matrix, fibronectin. The modular nature of some of the molecules involved is also indicated. Some recent work in this laboratory on some of these molecules is now discussed, with particular emphasis on CD2, a cell-surface molecule found on T lymphocytes, thymocytes and natural killer cells.

CD2

The basis of the cellular immune response is the circulation of cells of the T-lymphocyte lineage in the blood vessels and lymphatic system. These cells are poised to react to the presence of a foreign organism by congregating in the lymphoid organs and migrating through the endothelium and basement membrane, eventually to transfer to the site of infection. The specific response that leads to the killing of a virus-infected cell is the recognition, by the T-cell receptor (TCR), of an antigenic peptide in a class I MHC molecule on the antigen-presenting cell (APC). The recognition of the MHC/peptide complex by the

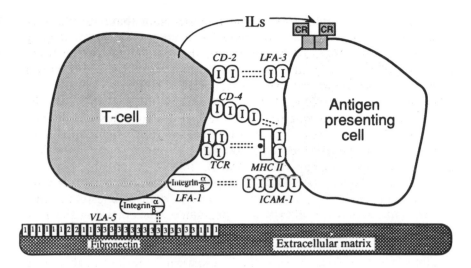

Figure 5.4 A schematic view of cell–cell and cell–matrix adhesion interactions, showing some of the modular proteins involved. The LFA-1/ICAM-1 and CD2/LFA-3 adhesion reactions that accompany the recognition of an MHC/peptide complex by a T-cell receptor are illustrated. The signals generated by T-cell activation lead to the production of cytokines such as IL-2, which bind to receptors which are also modular proteins. The integrin VLA-5 is involved in contacts between the T cell and the fibronectin in the extracellular matrix. For reasons of space the fibronectin modules F1, F2 and F3 are merely designated 1, 2 and 3

TCR/CD3 complex and CD8 leads to the production of biochemical signals (Ca^{2+}, IL-2) that then give rise to T-cell activation and proliferation (in the case of T-helper cells, class II MHC/peptide complexes associate with the TCR and CD4). Other intercellular protein–protein interactions between the T cell and the APC appear to give rise to biochemical signals that synergize with the primary activation signal from the TCR. The precise functional and structural mechanisms of these signalling processes is a subject of intensive investigation.

The regulation of the T cell between adherent and non-adherent states (and thereby the control of T-cell circulation) is a complex phenomenon. The adhesive state of a T cell is now known to depend on a number of specific intercellular protein-mediated interactions, as well as the precise state of activation of the T cell itself. Excellent reviews of our current understanding of lymphocyte adhesion reactions have recently been published (Springer, 1990; Dustin and Springer, 1991).

The analysis of monoclonal antibodies (mAbs) that are inhibitory for T-lymphocyte function has given rise to the identification of a number of cell-surface proteins that are involved in intercellular adhesion. Both the LFA-1/ICAM-1 and CD2/LFA-3 interactions, depicted schematically in Figure 5.4, were identified through screening mAbs for their ability to inhibit T-cell killing (LFA, lymphocyte function-related antigen; ICAM, intercellular adhesion molecule). These two adhesion interactions, which have been shown to be independent by examining the effect of combinations of mAbs, appear to have fundamentally different time-scales of action. Thus, the CD2/LFA-3 adhesion reaction becomes active over a period of hours after T-cell activation, whereas the adhesion mediated by the LFA-1/ICAM-1 pair is transiently increased over a period of minutes during T-cell activation (Dustin and Springer, 1991). The function of LFA-1 is known to be Mg^{2+} ion dependent, and this molecule is a member of the family of two chain proteins known as integrins (because these proteins often are involved in the link between the extracellular matrix and the cytoskeleton). ICAM-1 is confidently predicted to be a member of the immunoglobulin superfamily of cell-surface molecules, comprising five tandemly repeated Ig domains. The three-dimensional structure of the integrins and the structural basis of cation-dependent interaction of LFA-1 and ICAM-1 remains to be determined.

The expression of the CD2 antigen is limited to T lymphocytes and their precursors, while LFA-3 is expressed on virtually all cell types. A common assay for the interaction between these two molecules is the 'rosetting' of erythrocytes around activated T cells that express CD2 *in vitro*. Unactivated or 'resting' T cells do not form these rosettes, indicating that the adhesive state of the T cell is regulated, possibly by the charge state of the cell surface (Springer, 1990). Curiously, sheep erythrocytes have long been used as a clinical test for human T lymphocytes because of their ability to adhere strongly and induce rosetting. The rosetting in this case is mediated by a sheep homologue of LFA-3 which binds avidly to human CD2. In experiments with soluble forms of recombinant or purified CD2, the affinity for human LFA-3 is found to be low with $K_d \sim 10^{-6}$ M (Recny *et al.*, 1990).

CD2 and LFA-3 are transmembrane glycoproteins of around 50 kDa and 70 kDa molecular weight, respectively. LFA-3 can also be found in a 55 kDa glycolipid-linked form. CD2 is distinguished by a 117 amino acid cytoplasmic domain, rich in proline and histidine residues. It is thought that this domain may play a role in transmembrane signalling. There is evidence that CD2 may transduce a signal synergistic with signals derived from the TCR. Pairs of mAbs to CD2 or a single mAb in combination with LFA-3 can lead to antigen-independent T-cell stimulation. Recent work shows that this mechanism is dependent on the presence of the TCR or the FcγRIIIA (CD16) receptor (Spruyt *et al.*, 1991).

On the basis of amino acid sequence patterns, the extracellular portions of both CD2 and LFA-3 were each predicted to contain two immunoglobulin-like domains. On this basis both proteins were assigned membership of the immunoglobulin superfamily (IgSF) or I module class. The immunoglobulin superfamily is defined on the basis of statistical sequence pattern analysis (Williams and Barclay, 1988). This designation of CD2 and LFA-3 to I was despite the absence of the consensus cysteine residues in the first 100 amino acids (domain 1) of each sequence. These consensus cysteines usually form a disulphide link between strands B and F of the immunoglobulin fold. Rather, the assignment of the first domain of these proteins to the I family relied upon the recognition of certain short stretches of alternating hydrophobic and hydrophilic residues and a statistically significant sequence similarity match with other proteins of the family. In spite of this analysis, the assignment of CD2 to I has been disputed on the basis of *ab initio* secondary structure prediction and physical studies of soluble forms of the extracellular portion of the molecule. These studies included an analysis of circular dichroism spectra of one- and two-domain forms of human CD2, prepared by proteolytic digestion of full-length CD2 produced in a baculovirus expression system (Clayton *et al.*, 1987; Recny *et al.*, 1990). The conclusions of these studies were that the adhesive properties of CD2 are contained in the N-terminal domain (domain 1). It was predicted that domain 1 contained 22% α-helical character and, by direct comparison with a parallel analysis of immunoglobulin sequences, that the extracellular portion of the protein has a novel α/β protein fold.

Our studies of CD2 have focused on the identification of the three-dimensional solution structure of the N-terminal domain of *rat* CD2 by NMR spectroscopy. The construct used for the production of the first domain of this protein comprised residues 1–99. This domain has 43% sequence identity with residues 4–104 of human CD2. This rat CD2 domain was conveniently expressed in *E. coli* as a fusion with glutathione-*S*-tranferase (GST), using the p-GEX expression vector. The expressed fusion protein contains a Leu–Val–Pro–Arg–Gly–Ser thrombin cleavage site between the C terminus of the GST and the N terminus of the CD2 domain. The fusion protein was expressed at levels up to 40 mg/l. After recovery of the fusion protein from lysed *E. coli* cells on a glutathione agarose column, thrombin cleavage yielded a 101 amino acid protein Gly–Ser–CD2 (1–99) which was purified by gel filtration and used for the structure determination. While this protein yielded excellent 1D and 2D ^{1}H NMR spectra, uniformly ^{15}N-labelled CD2 domain 1 was produced, using the same expression system but with minimal growth medium supplemented with ^{15}N ammonium chloride. The ^{15}N-labelled material allowed the application of heteronuclear 2D and 3D NMR experiments that considerably enhanced the ease of the assessment of the sample purity, sequence-specific residue assignment,

assignment of long-range NOEs involving amide NH protons and the determination of $^3J_{HN\alpha}$ coupling constants. The spectra were recorded with a 3 mM CD2 sample at pH 4.2 and 23 °C or 30 °C.

As the assignment of the spectrum of rat CD2 domain 1 proceeded, it became clear that the secondary structure of the protein was dominated by β-sheet secondary structure. During further analysis it emerged that the secondary structure topology was very similar to that of an immunoglobulin V domain. Structures of CD2 domain 1 were computed on the basis of 713 NOE distance restraints (20 intraresidue, 356 sequential, 102 short-range, 60 medium-range and 185 long-range), 64 φ dihedral restraints and 68 H-bond distance restraints. The structure calculations were performed by dynamical simulated annealing starting from a random coil backbone configuration using the program X-PLOR (see above). The calculations were restricted to residues 3–99 of the domain, as both the Gly–Ser dipeptide leader and residues 1 and 2 showed no evidence of regular structure. In total, 16 independent structures of CD2 were obtained that satisfied the NOE distance restraints with no individual distance violation greater than 0.5 Å. While the structures cannot be considered to be high-resolution by modern NMR standards, they are of sufficient quality to determine the protein fold and the relative orientation of the internal side chains.

The structure of CD2 obtained from the NMR data clearly shows a fold that is very similar to that of an immunoglobulin V domain. The structure comprises nine extended strands constructed into two β sheets. The naming of these strands is given in Figure 5.5 in accordance with the naming of β strands in an Ig V-domain fold. Figure 5.5 shows that the folding topology of the β-strand segments of the CD2 domain is the same as that observed for CD4 domain 1 and two Ig V-domain structures (one from an Ig light chain and one from a heavy chain) obtained by X-ray crystallography. RMS differences for 28 core Cα atoms of CD2 and the three other structures are in the range 0.9–1.5 Å, similar in magnitude to the spread of the 16 converged CD2 structures (Driscoll *et al.*, 1991).

The structure obtained by NMR for rat CD2 domain 1 clearly puts CD2 in the immunoglobulin superfamily of cell surface antigens, confounding the secondary structure prediction and circular dichroism studies of Clayton *et al.* (1987). The absence of the consensus I module disuphide bond is one of the interesting features of the CD2 structure and may be one of the more important reasons for the failure of the predictive methods. In rat CD2 domain 1 residues Ile-18 and Val-78 are in the corresponding positions to the consensus disulphide. The side-chains of these two residues are both buried in the core of the structure and are in contact with one another, bridging the gap between the β sheets. In the NMR structures, the distance between the Cα atoms of

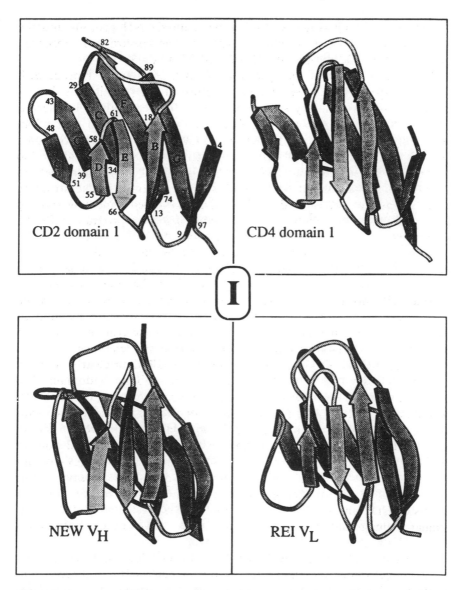

Figure 5.5 The structure of rat CD2 domain 1 is compared with other V-type I domains from immunoglobulin superfamily molecules: domain 1 of human CD4; a heavy chain V domain from the immunoglobulin Fab fragment NEW, and a domain from the REI light chain dimer (Bence-Jones protein). The standard β-strand nomenclature for I domains is illustrated for the CD2 domain

these residues is between 6.8 and 7.2 Å, well within the range observed for the separation between cysteine Cα atoms of Ig consensus disulphide bonds (5.6–7.4 Å). The major differences between the Ig V-type structures of Figure 5.5 are in the length and conformations of the loops at the 'top' of the domain, corresponding to the hypervariable complement-arity-determining regions (CDR) of Ig V domains. Thus, the loop connecting strands D and E of CD2 domain 1 is much shorter than that seen in Ig and CD4 domain 1 structures, whereas the loop con-necting strand B of one β sheet and the C strand on the other β sheet is much longer and more convoluted in conformation. We believe that it is significant that CD2 possesses a proline residue at the position immediately following the consensus disulphide position (Pro-19). A proline residue has not been observed in the corresponding position of any immunoglobulin molecule, while LFA-3, the adhesion receptor of CD2, does have such a proline residue. We therefore speculate that the LFA-3 domain may share some of these structural features of CD2, consistent with the notion that the heterotypic CD2/LFA-3 adhesion interaction may have evolved from a more primordial homotypic inter-action (Williams, 1987). Using a strategy of random mutagenesis, a number of point mutants of human CD2 have been identified which exhibit decreased binding to LFA-3 (Peterson and Seed, 1987). Origin-ally the residues involved had been assumed to lie in the CDR-type loops of the putative Ig V fold. However, on the basis of the three-dimensional structure of rat CD2 described here and sequence alignment of the human and rat CD2 sequences based on that structure, it became clear that these mutations mapped to the face of the major β sheet of the CD2 domain 1 structure, including residues in strands C' and F and the loops between strands C and C', and between strands F and G. The amino acid sequences of the rat and human CD2 sequences differ slightly in these regions (for example, there are two insertions in the C–C' loop in the human CD2 sequence). However, Figure 5.6 illustrates that the approximate locations of the side-chains of the corresponding amino acid side-chains in the rat CD2 structure are broadly distributed on the face of the β sheet. On the basis of this structural analysis, a further study of site-directed mutagenesis is being pursued in order to map more precisely the adhesion site in human CD2.

In summary, the NMR work on CD2 domain 1 clearly resolves the dispute over the nature of the fold of this important T-cell antigen. The structure provides a basis for mapping the residues of the human CD2 known to be important for adhesion to LFA-3. In addition, the structure is a prototype for predicted members of the IgSF that do not possess the consensus disulphide residues. Examples include LFA-3 domain 1, CEA domain 1, CD4 domain 3 and PDGF receptor domain 4, as well as predicted IgSF domains of myosin-binding muscle proteins. It is hoped to extend the NMR studies of T-cell adhesion receptors though the investigation of epitope mapping using NH exchange studies, protein dynamics and, in favourable cases, protein–protein complexes.

CD2 domain 1

Figure 5.6 A schematic illustrating the residues of the rat CD2 domain 1 structure that correspond to those residues of human CD2 that, when individually mutated, abrogate binding to human LFA-3. The residues are clustered on the major β-sheet face of the CD2 structure, the loops joining the F and G strands and the loop connecting strands C and C′

Fibronectin

Fibronectin is a large protein found in the extracellular matrix and serum. It is made up from three different kinds of structural unit, F1, F2 and F3 (Engel, 1991; Yamada, 1991). Among its diverse roles, fibronectin is involved in wound healing and embryonic development, since it binds to other components of the extracellular matrix and to integrin receptors on the cell surface. This is shown schematically in Figure 5.4. Two fibronectin chains of the kind shown in Figure 5.4 are linked by disulphide bonds and the F1 modules near the N termini are involved in interactions with collagen in the extracellular matrix, while the tenth F3 module interacts with integrins. The localization of these biological functions was achieved by dissecting the protein into fragments and then using the fragments to interfere with some biological process such as spreading of cells on a surface. The adhesion-recognition sequences have been further localized to short peptide fragments, including the Arg–Gly–Asp (RGD) sequence in the tenth F3 module and a Leu–Asp–Val (LDV) sequence which is found in an F3 module whose presence in intact fibronectin is made variable by alternative mRNA splicing (Yamada, 1991). Some of these short peptides show significant biological activity, and this has led to their use as lead compounds in drug development.

No structural information at atomic resolution was available about fibronectin until recently, but now structures for all three different types of module have been obtained by NMR (Baron *et al.*, 1990, 1992a,b;

Constantine *et al.*, 1991). The F1 module structure from fibronectin which was determined in this laboratory was the seventh type I module. We have also determined the structure of a homologous F1 module from tissue plasminogen activator which is shown in Figure 5.3 (Downing *et al.*, 1992).

The fibronectin type III module (F3) has been identified in numerous cell adhesion molecules and intracellular muscle related proteins. It is about 100 residues in length and recently we have achieved a complete assignment of a ^{15}N-labelled F3 module (100 residues) which contains an RGD sequence known to bind to integrins. This has allowed the secondary structure to be defined (Baron *et al.*, 1992a,b), and the complete structure is illustrated in Figure 5.3. The RGD sequence is located on a loop between the sixth and seventh β strands of the structure. Interestingly, the lack of ^{1}H–^{1}H NOEs and relatively low heteronuclear ^{15}N NOEs indicate that this loop is relatively flexible.

The topology of the F3 (Baron *et al.*, 1992a,b) module is compared in Figure 5.7 with the topologies of I modules (e.g. Williams and Barclay, 1988), the bacterial chaperonin protein PapD (Holmgren *et al.*, 1992), CD4 domain 2 (Ryu *et al.*, 1990; Wang *et al.*, 1990) and the CR module (de Vos *et al.*, 1992a). I modules in the V class (related to

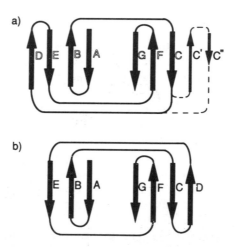

Figure 5.7 The β-strand topology of some modules discussed in the text. As drawn, neighbouring strands are hydrogen bonded to each other to form two β-sheets which sandwich together to form the core fold of the module. (a) The topology of most I modules is related to that seen for the variable (V-type) and constant (C-type) domains of immunoglobulin molecules. The V-type modules, which include CD2 domain 1, have the extra C'–C" strands. (b) The strand topology found in PapD (Holmgren *et al.*, 1992), CD4 domain 2 (Ryu *et al.*, 1990; Wang *et al.*, 1990), F3 (Baron *et al.* 1992a,b) and CR modules (de Vos *et al.*, 1992a). Note that strand D moves to the opposite β sheet compared with the majority of I modules

variable domains of Ig molecules) have two extra strands inserted between the C and D strands (strands C′ and C″). Often the C″ strand is irregular in V-type I module structures. C-type (related to constant domains from Ig molecules) have a core of a three-strand/four-strand β-sheet sandwich. F3 and CR modules and some I modules (e.g. the second domain from CD4), as well as domain 2 of PapD, share a common core of a three-strand/four strand β-sheet sandwich, but the two classes differ by 'sheet switching' of the D strand. Thus, the D strand is hydrogen bonded to the C strand rather than the E strand, putting it in a similar location to the C′ strand of the V-type I modules.

6 Blood Clot Homoeostasis

Most of the proteins in the enzyme cascades which control the rate of clot formation and dissolution are serine proteases whose protein–protein specificity is tuned by attaching various protein modules (Furie and Furie, 1988). A schematic view of some of the blood-clotting and fibrinolysis pathways, together with some of the modular proteins

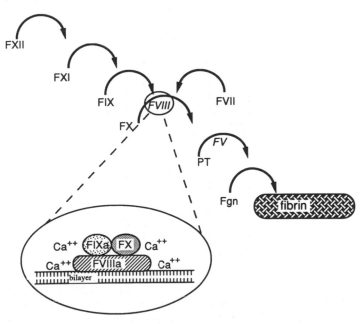

Figure 5.8 A schematic view of the blood clotting cascade—further details of the various factors involved can be found in Furie and Furie (1988) (PT, prothrombin; Fgn, fibrinogen). A representation of one of the complexes formed in the cascade is shown in the inset, where activated factor IX (FIX) cleaves factor X (FX) in the presence of cofactors factor VIIIa (FVIIIa), calcium ions and phospholipid

involved, is given in Figures 5.8 and 5.9. Since intact versions of the molecules involved have proved to be very difficult to crystallize, assessment of the function and atomic resolution structure of modules in the light of functional and low-resolution studies of the intact molecules is again probably the only feasible strategy. Most of the structures of the modules have now been determined and much progress has been made with studying the intact molecules and their mutants. The next phase of the work will be to define the 3D structures and interaction patches of the intact enzymes more precisely. Understanding the various protein–protein interactions in clotting homoeostasis is clearly of considerable medical and pharmaceutical interest. Two particular examples which we are working on will be briefly described here.

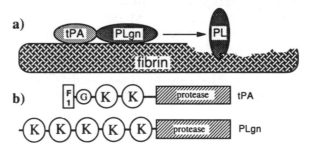

Figure 5.9 Fibrinolysis: (a) tissue plasminogen activator (tPA) activates plasminogen (PLgn) to plasmin (PL) which then dissolves a fibrin clot; (b) The modular structure of tPA and plasminogen

Tissue Plasminogen Activator (tPA)

There continues to be great medical and commercial interest in tPA as a selective fibrinolytic agent. No 3D structure of the intact molecule is available, but the structure of the five structural components of the molecule (see Figure 5.9) are becoming well understood. X-ray and NMR structures of the second kringle module have appeared (Byeon and Llinas, 1991; de Vos *et al.*, 1992b) and the structure of the F1 module of tPA has recently been determined in this laboratory by NMR (Downing *et al.*, 1992). The remaining modules can be modelled with a reasonable degree of precision because of the availability of structures for relatively close homologues. The serine protease domain is similar to kallikreins and other serine proteases with known X-ray structures; the first and second kringles are quite closely related and the G module can be modelled from the various NMR structures obtained here and in other laboratories.

    An important aspect, however, is to define the globular fold and the functional patches of the intact protein. Interestingly, tPA does not, unlike most serine proteases, require a proteolytic activation step but is active in both single-chain and two-chain forms. Some of the functional patches of the molecule have been well characterized; for example, there is a specific lysine-binding site in the second kringle which has been associated with fibrin binding (Byeon and Llinas, 1991; de Vos *et al.*, 1992b). Extensive site-specific mutagenesis experiments have, however, recently suggested that fibrin binding also involves the F1, G, K1 and protease domains (Haber *et al.*, 1989; Bennett *et al.*, 1991). Further clues about structure and function of intact tPA can be deduced from changes in function which result from changes in glycosylation (Witwer and Howard, 1990). Module deletion and insertion experiments, together with differential scanning calorimetry (Novokhatny *et al.*, 1991), have given evidence for relatively strong interactions between modules. All these data will only be interpretable in terms of information about the structure of combinations of modules as well as of isolated ones. One way of approaching this problem is to produce pairs of modules for structural studies, to see how the modules fit together. As a start in this direction, the F1–G module pair from tPA has been produced in a yeast expression system and NMR studies on the structure are well advanced in this laboratory.

Factor IX

Factor IX is made up of two G and one Gla module plus a serine protease domain. It is activated by the removal of a short peptide from a region just before the protease domain. In a similar way to tPA, information is now available about related structures of all the components. The NMR structure of the first G module has been determined here (Baron *et al.*, 1992b) and the structure of a Gla module is known (Soriano-Garcia *et al.*, 1992). The intact molecule can thus be modelled, but possible interactions between domains are still unclear. Mapping functional regions of the factor IX molecule has been aided by the location and characterization of a large number of amino acid substitutions found in patients with haemophilia B (Giannelli *et al.*, 1990).

The structure of the first G module of factor IX and the region which might bind calcium were predicted after the structure of EGF was determined (Cooke *et al.*, 1987). In a collaboration with George Brownlee of the Sir William Dunn School of Pathology, Oxford, we have subsequently studied the first G module extensively by NMR. It is relatively simple to observe specific calcium binding to the isolated module by monitoring the strong calcium-induced upfield shift of the 2,6 proton resonance of Tyr 69. Various substitutions of amino acids in the putative calcium binding site of the first G module have been made (see Figure 5.10). These have very little effect on the observed chemical shifts and the structure of the module in the absence of

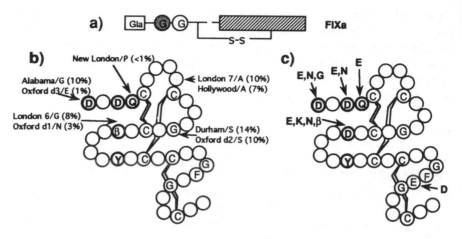

Figure 5.10 Mutations in the first G module of factor IX. (a) Shows the modular nature of factor IX, which is activated to factor IXa by release of a peptide from the region joining the second G module and the protease domain. The disulphide bridge remains intact. (b) Shows some of the mutations which have been found in haemophilia patients. (c) Shows different versions of the first G module which have been produced in Oxford to study its role in calcium binding and blood clotting. From Handford *et al.* (1991)

calcium, but there are large differences in calcium-binding affinity (Handford *et al.*, 1990, 1991). These changes in affinity correlate with changes observed in clotting activity of haemophiliacs. For example, the 'Alabama' substitution, D47G, reduces clotting activity to 10% of normal, while the K_d for calcium binding to the isolated module, under physiological conditions, is increased from 1.8 mM to 37 mM.

This approach, where a functional parameter on an isolated module is compared with a functional parameter of the intact molecule, is another aspect of the 'dissect and rebuild' strategy which is very powerful. It has begun to give information about important protein–protein and protein–metal ion interactions in a very large complex of factors IX, X, VIII, Ca^{2+} and phospholipid which will be extremely difficult to study by conventional physical methods. This research is also beginning to provide an explanation for how some mutations of factor IX lead to haemophilia.

7 Conclusions

We have tried to illustrate the power and scope of modern NMR methods applied to a few examples of specific protein–protein interactions. Although complete structural analysis of proteins by NMR is limited to those of relatively small size, the kind of dissect and rebuild approach outlined here allows useful information about quite complex structural problems to be obtained. We foresee a prosperous future for the use of NMR spectroscopy in the application to modular proteins, of which the study of how neighbouring modules interact in mosaic proteins is likely to be a major area of investigation. Already in our laboratory we are in the process of studying a number of module 'pairs' to elucidate both their structural and their dynamic properties.

References

Atkinson, A. and Williams, R. J. P. (1990). Solution structure of the kringle 4 domain from human plasminogen by ^{1}H NMR and distance geometry. *J. Mol. Biol.*, **212**, 541–552

Barlow, P. N., Norman, D. G., Steinkasserer, A., Horne, T. J., Pearce, J. M., Driscoll, P. C., Sim, R. B. and Campbell, I. D. (1992). Solution structure of the fifth repeat of factor H—a second example of the complement control protein module. *Biochemistry*, **31**, 3626–3634

Baron, M., Main, A. L., Driscoll, P. C., Mardon, H. J., Boyd, J. and Campbell I. D. (1992a). ^{1}H NMR assignment and secondary structure of the cell adhesion type III module of fibronectin. *Biochemistry*, **31**, 2068–2073

Baron, M. Norman, D. G. and Campbell, I. D. (1991) Protein modules. *TIBS*, **16**, 13–17

Baron, M. Norman, D. G., Harvey, T. S., Handford, P. A., Mayhew, M., Tse, A. G. D., Brownlee, G. G. and Campbell, I. D. (1992b). The 3D structure of the first EGF-like module of human factor IX: comparison with EGF and TGF-α. *Protein Sci.*, **1**, 81–90

Baron, M. Norman, D. G., Willis, A. and Campbell, I. D. (1990). Structure of the fibronectin type 1 module. *Nature*, **345**, 642–646

Bax, A. Experimental techniques for studies of biopolymers. *Curr. Topics Struct. Biol.*, **1**, 1030–1035

Bennett, W. A. F., Paoni, N. F., Keyt, B. A., Bolstein, D., Jones, A. J. S., Presta, L., Wurm, F. M. and Zoller, M. J. (1991). High resolution analysis of functional determinants on human tPA. *J. Biol. Chem.*, **266**, 5191–5201

Brown, S. C., Weber, P. L. and Mueller, L. (1987). Toward complete ^{1}H NMR spectra in proteins. *J. Magn. Reson.*, **77**, 166–169

Brünger, A. T. (1988). *XPLOR Manual.* Yale University, New Haven, CT

Brünger, A. T., Clore, G. M., Gronenborn, A. M. and Karplus, M. (1987). Solution conformations of human growth hormone releasing factor: comparison of the restrained molecular dynamics and distance geometry methods for a system without long-range distance data. *Protein Engng*, **1**, 399–406

Byeon, I. L. and Llinas, M. (1991). Solution structure of the tissue type plasminogen activator kringle 2 domain complexed to 6-aminohexanoic acid, an antifibrinolytic drug. *J. Mol. Biol.*, **222**, 1035–1051

Campbell, I. D. and Baron, M. (1991). The structure and function of protein modules. *Phil. Trans. Roy. Soc. Lond.*, **B332**, 165–170

Chothia, C. (1991). Antigen recognition *Curr. Opinion Struct. Biol.*, **1**, 53–59

Clayton, L. K., Sayre, P. H., Novotny, J. and Reinherz, E. L. (1987). Murine and human T11 (CD2) cDNA sequences suggest a common signal transduction mechanism. *Eur. J. Immunol.*, **17**, 1367–1370

Clore, G. M., Brünger, A. T., Karplus, M. and Gronenborn, A. M. (1986). Application of molecular dynamics with interproton distance restraints to three-dimensional protein structure determination: a model study of crambin. *J. Mol. Biol.*, **191**, 523–551

Clore, G. M. and Gronenborn A. M. (1991a). Structures of larger proteins in solution: three- and four-dimensional heteronuclear NMR spectroscopy. *Science*, **252**, 1390–1399

Clore, G. M. and Gronenborn, A. M. (1991b). Two-, three-, and four-dimensional NMR methods for obtaining larger and more precise three-dimensional structures of proteins in solution. *Ann. Rev. Biophys. Biophys. Chem.*, **20**, 29–63

Constantine, K. L., Ramesh, V., Banyai, L., Trexler, M., Patthy, L. and Llinas, M. (1991). Sequence-specific ^{1}H NMR assignments and structural characterisation of bovine seminal fluid protein PDC-109 domain b. *Biochemistry*, **30**, 1663–1672

Cooke, R. M., Wilkinson, A. J., Baron, M., Pastore, A., Tappin, M. J., Campbell, I. D., Gregory, H. and Sheard, B. (1987). The solution structure of human epidermal growth factor. *Nature*, **327**, 339–341

Doolittle, R. F. (1989). Similar amino-acids revisited. *TIBS*, **14**, 244–245

Dorit, R. L. and Gilbert, W. A. (1991). The limited universe of exons. *Curr. Opinion Struct. Biol.*, **1**, 973–977

Downing, A. K., Driscoll, P. C., Harvey, T. S., Dudgeon, T. J., Smith, B. O., Baron, M. and Campbell, I. D. (1993). The solution structure of the fibrin binding finger domain of tissue-type plasminogen activator determined by ^{1}H NMR. *J. Mol. Biol.* (in press)

Driscoll, P. C., Clore, G. M., Beress, L. and Gronenborn, A. M. (1989). A proton nuclear magnetic resonance study of the antihypertensive and antiviral protein BDS-I from the sea anemone *Anemonia sulcata*: Sequential and stereospecific resonance assignment and secondary structure. *Biochemistry*, **28**, 2178–2187

Driscoll, P. C., Cyster, J., Campbell, I. D. and Williams, A. F. (1991). Structure of domain 1 of rat T lymphocyte CD2 antigen. *Nature*, **353**, 762–765

Dustin, M. L. and Springer, T. A. (1991). Role of lymphocyte adhesion receptors in transient interactions and cell locomotion. *Ann. Rev. Immunol.*, **9**, 27–66

Engel, J. (1991). Common structural motifs in proteins of the extracellular matrix. *Curr. Opinion Cell Biol.*, **3**, 779–785

Furie, B. and Furie, B. C. (1988). The molecular basis of blood coagulation. *Cell*, **53**, 505–517

Haber, E., Quertermous T., Matsueda, G. R. and Runge, M. S. (1989). Innovative approaches to plasminogen activator therapy. *Science*, **243**, 51–56

Handford, P. A., Baron, M., Mayhew, M., Willis, A., Beesley, T., Brownlee, G. G. and Campbell, I. D. (1990). The first EGF-like domain of human factor IX has a high affinity Ca^{++} binding site. *EMBO Jl*, **9**, 475–480

Handford, P. A., Mayhew, M., Baron, M., Winship, P. R., Campbell, I. D. and Brownlee, G. G. (1991). Key residues involved in calcium-binding motifs in EGF-like domains. *Nature*, **351**, 164–167

Harrison, S. C. (1991). The structural taxonomy of DNA-binding domains. *Nature*, **353**, 715–719

Holmgren, A., Kuehn, M. J., Bränden, C.-I. and Hultgren, S. J. (1992). Conserved immuno-globulin-like features in a family of periplasmic pilus chaperones in bacteria. *EMBO Jl*, **16**, 1617–1622

Hommel, U., Dudgeon, T. J., Fallon, A., Edwards, R. M. and Campbell, I. D. (1991). Structure–function relationship in human epidermal growth factor studied by site-directed mutagenesis and ^{1}H NMR. *Biochemistry*, **30**, 8891–8

Hommel, U., Harvey, T. S., Driscoll, P. C. and Campbell, I. D. (1992). Human epidermal growth factor: high resolution solution structure and comparison with human TGF-α. *J. Mol. Biol.* (in press)

Kay L. E. and Bax A. (1990). New methods for the measurement of NH-CαH coupling constants in ^{15}N-labeled proteins. *J. Magn. Reson.*, **86**, 110–126

Kay, L. E., Torchia, D. A. and Bax, A. (1989). Backbone dynamics of proteins as studied by ^{15}N inverse detected heteronuclear NMR: application to staphylococcal nuclease. *Biochemistry*, **28**, 8972–8979

Kraulis, P. J. (1991). MOLSCRIPT: A program to produce both detailed and schematic plots of protein structures. *J. Appl. Crystallog.*, **24**, 946–950

Maliett, S., Fossum, S. and Barclay, A. N. (1990). Characterization of the MRC OX40 antigen of activated CD4 positive T lymphocytes—a molecule related to nerve growth factor receptor. *EMBO Jl*, **9**, 1063–1068

Marion, D. and Bax, A. (1988). Baseline distortion in real-Fourier-transform NMR spectra. *J. Magn. Reson.*, **79**, 352–356

Marion, D., Ikura, M. and Bax, A. (1989). Improved solvent suppression in one- and two-dimensional NMR spectra by convolution of time-domain data. *J. Magn. Reson.*, **84**, 425–430

Messerle, B. A., Wider, G., Otting, G., Weber, G. and Wüthrich, K. (1989). Solvent suppression using a spin lock in 2D and 3D NMR spectroscopy with H_2O solutions. *J. Magn. Reson.*, **85**, 608–613

Montelione, G. T., Wüthrich, K., Nice, E. C., Burgess, A. W. and Sheraga, H. A. (1987). Solution structure of murine EGF; determination of the polypeptide backbone chain-fold by NMR and distance geometry. *Proc. Natl Acad. Sci. USA*, **84**, 5226–5230

Müller, L. (1987). P. E. COSY, a simple alternative to E. COSY. *J. Magn Reson.*, **72**, 191–196

Nilges, M., Clore, G. M. and Gronenborn, A. M. (1990). ^{1}H-NMR stereospecific assignments by conformational data-base searches. *Biopolymers*, **29**, 813–822

Nilges, M., Gronenborn, A. M., Brünger, A. T. and Clore, G. M. (1988). Determination of three-dimensional structure of proteins by simulated annealing with interproton distance restraints: Application to crambin, potato carboxypeptidase inhibitor and barley serine proteinase inhibitor 2. *Protein Engng*, **2**, 27–38

Norman, D. G., Barlow, P. N., Baron, M., Day, A. J., Sim, R. B. and Campbell, I. D. (1991). Three-dimensional structure of a complement control protein module in solution. *J. Mol. Biol.*, **219**, 717–725

Norwood, T. J., Crawford, D. A., Stevenson, M. E., Driscoll, P. C. and Campbell, I. D. (1993). [1]H–[15]N nuclear magnetic resonance studies of the c subunit of the *E. coli* F_1F_0 ATP synthase: assignment and secondary structure. *Biochemistry* (in press)

Novokhatny, V. V., Ingham, K. C. and Medved, L. V. (1991). Domain structure and domain–domain interactions of recombinant tPA. *J. Biol. Chem.*, **266**, 12994–13002

Patthy L. (1985). Evolution of the proteases of the blood clotting and fibrinolysis by assembly from modules. *Cell*, **41**, 657–663

Patthy, L. (1987). Intron-dependent evolution: preferred types of exons and introns. *FEBS Lett.*, **214**, 1–7

Patthy, L. (1991). Modular exchange principles in proteins. *Curr. Opinion Struct. Biol.*, **1**, 351–361

Peterson, A. and Seed B. (1987). Monoclonal antibody and ligand binding sites of the T-cell erythrocyte receptor (CD2). *Nature*, **329**, 842–846

Plateau, P. and Guéron, M. (1982). Exchangeable proton NMR without base-line distortion, using new strong-pulse sequences. *J. Am. Chem. Soc.*, **104**, 7310–7311

Recny, M. A., Neidhart, E. A., Sayre, P. H., Ciardelli, T. L. and Reinherz, E. L. (1990). Structural and functional characterisation of the CD2 immunoadhesive domain. *J. Biol. Chem.*, **265**, 8542–8549

Redfield, C. and Dobson, C. M. (1990). ^{1}H NMR studies of human lysozyme:

Spectral assignment and comparison with hen lysozyme. *Biochemistry*, **29**, 7201–7214

Reid, K. B. M. and Day, A. J. (1989). Structure function relationships of the complement system. *Immunol. Today*, **10**, 177–180

Ryu, S. E., Kwong, P. D., Truneh, A., Porter, T. G., Arthos, J., Rosenberg, M., Dai, X., Xuong, N., Axel, R., Sweet, R. W. and Hendrickson, W. A. (1990). Crystal structure of HIV-binding recombinant fragment of human CD4. *Nature*, **348**, 419–426

Selander, M., Persson, E., Stenflo, J. and Drakenberg, T. (1990). [1]H assignment and secondary structure of the Ca^{2+} form of the amino-terminal epidermal growth factor like domain in coagulation factor X. *Biochemistry*, **29**, 8111–8

Soriano-Garcia, M., Padmanabhan, K., de Vos, A. M. and Tulinsky, A. (1992). The Ca^{++} ion and membrane binding structure of the gla domain of Ca-prothrombin fragment 1. *Biochemistry*, **31**, 2554–2566

Soriano-Garcia, M., Park, C. H., Tulinsky, A., Ravichandran, K. G. and Skrzpzcak-Jankun, E. (1989). Structure of Ca^{++} prothrombin fragment 1 including the conformation of the gla domain. *Biochemistry*, **28**, 6805–6810

Soutar, A. K. and Knight, B. L. (1990). Structure and regulation of the LDL-receptor and its gene. *Br. Med. Bull.*, **46**, 891–916

Springer, T. A. (1990). Adhesion receptors of the immune system. *Nature*, **346**, 425–434

Spruyt, L. L., Glennie, M. J., Beyers, A. D. and Williams, A. F. (1991). Signal transduction by the CD2 antigen T cells and natural killer cells: Requirement for expression of a functional T cell receptor or binding of antibody Fc to the receptor, FcγRIIIA (CD16). *J. Exptl Med.*, **174**, 1407–1415

Tappin, M. J. Cooke, R. M., Fitton, J. E. and Campbell, I. D. (1989). A high resolution ^{1}H NMR study of human transforming growth factor α: structure and pH dependent conformational interconversion. *Eur. J. Biochem.*, **179**, 629–637

de Vos, A. M., Ultsch, M. H., Kelley, R. F., Padmanabhan, K., Tulinsky, A., Westbrook, M. L. and Kossiakof, A. A. (1992a). Crystal structure of the kringle 2 domain of tissue plasminogen activator at 2.4Å resolution. *Biochemistry*, **31**, 270–279

de Vos, A. M., Ultsch, M. and Kossiakof, A. A. (1992b). Human growth hormone and extracellular domain of its receptor: crystal structure of the complex. *Science*, **255**, 306–312

Wang, J., Yan, Y., Garret, T. P. J., Liu, J., Rodgers, D. W., Garlick, R. L., Tarr, G. E., Hussain, Y., Reinherz, E. L. and Harrison, S. C. (1990). Atomic structure of a fragment of CD4 containing two immunoglobulin-like domains. *Nature*, **348**, 411–419

Weis, W. I., Kahn, R., Fourme, R., Drickamer, K. and Hendrickson, W. A. (1991). Structure of the calcium-dependent lectin domain from a rat mannose-binding protein determined by MAD phasing. *Science*, **254**, 1608–1615

Williams, A. F. (1987). A year in the life of the immunoglobulin superfamily. *Immunol. Today*, **8**, 298–303

Williams, A. F. and Barclay, A. N. (1988). The immunoglobulin superfamily—domains for cell surface recognition. *Ann. Rev. Immunol.*, **6**, 381–405

Witwer, A. J. and Howard, S. C. (1990). Glycosylation at Asn-184 inhibits the conversion of single chain to two-chain tPA by plasmin. *Biochemistry*, **29**, 4175–4180

Wüthrich, K. (1986). *NMR of Proteins and Nucleic Acids*. New York, Wiley

Wüthrich, K. (1989). Protein structure determination in solution by NMR. *Science*, **243**, 45–50

Yamada, K. M. (1991). Adhesive recognition sequences. *J. Biol. Chem.*, **266**, 12809–12812

6

NMR Structural Studies of Membrane Proteins

S. J. Opella and P. A. McDonnell

1 Introduction

Membrane proteins are responsible for many significant biological functions, including those of enzymes, channels and receptors. Unfortunately, membrane proteins present serious obstacles for applications of the most widely used methods of structural biology; in particular, membrane proteins are notoriously difficult to crystallize and the multidimensional NMR methods that are successful in determining the structures of globular proteins in solution are hampered by the slow reorientation rates of proteins complexed with lipids. As a consequence of these experimental problems, descriptions of the structures and dynamics of membrane proteins have lagged far behind those of globular proteins. The few structures of membrane proteins determined by diffraction (Deisenhofer et al., 1985; Rees et al., 1989; Weiss et al., 1991) and image reconstruction (Henderson et al., 1990) have proven to be exceptions, since many other interesting membrane proteins have not been amenable to crystallization. Although NMR spectroscopy is at an even earlier stage in its application to membrane proteins than X-ray crystallography, it is now emerging as an effective method for describing the dynamics of membrane proteins, and substantial progress is being made in its development as a method for determining the structures of membrane proteins.

This chapter describes how many of the technical difficulties associated with NMR studies of membrane proteins can be overcome with an approach that combines the methods of high-resolution solid-state NMR spectroscopy on samples of membrane proteins in phospholipid bilayers with those of multidimensional solution NMR spectroscopy on samples in detergent micelles (Shon *et al.*, 1991 a,b; Opella, 1993; McDonnell *et al.*, 1993b). Both high-resolution solid-state NMR spectroscopy (Opella *et al.*, 1987) and multidimensional solution NMR spectroscopy are independently capable of determining the backbone and side-chain structures of peptides (Dyson and Wright, 1991) and proteins (Wüthrich, 1986; Clore and Gronenborn, 1989), including membrane proteins. However, the combination of spectroscopic methods and extra steps in sample preparation are required to obtain results on membrane proteins at the current state of the art of NMR spectroscopy. As the instrumentation, especially magnets, and experimental techniques are improved, membrane protein studies by NMR spectroscopy will become much more feasible and informative.

NMR studies of membrane proteins are difficult only because of the characteristics of the samples rather than those of the proteins themselves. Aside from the fact that membrane proteins are generally insoluble in water, the medium in which globular proteins are studied, these proteins do not have unusual features. Membrane proteins are typically monomers in membrane environments and are characteristically organized around long rigid hydrophobic helices and shorter amphipathic helices separated by loops, some of which are mobile, and mobile N- and C-terminal segments.

Many, but not all, membrane peptides and proteins are soluble in various combinations of organic solvents, such as chloroform, methanol or trifluoroethanol. The overall reorientation rates of proteins in these solvents are similar to those in water, and multidimensional solution NMR experimental methods can be applied to peptides and proteins in these solvents. Although the use of these solvents does make some NMR studies feasible, there is always serious concern about the state of the protein in this type of environment. Therefore, more appropriate membrane environments are of interest, in spite of their limitations for experimental NMR spectroscopy.

The two well-accepted model membrane environments for membrane proteins, (A) detergent micelles and (B) phospholipid bilayers, are shown schematically in Figure 6.1. Neither of these systems is very well

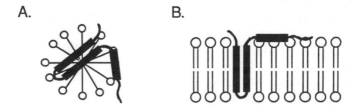

Figure 6.1 Schematic drawings of a protein in membrane environments: (A) detergent micelle; (B) phospholipid bilayer

suited for multidimensional solution NMR spectroscopy, because protein–lipid complexes reorient much more slowly than proteins alone in solution. Although a protein in a detergent micelle in aqueous solution reorients much faster than a protein in a fully hydrated phospholipid bilayer, its reorientation is still slow enough to cause problems for multidimensional solution NMR experiments. A small 5000 dalton protein in a micelle behaves like a large 30 000 dalton protein in aqueous solution. Although spectral complexity is minimal in this situation, the most serious spectroscopic difficulties of large proteins, broad lines and rapid spin-diffusion exist even with the best micelle preparations. Isolated membrane proteins, whether prepared by purification from cells or automated solid-state peptide synthesis, can be reconstituted into phospholipid bilayers. Protein-containing bilayers can be prepared as oriented or unoriented multilamellar liposome samples where the proteins do not reorient on NMR time-scales as slow as 10^{-4} s, which enables them to be investigated by the methods of solid-state NMR spectroscopy. It is also possible to sonicate the bilayers to form relatively small spherical vesicles. However, the overall correlation times of proteins in vesicles (10^{-6} s) are inconvenient for nearly all types of NMR experiments, since the reorientation is too rapid for solid-state NMR techniques and too slow for solution NMR techniques. As a result, only a limited amount of information has been derived from solution NMR studies of protein in sonicated vesicles. In contrast, a great deal of valuable information about the structure and dynamics of membrane proteins is being derived from both multidimensional solution NMR studies of proteins in micelles and solid-state NMR studies of proteins in oriented and unoriented bilayers. These two approaches are synergistic, and applying them in combination has made it possible to characterize several membrane peptides and proteins in some detail.

2 Some Pertinent Aspects of NMR Spectroscopy

In order to appreciate the design and implementation of experimental NMR studies of membrane proteins, which involve substantial compromises compared with the thorough studies that are now almost routinely performed on globular proteins in solution, it is essential to recognize that experimental NMR studies of membrane proteins are near the limits of feasibility with currently available technology. Membrane proteins require that the most powerful spectroscopic experiments, spectrometers with the highest magnetic fields and comprehensive isotopic labelling schemes be fully utilized. This is the case for both multidimensional solution NMR studies of detergent micelle samples and solid-state

NMR studies of phospholipid bilayer samples. The combination of these methods, augmented by molecular dynamics calculations, fills in gaps left by each of them, yielding models of membrane proteins that demonstrate the overall feasibility of determining their structures and describing their dynamics in sufficient detail to derive models capable of giving some understanding of their functions.

Multidimensional Solution NMR Spectroscopy of Membrane Proteins in Micelles

Multidimensional solution NMR methods are applicable to peptides and proteins in micelles, because the reorientation of the protein–micelle complex in solution is rapid enough (10^{-8} s) to average the spectral parameters from the various nuclear spin interactions to their isotropic values and to enable the indirect effects of their fluctuations, as monitored by relaxation phenomena, to describe local dynamics as well as provide spatial information in the form of relative proximities. However, the methods of solution NMR spectroscopy are limited by the broad linewidths and efficient spin diffusion that accompany the relatively slow reorientation of protein–micelle complexes in solution. These limitations can be reduced substantially with careful sample preparation, especially with isotopic labelling of the protein and with sufficient detergent concentrations to form uniform micelles.

Protein structure determination by multidimensional NMR spectroscopy (Wüthrich, 1986; Clore and Gronenborn, 1989) is straightforward for relatively small globular proteins in solution, especially if uniformly ^{15}N- and ^{13}C-labelled samples are available. There are several basic steps in the multidimensional solution NMR approach to protein structure determination that are described in considerable detail in other chapters of this volume. These steps include: the resolution and assignment of backbone and side-chain resonances based on through-bond and through-space interactions observed in multidimensional NMR spectra; the measurement of many homonuclear ^{1}H/^{1}H NOEs among the assigned resonances, supplementing these short-range distances with other structural constraints including spin–spin coupling constants; the interpretation of all of the structural constraints in terms of secondary structure; and the generation of a family of three-dimensional protein structures using distance geometry, molecular dynamics and other calculations (Havel and Wüthrich, 1985; Brünger *et al.*, 1986; Braun, 1987; Scheek *et al.*, 1989). The availability of high-field spectrometers, cloned proteins that can be expressed in organisms grown on chemically and isotopically defined media, and software for data processing and protein structure determination have led to the widespread application of NMR to the structural biology of relatively small globular proteins.

There are two main reasons that larger proteins present difficulties for multidimensional solution NMR spectroscopy. First, they are more complex than small proteins, with many more resonances to resolve and assign. This problem is addressed successfully through the use of experiments with high dimensionality. Second, larger proteins reorient more slowly in solution, presenting the same spectroscopic problems seen in proteins in micelles due to broad lines and spin diffusion. These problems are quite difficult to address. The most direct approaches involve reducing the overall correlation time by using the smallest functional polypeptide chain or raising the temperature. However, additional development is needed in order to routinely apply this approach to larger proteins and proteins in complexes with DNA and micelles. These problems are directly confronted in studies of membrane proteins in micelles because of their slow reorientation rates.

Multidimensional solution NMR studies of membrane proteins in micelles are deceptively difficult. While one-dimensional homonuclear spectra and two-dimensional heteronuclear correlation spectra of these samples often appear reasonably well resolved, the devastating effects of spin diffusion on homonuclear NOE measurements compound the sensitivity and resolution problems caused by the broad linewidths of the resonances and their limited chemical shift dispersion of these highly helical proteins. Extensive efforts to optimize the samples, experiments and instrumentation have been required in order to resolve and assign all backbone ^{1}H and ^{15}N amide resonances of both Pf1 and fd coat proteins in micelles (Shon *et al.*, 1991a,b; McDonnell *et al.*, 1993b). Since these are typical membrane proteins, the methods worked out with these examples should be applicable to many other membrane proteins.

The experimental conditions for solubilizing these proteins in both perdeuterated dodecylphosphocholine (DPC) and sodium dodecyl sulphate (SDS) micelles had to be established. In favourable cases DPC yields narrower and more intense protein resonances than other detergents and eliminates the chemical nature of the head group as a potential source of criticism (Brown, 1979). However, in some cases SDS remains the detergent of choice because it gives the highest-quality multidimensional NMR spectra. The optimization of the protein-containing micelle is at once the simplest and most subtle aspect of the research. All aspects of the sample, including the choice of detergent, concentration of polypeptides and detergent, pH, temperature and counterions, must be taken into account in order to prepare samples capable of yielding reproducible, resolved spectra that can be reliably interpreted. The chemical purity of the detergent also has a significant role in determining the stability of the protein and the ultimate quality of the spectra.

The single most important aspect of sample preparation is to ensure that adequate concentrations of highly purified detergent are present.

This is the only way to obtain reliable, well-resolved, single-line spectra with good sensitivity of membrane proteins in micelles. Inadequate detergent concentrations lead to spectral artefacts. This is illustrated in Figure 6.2 with two-dimensional heteronuclear correlation spectra from uniformly ^{15}N-labelled fd coat protein as a function of SDS concentration (McDonnell *et al.*, 1993b). The uniformly ^{15}N-labelled coat protein spectrum is expected to have 48 correlation peaks, one for each amino acid in the protein, with the exception of the N terminus and the proline residue. However, the number of resonances observed in the spectra depends on the concentration of the detergent, indicating that the protein spectra reflect the state of the micelles in the solution. At 40 mM SDS, the spectrum of uniformly ^{15}N-labelled coat protein in Figure 6.2(A) has broad, weak, poorly resolved and in some cases missing resonances when compared with a spectrum obtained on samples with ten or more times the concentration of detergent, even though 40 mM is about five times higher than the critical micelle concentration of SDS (Helenius *et al.*, 1979). When the detergent concentration is increased to 115 mM, the linewidths narrow in both the ^{1}H and ^{15}N dimensions and many extra peaks appear in the spectrum in Figure 6.2(B). The extra correlation peaks are not present for all residues and are not true doublings of resonances, since they generally are not of equal intensity. These extra peaks have been interpreted as arising from protein dimers (Henry and Sykes, 1990, 1992). However, when the concentration of SDS is above 480 mM, the extra resonances disappear, as shown in the spectrum in Figure 6.2(C), which is identical with that obtained with higher SDS concentrations. All of the spectra obtained with SDS concentrations greater than 300 mM have one resonance for each residue, as expected for the monomeric membrane-bound form of fd coat protein (Shon *et al.*, 1991b; McDonnell *et al.*, 1993b). The extra resonance peaks observed at low detergent concentrations, as in Figure 6.2(B), probably result from complex equilibria of the detergent in micelles. The type, purity and amount of detergent used to solubilize a membrane protein must be carefully adjusted for each protein before performing NMR structural studies.

The technical problems associated with applying solution NMR methods to membrane proteins, regardless of the environment of the sample, are compounded by the structural features of the major class of membrane proteins. These proteins have their structures dominated by helical secondary structure, in the form of hydrophobic membrane-spanning helices and amphipathic bridging helices, which have limited chemical shift dispersion when compared with other types of secondary and tertiary structural elements. The loop and end portions of the proteins are often mobile, which serves to limit their chemical shift dispersion, as well as their interresidue homonuclear NOEs.

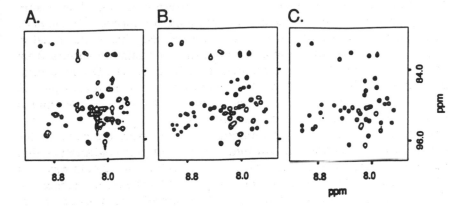

Figure 6.2 Amide resonance region of two-dimensional ^{1}H–^{15}N heteronuclear correlation spectra of uniformly ^{15}N-labelled fd coat protein at different sodium dodecylsulphate (SDS) concentrations. All samples were 1.9 mM coat protein, 40 mM NaCl, in H$_2$O, pH 4.0, 50 °C. (A) 40 mM SDS; (B) 115 mM SDS; (C) 180 mM SDS. From McDonnell *et al.* (1992b)

Solid-state NMR Spectroscopy of Membrane Proteins in Lipid Bilayers

Immobile proteins, such as those strongly associated with phospholipid bilayers, can be studied by solid-state NMR spectroscopy. Solid-state NMR spectroscopy enables both protein structure and dynamics to be described in parallel, because of the absence of the effects of overall reorientation. This is of particular interest for membrane proteins, which characteristically have mobile loops and termini juxtaposed with rigid, structured helical regions.

All NMR spectral features are influenced by molecular motions. This occurs directly through the motional averaging of line shapes and indirectly through the induction of relaxation. In cases where the overall molecular reorientation is slow compared with the time scales determined by the nuclear spin interactions, then local, internal molecular motions can be analysed from their effects on the line shapes. The slowest time-scales available from the solid-state NMR experiments are around 10^{-4}s; therefore, as long as the molecule of interest reorients slowly compared with this, the spectral properties of solids rather than solutions are observed in the spectra. As a result, solid-state NMR methods are appropriate for many complexes of biopolymers, such as virus particles in solution or fully hydrated membrane bilayers, even though these samples are not physically solids. It is also the reason solid-state NMR spectroscopy gives such favourable opportunities for distinguishing mobile segments of membrane proteins against the background of rigid regions of secondary structure (Bogusky *et al.*, 1987). In studies of membrane proteins, one of the key experiments is to identify the structured regions on the basis of their dynamics. This simplifies the process of structure determination and assists by locating the boundaries to the regions of stable secondary structure. After the backbone structure of the protein is determined, then it is essential to analyse the side-chain dynamics (Opella, 1986).

Although solid-state NMR spectroscopy has long been established as an independent method for determining molecular structure at atomic resolution, only recently has it become possible to obtain detailed structural information on proteins by utilizing high-field spectrometers; multiple-pulse, multiple-resonance, multidimensional experiments; and specific, selective and uniform isotopic labelling schemes. The basic physical properties of the relevant nuclear spin interactions are so well understood that spectroscopic interpretations have not been a limiting factor in applying solid-state NMR spectroscopy to the study of proteins. The solid-state NMR approach that we are developing for determining the structures of proteins in oriented samples relies on the spectral simplifications that result from uniaxial sample orientation parallel to the direction of the applied magnetic field (Opella *et al.*, 1987). The spin interactions at individual sites yield single resonance lines characterized by their resonance frequencies or doublets (or triplets) characterized by the magnitudes (and asymmetry) of their splittings. The observed values of the frequencies and splittings depend on the orientations of the principal axes of the spin-interaction tensors present at each site relative to the direction of the applied magnetic field. Since the direction of sample orientation and the applied magnetic field of the NMR spectrometer is the same, it defines a frame of reference for the evaluation of the orientational information. The orientations of many spin-interaction tensors have been established in their molecular frames of reference, enabling angular factors to be determined from the experimental data. Molecular structures can be determined on the basis of angles alone, given standard bond lengths and geometries; therefore, it is possible to determine the structure of a protein with a sufficient number of orientationally dependent spectroscopic measurements. Protein structures can also be determined from many short-range internuclear distance measurements, which can be made with both solution NMR experiments, through homonuclear NOEs, and with several different solid-state NMR experiments. We have performed experiments that yield distance information on both oriented (Cross *et al.*, 1983a) and unoriented (Frey and Opella, 1984) samples, and others are developing spectroscopic methods for making quantitative distance measurements between isotopically labelled sites in unoriented samples (Raleigh *et al.*, 1988; Gullion and Schaefer, 1989).

In principle, solid-state NMR spectroscopy is fully capable of determining the structures of immobile, oriented proteins (Opella *et al.*, 1987; Opella and Stewart, 1989), including membrane proteins embedded in lipid bilayers oriented between glass plates (Bechinger *et al.*, 1991; Shon *et al.*, 1991a,b; McDonnell *et al.*, 1992b; Opella, 1993; Opella *et al.*, 1993). A very high degree of orientation can be achieved when small amounts of lipids are placed between glass plates and partially dehydrated under controlled conditions. This sample arrangement compromises the orientation of the glass plates, gives large edge effects and gives a poor filling factor, making the spectroscopic experiments very inefficient. The feasibility of NMR experiments on oriented membrane samples is greatly improved by using a single pair of square glass plates (as large as 22×22 mm) with the rf coil wrapped directly around the plates. This flat-coil arrangement, as shown in Figure 6.3, enables solid-state NMR spectra of oriented proteins to be obtained efficiently on a minimum amount of material (Bechinger and Opella, 1991).

Complete polypeptide backbone structures require that several spectral parameters be measured to characterize the angles between each of the peptide planes and the direction of sample orientation (Opella *et al.*, 1987; Opella and Stewart, 1989). However, once the secondary structure of a membrane-bound form of a protein is established on the basis of NOE measurements in micelle samples, a single spectral

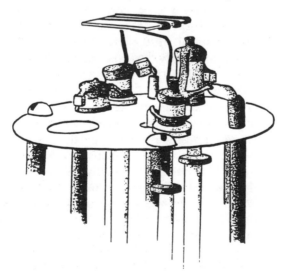

Figure 6.3 Flat coil probe for solid-state NMR experiments on proteins in lipid bilayers oriented between a single pair of large square glass plates. From Bechinger and Opella (1991)

parameter — for example, the ^{15}N chemical shift — is sufficient to establish the orientations of the helices relative to the plane of the bilayer (Bechinger *et al.*, 1991; Shon *et al.*, 1991a,b; McDonnell *et al.*, 1993b; Opella, 1993; Opella *et al.*, 1993). Specific, selective and uniform labelling strategies are feasible for ^{15}N NMR studies of proteins, and solid-state ^{15}N NMR spectra of oriented peptides and protein in lipid bilayers can be simply and qualitatively interpreted by using the nearly axially symmetric ^{15}N amide chemical shift tensor as a guide to the orientation of the peptide planes within the helices. The determination of helix orientations with respect to the lipid bilayer normal provides tertiary structural information that is essential for understanding the structure and function relationships of membrane proteins. For membrane proteins with structures dominated by helices, a description of the secondary structure of the backbone and its orientations is of great value and goes a long way towards characterizing the structure of the protein in the membrane.

Isotopic Labelling

Isotopic labelling is an essential part of investigations of all but the smallest and most tractable peptides and proteins. There are basically three different labelling strategies: specific, selective and uniform. Uniform labelling occurs when all of the nuclei of one type are labelled. This is usually accomplished by manipulating the growth media so that all of the nuclei of a certain type come from a single source. For example, all of the nitrogens in proteins can be derived from ^{15}N ammonium sulphate supplied in the growth media. The spectrum in Figure 6.4(A) is from a uniformly ^{15}N-labelled sample of fd coat protein, showing resonances from all of the amide sites. The spectrum in Figure 6.4(B) is from a selectively [^{15}N]-Lys-labelled sample of fd coat protein, and, in contrast to the spectrum of the uniformly ^{15}N-labelled sample in Figure 6.4(A), it has only five resonances, each of which corresponds to one of the five lysine residues in the protein. Specific labelling is where isotopic labels are placed at predetermined locations such that resolution or assignment results. It is usually carried out by automated solid-phase peptide synthesis. The spectrum in Figure 6.4(C) has one correlation resonance from the one specifically [^{15}N]-Ala-labelled amide site in a 21 residue peptide with a total of 8 alanine residues.

The various labelling strategies are complementary and, in practice, it is frequently the case that a labelled protein is prepared to solve a particular spectroscopic problem of overlap or assignment, rather than as part of a general plan. However, there are several reasons why it is preferable to devise experiments that utilize uniformly labelled proteins. Most importantly, they enable data to be obtained on all of the residues

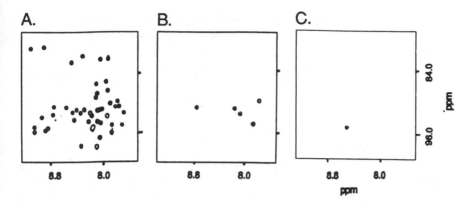

Figure 6.4 Amide resonance region of two-dimensional ^{1}H–^{15}N heteronuclear correlation spectra illustrating different labelling strategies. (A) Uniformly ^{15}N-labelled fd coat protein in micelles. There are 48 labelled amide sites in the protein. (B) Selectively [^{15}N]-Lys-labelled fd coat protein in micelles. There are 5 labelled lysine amide sites in the protein. (C) Specifically [^{15}N]-Ala-labelled PGLa peptide. One out of the 8 alanine residues is labelled. The peptide has a total of 21 residues

at once by utilizing the various spin interactions to provide resolution in several dimensions. In addition, it is generally quicker and less expensive to prepare one uniformly labelled sample than the multiple selectively labelled sample needed to obtain data from many residues. Overall, one would like to obtain the basic data in the form of frequencies, splittings and other evidence of interactions for all of the sites at once. And if specific or selective labels are needed for resolution or assignment, then they should be used solely for that purpose rather than for the basic measurements, which are tedious and error-prone.

It is possible to adjust the isotopic composition of the proteins to narrow the resonances, attenuate spin diffusion and retain reasonable receptivity for the nuclear spin interactions that occur both through-bond and through-space. We have found that some experiments, especially detection of ^{1}H–^{1}H homonuclear NOEs, can be performed on proteins uniformly labelled with ^{15}N (98%), ^{2}H on all carbon sites (80%) and ^{2}H (50%) on amide nitrogen sites (Shon *et al.*, 1991a,b; McDonnell *et al.*, 1993b) that can not be performed on samples that are only uniformly ^{15}N-labelled.

Molecular Dynamics Simulations of Membrane Proteins

The theoretical methods of structural biology are also much better suited to globular proteins than to membrane proteins. However, classical molecular dynamics simulations, which consist of numerically solving Newton's equations of motion for all of the atoms in the system, have profoundly influenced our understanding of the structure, dynamics and functions of globular proteins (Brooks *et al.*, 1988) and even qualitative information about membrane proteins will contribute to NMR invest-igations. Results for Pf1 coat protein show that vacuum MD simulations can provide information directly relevant to the structure and dynamics of membrane-bound proteins (Tobias *et al.*, 1993). The agreement be-tween the results of these calculations and NMR experiments is reas-suring. The simulations identified correctly the locations of the helix and loop regions in the coat proteins. Furthermore, the relative mobility of the residues in the helical and loop regions were reproduced qualit-atively in the simulations. There are several potential ways of combining and comparing results from experimental NMR spectroscopy with those from MD simulations for membrane proteins. Since membrane proteins are typically constructed with mobile and rigid sections interspersed, the MD and NMR results are highly complementary in identifying the boundaries of the secondary structure elements.

3   Examples of NMR Studies of Membrane Proteins

Filamentous Bacteriophages Coat Proteins

The major coat proteins from filamentous bacteriophages provide tract-able experimental systems for investigating processes that occur at membranes, including the roles of signal sequences, protein translocation across a membrane and membrane-mediated virus assembly (Marvin and Hohn, 1969; Makowski, 1984; Russel, 1991). These proteins are small (46–50 residues) and can be readily prepared in large quantities labelled with stable isotopes. The major coat protein is synthesized in the cytoplasm of its host during viral infection as a precursor, procoat protein, with an N-terminal signal sequence. After synthesis, procoat protein is inserted into the inner membrane, where the signal sequence is rapidly removed by bacterial leader peptidase, yielding the mature coat protein (Zimmermann *et al.*, 1982; Dalbey and Wickner, 1985). Many copies of the coat protein are stored in the membrane prior to phage assembly, during which the coat protein undergoes a structural transition as it is transferred from the membrane to a newly extruded virus particle. Comparisons of the structural and dynamic properties of the various coat protein species, especially between the precursor and mature coat protein, the membrane-bound form and the structural form in virus particles, and the coat proteins from class I (fd, f1, M13) and class II (Pf1) bacteriophages, are of interest.

Biophysical techniques, including solid-state NMR spectroscopy (Cross *et al.*, 1983b; Opella *et al.*, 1987), neutron (Nambudripad *et al.*, 1991) and X-ray fibre diffraction (Marvin *et al.*, 1974; Glucksman *et al.*, 1992), have been successful in describing the structure and dynamics of the structural forms of the coat proteins in the bacteriophages in the virus particles. The coat proteins in the virus are almost entirely α-helical, except for unstructured and mobile residues at the amino terminus of fd coat protein (Opella *et al.*, 1987; Glucksman *et al.*, 1992) and a loop connecting two helical segments of Pf1 coat protein (Nambudripad *et al.*, 1991; Shon *et al.*, 1991a). The membrane-bound forms of both coat proteins are more complex than the structural forms in the virus particles.

In our studies, solution and solid-state NMR spectroscopy were combined to determine the structure and describe the dynamics of the membrane-bound forms of both Pf1 (Shon *et al.*, 1991a) and fd (Shon *et al.*, 1991b; McDonnell *et al.*, 1993b) coat proteins. The secondary structures of the membrane-bound forms of the proteins in micelles were established by ^{1}H–^{1}H nuclear Overhauser effect (NOE) measurements, and the arrangements of the helical segments were determined by measurements of angular parameters of the proteins in oriented bilayers using solid-state NMR experiments. The distance measured with ^{1}H–^{1}H homonuclear NOE measurements can be used to determine secondary and tertiary structures of proteins in solution (Wüthrich, 1986). Amide ^{1}Hs on adjacent residues in an α helix are separated by less than 3 Å and give strong cross-peaks in NOE spectra that are particularly useful for identifying the secondary structure of individual residues in membrane proteins. In order to observe homonuclear NOEs for resonances from residues throughout the coat proteins, it was essential to reduce the problems resulting from the broad lines and efficient spin diffusion that accompany the slow reorientation of a protein in micelles by labelling the protein uniformly with ^{2}H on all carbon sites and with ^{15}N on all nitrogen sites (Shon and Opella, 1989).

The correlation resonances in the spectrum in Figure 6.5A have been assigned to specific residues, as noted, based on the identification of many resonances by residue type using selectively ^{15}N-labelled samples and the secondary-structure short-range NOEs observed throughout the protein sequence (Shon *et al.*, 1991b; McDonnell *et al.*, 1993a). The combined two-dimensional ^{1}H–^{15}N heteronuclear correlation and ^{1}H NOE spectrum of fd coat protein in micelles in Figure 6.5B has NOE cross-peaks between many of the amide correlation peaks. There is a continuous pattern of amide ^{1}H NOEs between amide correlation resonances from adjacent residues between residues 7 and 50. These NOEs demonstrate the extensive helical secondary structure of the membrane-bound form of the protein. The observation of additional NOE cross-peaks

between amide ¹Hs and the appropriate alpha carbon ¹Hs strongly
supports the presence of an α helix. Residues at both the amino and
carboxy terminal ends of fd coat protein in micelles are mobile (Bogusky
et al., 1987, 1988; Leo *et al.*, 1987). However, despite evidence of flexibility of the carboxy terminus by both solution and solid-state NMR
parameters, the observation of NOE cross-peaks indicates the presence
of helical secondary structure all the way through the carboxyl terminus.
In contrast, NOE cross-peaks are not observed among the N-terminal
residues; therefore, these residues are mobile and unstructured by
all available NMR spectral criteria. The remaining residues of the fd
coat protein are rigid on all time-scales except for the evidence of
mobility at tyrosine 21 on the slower 10^{-4} s time-scale (McDonnell *et
al.*, 1993b).

Solid-state NMR spectroscopy can be used to determine protein
structure by utilizing the anisotropy of nuclear spin interactions of
immobile and uniaxially oriented sites. Several different spectral parameters can be measured for the nuclei in a peptide plane, giving direct

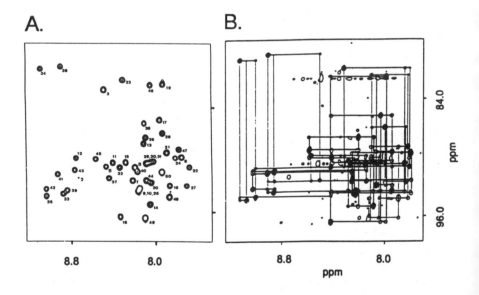

Figure 6.5 (A) Amide resonance region of two-dimensional ¹H–¹⁵N correlation
spectrum of uniformly ¹⁵N-labelled coat protein in 480 mM sodium dodecylsulphate (SDS), pH 4.0, 40 mM NaCl, 50 °C, indicating the resonance assignments.
(B) Amide resonance region of combined ¹H–¹⁵N correlation and ¹H–¹H NOE
spectrum of uniformly ²H and ¹⁵N-labelled fd coat protein in (SDS) micelles at
40 mM NaCl, pH 4.0, 50 °C. The lines connect the NOE cross-peaks between
correlation peaks from adjacent residues. From Shon *et al.* (1991b); McDonnell
et al. (1993b)

measurements of the angles between the groups and bonds in the planes and the direction of sample orientation. The ^{15}N chemical shift powder patterns from unoriented samples of uniformly and selectively ^{15}N-labelled fd coat protein in bilayers establish that the membrane-bound form has both mobile and rigid backbone sites. Comparison of the spectra in Figure 6.6 shows the dramatic effects of macroscopic sample orientation, since ^{15}N resonances change from broad, characteristically shaped powder patterns to narrow, single lines with frequencies determined by the orientation of the peptide planes with respect to the direction of the magnetic field.

^{15}N solid-state NMR spectra of uniformly ^{15}N-labelled and selectively [^{15}N]-Leu- and selectively [^{15}N]-Tyr-labelled coat proteins in oriented phospholipid bilayers are shown in Figure 6.6(A, D, F, respectively). Comparison of the spectra in Figure 6.6(D, E) clearly shows the effect of sample orientation: the two ^{15}N resonances go from a broad, characteristically shaped powder pattern to narrow, single lines with frequencies determined by the orientation of the peptide planes with respect to the direction of the magnetic field (Shon *et al.*, 1991b). Since the secondary structure for the two leucine residues is established to be α-helical by the homonuclear $^1H-^1H$ NOE cross-peaks in Figure 6.5(B), a single spectral parameter, in this case the amide ^{15}N chemical shift, is sufficient to establish the orientations of the α helices relative to the plane of the bilayer (Bechinger *et al.*, 1991; Shon *et al.*, 1991a, b; Opella, 1993). The spectrum from an oriented sample of uniformly ^{15}N-labelled coat protein in Figure 6.6(A) has two bands of intensity corresponding approximately to the two frequencies seen in the spectrum of selectively [^{15}N]-Leu-labelled coat protein in the spectrum in Figure 6.6(D). In contrast, the spectrum of selectively [^{15}N]-Tyr-labelled coat protein in Figure 6.6(F) has one resonance line at one of the positions observed in Figure 6.6(A, D); the other Tyr resonance is barely discernible at the isotropic resonance frequency. These oriented solid-state NMR spectra can be interpreted with considerable confidence, since the observed resonance frequencies are near the discontinuities of the powder pattern for the amide group, which allows the nearly axially symmetric ^{15}N amide chemical shift tensor to be used as a qualitative guide to the orientation of the peptide groups within the helices. An amide N–H bond approximately parallel to the direction of the applied magnetic field has a ^{15}N resonance frequency near that of the principal-element $\sigma_{\parallel}$ and an amide N–H bond perpendicular to the field has a ^{15}N resonance frequency near that of the principal-element $\sigma_{\perp}$ (Harbison *et al.*, 1984a). The two resonance bands in the spectrum in Figure 6.6(A) of uniformly ^{15}N occur near the resonance frequencies associated with $\sigma_{\parallel}$ and $\sigma_{\perp}$ of the amide chemical shift tensor. The spectrum in Figure 6.6(D) of [^{15}N]-Leu-labelled coat protein in oriented bilayers has one line with a resonance frequency near $\sigma_\parallel$ and a second line with a resonance frequency near $\sigma_\perp$, suggesting that the hydrophobic helix spans the bilayer because the NH group of one of the two leucine residues is parallel to the direction of orientation, and that the amphipathic helix is in the plane of the bilayer because the NH group of the other leucine residue is perpendicular to the direction of orientation. The spectrum in Figure 6.6(F) has one line with a resonance frequency near $\sigma_\parallel$, which also suggests that the long hydrophobic helix spans the bilayer. Isotropic resonance intensity can be difficult to observe in the spectra from oriented samples, as seen in Figure 6.6(A) for uniformly ^{15}N-labelled coat protein and in Figure 6.6(G) for selectively [^{15}N]-Tyr-labelled coat protein. Generally, we have found that spectra from unoriented samples provide more reliable evidence for the presence of mobile residues.

The results of our solution and solid-state NMR experiments on micelle and bilayer samples of fd coat protein are summarized by the model in Figure 6.7. Other investigations of the membrane-bound form of fd coat protein have arrived at quite different conclusions about the structure of the protein in membranes, primarily because of two incorrect proposals: the CD spectrum of the membrane-bound form of the protein has been interpreted to indicate the presence of significant

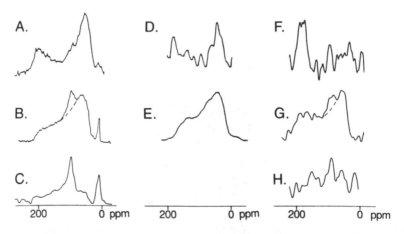

Figure 6.6 Solid-state NMR spectra of uniformly and selectively ^{15}N-labelled fd coat protein in phospholipid bilayer samples. (A) Uniformly ^{15}N-labelled fd coat protein in oriented bilayers. (B) Uniformly ^{15}N-labelled fd coat protein in unoriented bilayers obtained with cross-polarization. (C) Same sample as B with the spectrum obtained with direct pulsed excitation at the ^{15}N resonance frequency. (D) [^{15}N]-leucine-labelled fd coat protein in oriented bilayers. (E) [^{15}N]-leucine-labelled fd coat protein in unoriented bilayers. (F) [^{15}N]-tyrosine labelled fd coat protein in oriented bilayers. (G) [^{15}N]-tyrosine labelled fd coat protein in unoriented bilayers obtained with cross-polarization. (H) Same as G, with the spectrum obtained with direct pulse excitation at the ^{15}N resonance frequency. From McDonnell *et al.* (1993b)

amounts of β sheet instead of α helix (Nozaki *et al.*, 1976; Williams and Dunker, 1977; Spruijt *et al.*, 1989), and NMR spectra showing apparent doubling of resonances due to inadequate amounts of detergent have been interpreted as showing the protein to exist as an asymmetric dimer (Henry and Sykes, 1990, 1992). Figure 6.7 shows that fd coat protein is a typical membrane protein consisting of exclusively α-helical secondary structure interspersed with mobile regions (Shon *et al.*, 1991b; McDonnell *et al.*, 1993b). The secondary structure of the protein consists of two α helices based on the observation of homonuclear NOEs between amide resonances from adjacent residues. Solid-state NMR of oriented coat protein samples in bilayers determined the orientation of the helices with the amphipathic helix parallel to the bilayer and the hydrophobic helix perpendicular to the bilayer and membrane spanning. A turn that appears structured on the 10^{-9} s time-scale connects the two helical regions, although it shows evidence of mobility on the 10^{-4} s time-scale. Residues 1–5 at the amino terminus are highly mobile, on the basis of relaxation data in micelle samples and motional averaging of line shapes in bilayer samples. The C-terminus is mobile, but possesses stable, secondary structure, because homonuclear NOEs extend to the C-terminal residue. The presence of one resonance for each amide site demonstrates that the protein is a monomer.

Proteins that insert into or translocate across a membrane are usually synthesized with a 15–30 residue amino terminal polypeptide essential for proper secretion and membrane assembly. Detailed mechanisms by which signal sequences translocate proteins is not known; however, the

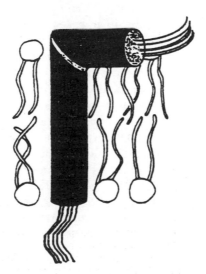

Figure 6.7 Model of membrane-bound fd coat protein. From McDonnell *et al.* (1992b)

signal sequence must influence part or all of the structure of a precursor
protein, because one of its peptide bonds is a specific substrate for leader
peptidase, an enzyme capable of processing a broad range of substrates.
The determination of the structure and dynamics of a precursor protein
and the comparison of a precursor protein with its mature protein's
structure and dynamics will help elucidate the role of the signal sequence
in protein translocation.

The dynamics of membrane-bound M13 procoat protein, which is
probably the smallest and most intensively studied precursor protein,
were characterized by the methods of ^{15}N solution and solid-state NMR
spectroscopy. Protein backbone dynamics on the 10^{-9} s time-scale can
readily be characterized by solution NMR spectroscopy (Leo *et al.*,
1987; Bogusky *et al.*, 1987, 1988), because the nuclear spin relaxation
at amide nitrogen sites is dominated by the fluctuations of the heteronuclear dipole–dipole interactions between the nitrogen and its bonded
hydrogen. Residues are identified as mobile on the rapid 10^{-9} s time-
scale when their amide resonances have large negative heteronuclear
^{1}H–^{15}N NOEs. The ^{15}N chemical shift powder pattern from unoriented
fd coat protein samples in bilayers established that the membrane-bound
form of the protein has both mobile and rigid backbone sites on the
10^{-4} s time-scale. Solid-state ^{15}N NMR can evaluate backbone motions
of immobile proteins in unoriented samples because powder pattern line
shapes are strongly affected by motional averaging. Amide sites undergoing large-amplitude fluctuations with correlation times shorter than
about 10^{-4} s have relatively narrow resonance intensity at the isotropic
position. The spectra in Figure 6.8 compare the effects of internal
mobility on the procoat protein in micelles and bilayers.

M13 procoat protein samples were prepared by solid-phase peptide
synthesis, because leader peptidase removes signal sequences efficiently
in vivo and attempts to increase the production of mutant, unprocessed
M13 procoat proteins have not yielded sufficient materials for structural
studies as yet. Solid-phase peptide synthesis enabled ^{15}N amide labels
to be incorporated specifically into two different residues in each protein
sample, allowing each spectroscopic experiment to be controlled internally.
^{15}N solution and solid-state NMR spectroscopy of M13 procoat protein
in micelles and in bilayers determined both mobile and rigid sections in
the precursor protein (McDonnell *et al.*, 1993a), as shown by the spectra
in Figure 6.8. Residues at both the amino and carboxy termini are
mobile on both solution and solid-state NMR time-scales. In addition,
a mobile section around the cleavage site for leader peptidase, residues
−1/+1, exists in M13 procoat protein. Residues in the hydrophobic
section of the signal sequence expected to have helical secondary structure
on the basis of studies of isolated signal sequence (Shinnar and Kaiser,
1984; Bruch *et al.*, 1989; Endo *et al.*, 1989; Karslake *et al.*, 1990;

Yamamoto *et al.*, 1990) are immobile on both solution and solid-state NMR time-scales. M13 procoat protein's dynamics is typical of a membrane protein because of the presence of mobile and rigid backbone sites.

Using the same approach that determined the structure and dynamics of membrane-bound fd coat protein, the structure and dynamics of membrane-bound Pf1 coat protein were characterized (Shon *et al.*, 1991a). As with fd coat protein, spectra were obtained on isotopically labelled coat protein samples in micelles and in bilayers, and results from solution and solid-state NMR were combined. Solution NMR experiments characterized the secondary structural elements in micelle samples and solid-state NMR experiments identified the orientation of the helix with respect to the bilayer. Backbone dynamics were analysed on the 10^{-9} s and 10^{-4} s time-scales by both solution and solid-state NMR. Even though it is a small 46 residue protein, membrane-bound Pf1 coat protein consists of five distinct regions determined by results from solution and solid-state NMR (Shon *et al.*, 1991a). Homonuclear ^{1}H–^{1}H NOEs in micelles using solution NMR defined residues 6–13 and residues 19–42 to form stable helical structures. The existence of the two α helices is strongly supported by the observation of many of the

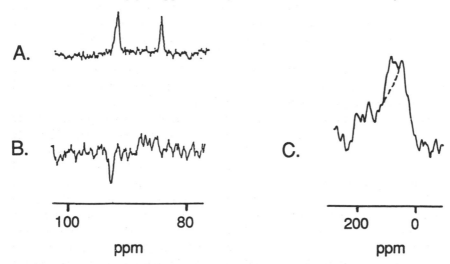

Figure 6.8 Comparison of backbone dynamics of M13 procoat protein specifically [^{15}N]-Ala$_{-1}$- and ^{15}N-Gly$_{+23}$-labelled. The residue at the −1 position of the signal sequence is mobile and the residue at the +23 position of the protein is rigidly held within the protein structure. (A) Procoat protein in micelles. ^{15}N NMR spectrum obtained by direct polarization transfer (DEPT). (B) Procoat protein in micelles. ^{15}N NMR spectrum with the heteronuclear ^{1}H–^{15}N NOE expressed obtained by direct pulsed ^{15}N excitation with continuous ^{1}H irradiation. (C) Procoat protein in unoriented phospholipid bilayers. Solid-state ^{15}N NMR spectrum obtained by cross-polarization. From McDonnell *et al.* (1993a)

appropriate cross-peaks between amide ^{1}H and C_α ^{1}H resonances in both two- and three-dimensional spectra. ^{15}N solid-state NMR spectroscopy of oriented coat protein samples in bilayers determined the orientation of the two helices with the amphipathic helix parallel to the bilayer plane and the hydrophobic helix perpendicular to the bilayer plane. Both solution and solid-state NMR spectroscopy identified mobile residues at the amino and carboxy termini and a mobile loop of five residues joining the two helices in the protein. The structure of membrane-bound Pf1 coat protein consists of α helices and mobile termini and loops, the same structural motifs present in fd coat protein and other membrane proteins.

Peptides

A wide variety of membrane-associated peptides, including small synthetic model peptides (Mueller *et al.*, 1986), glucagon (Braun *et al.*, 1983) and gramicidin (Nicholson *et al.*, 1987; Bystrov and Arseniev, 1988; Cornell *et al.*, 1988; Davis, 1988; Hing *et al.*, 1990) have been investigated by NMR spectroscopy. The results of NMR studies on hydrophobic and amphipathic helical peptides with more than 20 residues are directly applicable to the development and application of NMR methods for studying membrane proteins and for dissecting the factors contributing to the structural and dynamic properties of membrane proteins.

Mellitin has been studied extensively by NMR spectroscopy (Brown *et al.*, 1982; Bazzo *et al.*, 1988; Ikura *et al.*, 1991). This 26 residue peptide disrupts membranes. Solution NMR studies of mellitin in detergent micelles show that the N-terminal and C-terminal regions are alpha-helical and separated by a kink. The NMR data from mellitin have been subjected to considerable refinement and provide a model for other amphipathic helical peptides that interact with membranes. δ-Haemolysin is another 26 residue peptide that interacts with membranes. NMR studies of the peptide in micelles indicate that it forms an extended helix (Lee *et al.*, 1987). Typical of many membrane-bound amphipathic helical peptides, δ-haemolysin gives a quite well-resolved one-dimensional ^{1}H NMR spectrum, with some limitations in the two-dimensional spectrum, resulting from the broad resonance linewidths and efficient spin diffusion. Galanin is a 29 residue peptide found in tissues of the nervous system. NMR studies of the peptide show that it adopts a highly helical structure in trifluoroethanol/water (Wennerberg *et al.*, 1990). The 32 residue polypeptide hormone calcitonin was characterized in SDS micelles and found to have an amphipathic helix running into the N-terminal disulphide-bridged loop (Motta *et al.*, 1991).

The major functional channels in membranes, such as the nicotinic acetylcholine receptor and cation channels, are large oligomeric proteins.

Structural studies of these proteins are limited by their overall size and the difficulty in crystallizing them. A promising approach to their analysis is to synthesize and study peptide sequences corresponding to segments of the proteins (Montal, 1990). We have studied the peptide M2 selected from the δ subunit of the torpedo acetylcholine receptor, because homology and model studies suggest that this sequence-specific motif is responsible for specific functions in the channel activity of the receptor (Oiki *et al.*, 1988). The 23 residue M2δ peptide forms channels by self-association when added to membranes. This peptide closely mimics the ion channel properties of the receptor in model membranes.

Solution NMR experiments demonstrate that there is substantial helical secondary structure in the M2δ peptide in micelles (Opella *et al.*, 1993) and solid-state NMR studies have shown that the M2δ is immobilized by its interactions with phospholipid bilayers and is a trans-membrane helix (Bechinger *et al.*, 1991). For these experiments, the 23 residue M2δ peptide was synthesized with Ala 12 specifically labelled with ^{15}N. An experimental solid state ^{15}N NMR spectrum of the specifically labelled M2δ peptide in oriented phospholipid bilayers is presented in Figure 6.9(B). The anisotropies of both the chemical shift and heteronuclear ^{1}H–^{15}N dipole–dipole interactions are potentially present in this immobile sample. However, the heteronuclear (^{1}H–^{15}N) dipole–dipole interactions are decoupled by high-power irradiation of the ^{1}H resonances in the spectra in Figure 6.9; therefore, the spectral features reflect only the ^{15}N chemical shift interaction of the labelled site in the peptide. The experimental spectrum in Figure 6.9(B) of the oriented peptide in lipid bilayers has a relatively narrow single-line resonance, demonstrating that the sample is oriented along the direction of the applied magnetic field. The interpretation of the spectrum in Figure 6.2(B) is particularly interesting in comparison with the experimental results for magainin oriented in bilayers.

The S4 segment of the first internal repeat of rat brain sodium channels has been synthesized and studied in trifluoroethanol/water by multidimensional solution NMR spectroscopy (Mulvey *et al.*, 1989). The voltage-gated sodium channel is a large multi-subunit protein that resides in membranes. The S4 segment is believed to be part of the voltage sensor of the sodium channel. Two-dimensional homonuclear NOE spectra showed sequential NOEs between amide N–Hs from residues 2 through 17. Overall, 73 interresidue and 59 intraresidue restraints were observed and used to determine a structure with distance geometry and molecular dynamics calculations. The structure derived from the NMR data has a well-defined α-helical region for the first 14 residues of the peptide. No evidence was seen for a well-defined structure for the C-terminal regions of the peptide.

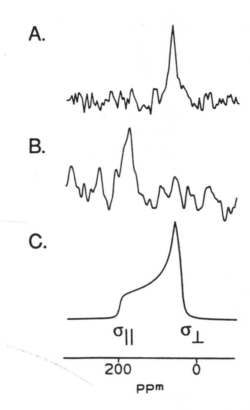

Figure 6.9 (A) and (B) are experimental solid-state ^{15}N NMR spectra of 23 residue amphipathic peptides in oriented lipid bilayers. (A) [^{15}N]- Ala$_{15}$-labelled magainin2. (B) ^{15}N-Ala$_{12}$-labelled M2δ. (C) Simulation of powder pattern for an immobile ^{15}N amide site. From Bechinger *et al.* (1991)

Many membrane-interactive peptides have antibiotic activities. Cecropin A, a 37 residue peptide found in *Hyalophora cecropia* and other insects, is part of their defence mechanism against bacteria. NMR studies show that this peptide is essentially a random coil in aqueous solution. This peptide has also been studied in hexafluoroisopropyl alcohol solutions by NMR spectroscopy (Holak *et al.*, 1988). Interproton distance restraints determined by nuclear Overhauser enhancement measurements and distance restraints hydrogen bonds were used as a basis for three-dimensional structure determination by simulated annealing. The resulting converged structures indicate that there are two α-helical regions extending from residues 5–21 and 24–37. Magainins are a family of 21–26 residue peptides that protect frogs from infections, even after severe injury, with a broad spectrum of antibacterial and antifungal activities (Bevins and Zasloff, 1990). These peptides interact strongly with bacterial and model membranes. Their antibiotic activity appears to be associated with the disruption of the electrochemical ionic gradient across cell membranes. Multidimensional solution NMR experiments indicate that magainins are unstructured in aqueous solution, but completely α-helical in detergent micelles and trifluoroethanol/water solutions (Marion *et al.*, 1989). Our solid-state NMR experiments on oriented peptide-containing bilayers show that magainin peptides lie parallel to the surface of membrane bilayers, in contrast to genuine channel-forming peptides that span bilayers (Bechinger *et al.*, 1991).

The ^{15}N NMR spectra of specifically labelled magainin and M2δ peptides in oriented phospholipid bilayers are compared in Figure 6.9. The resonance frequencies observed in the samples of M2δ and magainin are quite different, varying by nearly the full breadth of the ^{15}N amide chemical shift powder pattern, indicating that the planes containing the labelled peptide groups have very different orientations relative to the direction of the applied magnetic field. In order to fully determine the orientation of a labelled peptide plane, it would be necessary to measure at least one other spectral parameter associated with that site. However, since the secondary structure has been determined independently for both of the peptides, we can take advantage of the structural regularity of the α helix to extend the results from the specifically labelled site to the entire peptide. Both qualitative and quantitative analyses of the oriented ^{15}N chemical shifts are in agreement in describing the orientations of these two helical peptides in lipid bilayers. Magainin 2 is oriented approximately parallel to the plane of the bilayer, while M2δ is clearly a trans-membrane peptide oriented perpendicular to the plane of the bilayer. These orientations are illustrated in Figure 6.10. These results demonstrate that the combination of determining the secondary structure of proteins in micelles and the orientation of helices in bilayers that has been used successfully with the filamentous bacteriophage coat proteins (Shon *et al.*, 1991a,b; McDonnell *et al.*, 1993b) is applicable to amphipathic helical peptides as well (Bechinger *et al.*, 1991). Solid-state NMR spectroscopy has also been used to measure distances for structural studies of unoriented samples (Smith and Griffin, 1988), including magainins (Blazyk *et al.*, 1991).

Bacteriorhodopsin

Bacteriorhodopsin, the functional purple membrane protein of *Halobacterium halobium*, is one of the most intensively studied membrane proteins. Its structure is dominated by seven transmembrane hydrophobic helices (Henderson *et al.*, 1990). The protein utilizes a retinal chromophore to absorb light and translocate protons across the membrane. Extensive mutational studies indentifed amino acid residues important for ^{1}H pumping and for the retinal binding site (Khorana, 1988).

Both solution and solid-state NMR experiments have contributed to the descriptions of the structure, dynamics and mechanism of action of bacteriorhodopsin. Solid-state and solution NMR have been employed to study side-chain molecular dynamics of amino acids and to define mobile and rigid backbone sites in bacteriorhodopsin. Solution NMR on fragments of bacteriorhodopsin corresponding to transmembrane helices were determined in detergent micelles and in organic solvents, and compared with the single-crystal structures obtained from diffraction methods. NMR spectroscopy has also characterized the retinal chromophore and its interactions with the protein.

Solid-state NMR spectroscopy has been used to describe the backbone and side-chain dynamics of bacteriorhodopsin (Kinsey *et al.*, 1981; Rice *et al.*, 1987; Keniry *et al.*, 1984a,b; Bowers and Oldfield, 1988). ^{2}H solid-state NMR of bacteriorhodopsin found side-chains of surface residues to be more mobile than residues in the interior of the protein (Keniry *et al.*, 1984a,b). Most of the valine amino acid residues in bacteriorhodopsin are buried in the transmembrane helices and had side-chains that had limited motions. In contrast, residues near the termini and in the loops of the protein contributed to a central isotropic component superimposed on the powder pattern line shape, and indicated that some protein residues have large-scale rapid motions (Keniry *et al.*, 1984a). Backbone residues on the surface of bacteriorhodopsin, especially near the carboxy terminus, were found to be mobile, using ^{13}C solution NMR spectroscopy. Quantification of valine ^{13}C-labelled bacteriorhodopsin indicated motion in more than one residue, suggesting that one valine in the surface loops is also mobile (Bowers and Oldfield, 1988). These findings are in agreement with those for other membrane proteins in finding mobile residues near the termini and potentially in the loops connecting transmembrane helices.

Solution NMR spectroscopy characterized the structure of fragments in detergent micelles and in organic solvents obtained from proteolytic cleavage or solid-phase peptide synthesis corresponding to three of the transmembrane helices in bacteriorhodopsin. By two-dimensional homonuclear solution NMR, segment B (residues 34–65) of bacteriorhodopsin was found to consist of a right-handed α helix from residues 41–62, an unordered N-terminal region (residues 34–40) and a kink at Pro50 (Arseniev *et al.*, 1988; Pervushin *et al.*, 1991). The secondary structure of a fragment corresponding to segments F and G was also dominated with regions of extended right-handed α helix and had a flexible amino terminal segment (Barsukov *et al.*, 1990). Slight differences in the extent of residues involved in helical structure were observed for fragments in both micelles and organic solvents, with the anisotropic medium of micelles better simulating the environment of the membrane-spanning segments. In addition to determining the structure of individual helices in bacteriorhodopsin and backbone and side-chain dynamics, NMR spectroscopy was used to characterize the retinal chromophore and to determine the chemical environment of Asp residues necessary for the function of bacteriorhodopsin (Harbison *et al.*, 1984b; Engelhard *et al.*, 1989; Smith *et al.*, 1989).

Subunit c of the $F_1 F_0$ ATPase

Subunit c of the F_0 sector of the enzyme from *E. coli* is a small (79 residues), hydrophobic membrane-associated protein believed to play a key role in H^+ transport (Fillingame, 1990). This protein was studied in trifluoroethanol by multidimensional solution NMR spectroscopy (Norwood *et al.*, 1992). Its structure is predominantly helical, with two long segments connected by a mobile, unstructured loop with eight residues. Overall, its organization is like that of the other membrane proteins, with mobile N- and C-terminal regions and a mobile loop connecting two long hydrophobic helices.

4 Future Prospects

The major class of membrane proteins, exemplified by the photosynthetic reaction centre, bacteriorhodopsin and the filamentous bacteriophage coat proteins, is constructed with a limited repertoire of structural and dynamic elements. Although porins represent an entirely different class, and perhaps there are other classes of membrane proteins remaining to be discovered, it is likely that the majority of membrane proteins can have their structures analysed in the light of the few structures that have been determined. The extracellular and intracellular domains of the proteins of the photosynthetic reaction centre are, as expected for many large membrane proteins, constructed very much like globular proteins.

The membrane-associated domains are made up of long hydrophobic membrane-spanning helices, shorter amphipathic bridging helices in the plane of the bilayer, connecting loops with varying degrees of mobility, and mobile N and C terminal sections. NMR studies have been successful in identifying all of these elements and their orientations relative to each other and the membrane bilayer. The models in Figure 6.10 summarize the results of our NMR studies (Bechinger *et al.*, 1991; Shon *et al.*, 1991a,b; McDonnell *et al.*, 1993a,b; Opella, 1993; Opella *et al.*, 1993). These models demonstrate the feasibility of determining the structures of membrane proteins by NMR spectroscopy.

The instruments and methods for NMR spectroscopy are being improved rapidly, and the study of membrane proteins should become widespread in the near future. The most important advances are in the

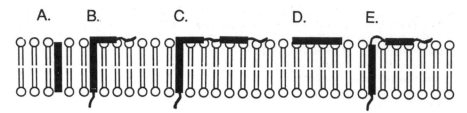

Figure 6.10 Models of membrane proteins in bilayers as determined by NMR spectroscopy: (A) M2δ peptide (Bechinger *et al.*, 1991); (B) fd coat protein (McDonnell *et al.*, 1993b); (C) M13 procoat protein (McDonnell *et al.*, 1993a); (D) magainin2 peptide (Bechinger *et al.*, 1991); (E) Pf1 coat protein (Shon *et al.*, 1991a)

use of high-field magnets, since they provide improved sensitivity and resolution for individual resonances and greater flexibility in the display and measurement of spectral parameters. NMR structural studies of globular proteins are enjoying extraordinary success and acceptance. However, multidimensional solution NMR spectroscopy and X-ray crystallography tend to work best on the same proteins, which have small to medium-size polypeptide chains and are water-soluble. The motivation for studying membrane proteins by NMR spectroscopy is clear, as is the promise of the approach. By characterizing the structure and dynamics of membrane proteins in model membrane environments on several different time-scales, it will be possible to arrive at detailed structural and functional models for this important class of proteins that is vastly underrepresented within structural biology.

Acknowledgments

This research is being supported by grants RO1 AI20770-09, RO1 GM24266-16, RO1 GM297541-10 and R24 RR05976-02 from the National Institutes of Health. P.A.M. was supported by a Biophysical Spectroscopy Training Grant to the University of Pennsylvania.

References

Arseniev, A. S., Maslennikov, I. V., Bystrov, V. F., Kozhich, A. T., Ivanov, V. T. and Ovchinnikov, Y. A. (1988). Two-dimensional ¹H-NMR study of bacterioopsin (34–65) polypeptide conformation. *FEBS Lett.*, **231**, 283–288

Barsukov, I. L., Abdulaeva, G. V., Arseniev, A. S. and Bystrov, V. F. (1990). Sequence-specific ¹H-NMR assignment and conformation of proteolytic fragment 163–231 of bacterioopsin. *Eur. J. Biochem.*, **192**, 321–327

Bazzo, R., Tappin, M. J., Pastore, A., Harvery, T. S., Carver, J. A. and Campbell, I. D. (1988). The structure of melittin. A ¹H NMR study in methanol. *Eur. J. Biochem.*, **173**, 139–146

Bechinger, B., Kim, Y., Chirlian, L. E., Gesell, J., Neumann, J. M., Montal, M., Tomich, J., Zasloff, M. and Opella, S. J. (1991). Orientations of amphipathic helical peptides in membrane bilayers determined by solid state NMR spectroscopy. *J. Biomol. NMR*, **1**, 167–173

Bechinger, B. and Opella, S. J. (1991). Flat coil probe for NMR spectroscopy of oriented membrane samples. *J. Magn. Reson.*, **95**, 585–588

Bevins, C. and Zasloff, M. (1990). Peptides from frog skin. *Ann. Rev. Biochem.*, **59**, 395–414

Blazyk, J., Hing, A. W., Schaefer, J. and Ferguson, M. (1991). Rotational-echo double resonance NMR spectroscopy of the antimicrobial peptide magainin2. *Cell Biol. Suppl.*, **15G**, 73

Bogusky, M. J., Leo, G. C. and Opella, S. J. (1988). Comparison of the dynamics of the membrane-bound form of fd coat protein in micelles and in bilayers by solution and solid-state nitrogen-15 nuclear magnetic spectroscopy. *Proteins Struct. Funct. Genet.*, **4**, 123–130

Bogusky, M. J., Schiksnis, R. A., Leo, G. C. and Opella, S. J. (1987). Protein backbone dynamics by solid state and solution ^{15}N NMR spectroscopy. *J. Magn. Reson.*, **72**, 186–190

Bowers, J. L. and Oldfield, E. (1988). Quantitative carbon-13 nuclear magnetic resonance spectroscopic study of mobile residue in bacteriorhodopsin. *Biochemistry*, **27**, 5156–5161

Braun, W. (1987). Distance geometry and related methods for protein structure determination from NMR data. *Quart. Rev. Biophys.*, **19**, 115–137

Braun, W., Wider, G., Lee, K. H. and Wüthrich, K. (1983). Conformation of glucagon in a lipid–water interphase by ^{1}H nuclear magnetic resonance. *J. Mol. Biol.*, **169**, 921–948

Brooks, C. L., Karplus, M. and Pettitt, B. M. (1988). Proteins: A theroetical perspective of dynamics, structure, and thermodynamics. *Adv. Chem. Phys.*, **71**, 1–259

Brown, L. (1979). Use of fully deuterated micelles for conformational studies of membrane proteins by high resolution ^{1}H nuclear magnetic resonance. *Biochim. Biophys. Acta*, **557**, 135–148

Brown, L. R., Braun, W., Kumar, A. and Wüthrich, K. (1982). High-resolution nuclear magnetic resonance studies of the conformation and orientation of melittin bound to a lipid–water interface. *Biophys. J.*, **37**, 319–328

Bruch, M. D., McKnight, J. and Gierasch, L. M. (1989). Helix formation and stability in a signal sequence. *Biochemistry*, **28**, 8554–8561

Brünger, A. T., Clore, G. M., Gronenborn, A. M. and Karplus, M. (1986). Three-dimensional structure of proteins determined by molecular dynamics with interproton distance restraints: application to crambin. *Proc. Natl Acad. Sci. USA*, **83**, 3801–3805

Bystrov, V. F. and Arseniev, A. S. (1988). Diversity of the gramicidin A spatial structure: Two-dimensional ^{1}H NMR study in solution. *Tetrahedron*, **44**, 925–940

Clore, G. and Gronenborn, A. (1989). Determination of three-dimensional structures of proteins and nucleic acids in solution by nuclear magnetic resonance spectroscopy. *CRC Crit. Rev. Biochem. Mol. Biol.*, **24**, 479–564

Cornell, B.A., Separovic, F., Baldassi, A.J. and Smith, R. (1988). Conformation and orientation of Gramicidin A in oriented phospholipid bilayers measured by solid-state carbon-13 NMR. *Biophys. J.*, **53**, 67–76

Cross, T.A., Frey, M.H. and Opella, S. J. (1983a). ^{15}N spin exchange in a protein. *J. Am. Chem. Soc.*, **105**, 7741–7743

Cross, T. A., Tsang, P. and Opella, S. J. (1983b). Comparison of protein and deoxyribonucleic acid backbone structure in fd and Pf1 bacteriophages. *Biochemistry*, **22**, 721–726

Dalbey, R. E. and Wickner, W. (1985). Leader peptidase catalyzes the release of exported proteins from the outer surface of the *Escherichia coli* plasma membrane. *J. Biol. Chem.*, **260**, 15925–15931

Davis, J. H. (1988). ^{2}H nuclear magnetic resonance of exchange labeled gramicidin in an oriented lyotropic nematic phase. *Biochemistry*, **27**, 428–436

Deisenhofer, J., Epp, O., Miki, K., Huber, R. and Michel, H. (1985). Structure of the protein subunits in the photosynthetic reaction centre of *Rhodopseudomonas viridis* at 3 Å resolution. *Nature*, **318**, 618–624

Dyson, J. and Wright, P. (1991). Defining solution conformations of small linear peptides. *Ann. Rev. Biophys. Biophys. Chem.*, **20**, 519–538

Endo, T., Shimada, I., Roise, D. and Inagaki, F. (1989). N-terminal half of a mitochondrial presequence peptide takes a helical conformation when bound to dodecylphosphocholine micelles: A proton nuclear magnetic resonance study. *J. Biochem.* **106**, 396–400

Engelhard, M., Hess, B., Emeis, D., Metz, G., Kreutz, W. and Siebert, F. (1989). Magic angle sample spinning ^{13}C nuclear magnetic resonance of isotopically labeled bacteriorhodopsin. *Biochemistry*, **28**, 3967–3975

Fillingame, R. (1990). Molecular mechanics of ATP synthesis by $F_1 F_0$ p type H^+-transporting syntases. In *The Bacteria*, Vol XII. Academic Press, 345–391

Frey, M.H. and Opella, S.J. (1984). ^{13}C spin exchange in amino acids and peptides. *J. Am. Chem. Soc.*, **106**, 4942–4945

Glucksman, M., Bhattacharjee, S. and Makowski, L. (1992). Three-dimensional structure of a cloning vector. *J. Mol. Biol.*, **226**, 455–470

Gullion, T. and Schaefer, J. (1989) Rotational-echo double resonance NMR. *J Magn. Reson.*, **81**, 196–200

Harbison, G., Jelinski, L., Stark, R., Torchia, D., Herfeld, J. and Griffen, R. (1984a). ^{15}N chemical shift and ^{15}N–^{13}C dipolar tensors for the peptide bond in [1-^{13}C] Glycyl[^{15}N]glycine hydrochloride monohydrate. *J. Magn. Reson.*, **60**, 79–82

Harbison, G. S., Smith, S. O., Pardoen, J. A., Mulder, P.P.J., Lugtenburg, J., Herzfeld, J., Mathies, R. and Griffin, R. G. (1984b). Solid-state ^{13}C NMR studies of retinal in bacteriorhodopsin. *Biochemistry*, **23**, 2662–2667

Havel, T. F. and Wüthrich, K. (1985). An evaluation of the combined use of nuclear magnetic resonance and distance geometry for the determination of protein conformations in solution. *J. Mol. Biol.*, **182**, 281–298

Helenius, A., McDaslin, D. R., Fries, E. and Tanford, C. (1979). Properties of detergents. *Meth. Enzymol.*, **56**, 734–749

Henderson, R., Baldwin, J. M., Ceska, R. A., Zemlin, F., Beckmann, E. and Downing, K. H. (1990). Model for the structure of bacteriorhodopsin based on high-resolution electron cryo-microscopy. *J. Mol. Biol.*, **213**, 899–929

Henry, G. D. and Sykes B. D. (1990). Detergent-solubilized M13 coat protein exists as an asymmetric dimer. *J. Mol. Biol.*, **212**, 11–14

Henry, G. D. and Sykes, B. D. (1992). Assignment of amide ^{1}H and ^{15}N NMR resonances in detergent solubilized M13 coat protein. *Biochemistry*, **31**, 5284–5297

Hing, A. W., Adams, S. P., Silbert, D. F. and Norberg, R. E. (1990). Deuterium NMR of Val1–(2-^{2}H)Ala3–Gramicidin A in oriented DMPC bilayers. *Biochemistry*, **29**, 4144–4156

Holak, T., Engstrom, A., Kraulis, P., Lindeberg, G., Bennich, H., Jones, T., Gronenborn, A. and Clore, G. M. (1988). The solution conformation of the antibacterial peptide Cecropin A: A nuclear magnetic resonance and dynamical simulated annealing study. *Biochemistry*, **27**, 7620–7629

Ikura, T., Go, N. and Inagaki, F. (1991). Refined structure of mellitin bound to perdeuterated dodecylphosphocholine micelles as studied by 2D-NMR and distance geometry calculations. *Proteins*, **9**, 81–89

Karslake, C., Piotto, M. E., Pak, Y. K., Weiner, H. and Gorenstein, D. G. (1990). 2D NMR and structural model for a mitochondrial signal peptide bound to a micelle. *Biochemistry*, **29**, 9872–9878

Keniry, M. A., Gutowsky, H. S. and Oldfield, E. (1984a). Surface dynamics of the integral membrane protein bacteriorhodopsin. *Nature*, **307**, 383–386

Keniry, M. A., Kintanar, A., Smith, R. L., Gutowsky, H. S. and Oldfield, E. (1984b). Nuclear magnetic resonance studies of amino acids and proteins. Deuterium nuclear magnetic resonance relaxation of deuteriomethyl-labeled amino acids in crystals and in *Halobacterium halobium* and *Escherichia coli* cell membranes. *Biochemistry*, **23**, 288–298

Khorana, H. G. (1988). Bacteriorhodopsin, a membrane protein that uses light to translocate protons. *J. Biol. Chem.*, **263**, 7439–7442

Kinsey, R. A., Kintanar, A. and Oldfield, E. (1981). Dynamics of amino acid side chains in membrane proteins by high field solid state deuterium nuclear magnetic resonance spectroscopy. *J. Biol. Chem.*, **256**, 9028–9036

Lee, K. H., Fitton, J. E. and Wüthrich, K. (1987). Nuclear magnetic resonance investigation of the conformation of δ-haemolysin bound to dodecylphosphocholine micelles. *Biochim. Biophys. Acta*, **911**, 144–153

Leo, G. C., Colnago, L. A., Valentine, K. G. and Opella, S. J. (1987). Dynamics of fd coat protein in lipid bilayers. *Biochemistry*, **26**, 854–862

McDonnell, P. A., Kim, Y., Tomich, J., Richards, J. and Opella, S. J. (1993a). Dynamics and orientation of the signal sequence of M13 procoat protein in membrane environments (unpublished results)

McDonnell, P. A., Shon, K., Kim, Y. and Opella, S. J. (1992b). fd coat protein structure in membrane environments (unpublished results)

Makowski, L. (1984). Structural diversity in filamentous bacteriophages. In McPherson, A. (Ed.), *Biological Macromolecules and Assemblies*, Vol. I, *The Viruses*. Wiley, New York, pp. 203–253

Marion, D., Zasloff, M. and Bax, A. (1989). A two-dimensional study of the antimicrobial peptide magainin2. *FEBS Lett.*, **227**, 21–26

Marvin, D. A. and Hohn, B. (1969). Filimentous bacterial viruses. *Bacteriol. Rev.*, **33**, 172–209

Marvin, D. A., Pigram, W. J., Wiseman, R. L., Wachtel, E. J. and Marvin, F. J. (1974). Filamentous bacterial viruses. XII. Molecular architecture of the class I (fd, Ifl, Ike) viron. *J. Mol. Biol.*, **88**, 581–598

Montal, M. (1990). Molecular anatomy and molecular design of channel proteins. *FASEB Jl*, **9**, 2623–2635

Motta, A., Pastore, A., Goud, N. A. and Morelli, M. A. C. (1991). Solution conformation of salmon calcitonin in sodium dodecyl sulfate micelles as determined by two-dimensional NMR and distance geometry calculations. *Biochemistry*, **30**, 10444–10450

Mueller, L. M., Frey, M. H., Rockwell, A. L., Gierasch, L. M. and Opella, S. J. (1986). Dynamics of a hydrophobic peptide in membrane bilayers by solid state NMR. *Biochemistry*, **25**, 557–561

Mulvey, D., King, G. F., Cooke, R. M., Doak, D. G., Harvery, T. S. and Campbell, I. D. (1989). High resolution [1]N NMR study of the solution structure of the S4 segment of the sodium channel protein. *FEBS Lett.*, **257**, 113–117

Nambudripad, R., Stark, W., Opella, S. and Makowski, L. (1991). Membrane mediated assembly of filamentous bacteriophage Pf1. *Science*, **252**, 1305–1308

Nicholson, L. K., Moll, F., Mixon, T. E., LoGrasso, P. V., Lay, J. C. and Cross, T. A. (1987). Solid state ^{15}N NMR of oriented lipid bilayer bound Gramicidin A. *Biochemistry*, **26**, 6621–6626

Norwood, T. S., Crawford, P. A., Stevenson, M. E., Driscoll, P. C. and Campbell, F. D. (1992). Heteronuclear ^{1}N–^{15}N nuclear magnetic resonance studies of the c subunit of the *Escherichia coli* F_1F_0 ATP synthase assignment and secondary structure. *Biochemistry*, **31**, 6285–6290

Nozaki, Y., Reynolds, J. and Tanford, C. (1978). Conformational states of a hydrophobic protein. The coat protein of fd bacteriophage. *Biochemistry*, **17**, 1239–1240

Oiki, S., Danho, W., Madison, V. and Montal, M. (1988). M2δ, a candidate for the structure lining the ionic channel of the nicotinic cholinergic receptor. *Proc. Natl Acad. Sci. USA*, **85**, 8703–8707

Opella, S. J. (1986). Protein dynamics by solid state NMR. *Meth. Enzymol.*, **17**, 327–361

Opella, S. J. (1993). NMR approaches to membrane protein structure. In White, S. (Ed.), *Membrane Protein Structure: Experimental Approaches*, Oxford (in press)

Opella, S. J., Gesell, J. and Bechinger, B. (1993). NMR spectroscopy of amphipathic helical peptides in membrane environments. In Epand, R. (Ed.), *The Amphipathic Helix*. CRC Press (in press)

Opella, S. J. and Stewart, P. L. (1989). Solid-state nuclear magnetic resonance structural studies of proteins. *Meth. Enzymol*, **176**, 242–275

Opella, S. J., Stewart, P. L. and Valentine, K. G. (1987). Protein structure by solid state NMR spectroscopy. *Quart. Rev. Biophys.*, **19**, 7–49

Pervushin, K. V., Arseniev, A. S., Kozhich, A. T. and Ivanov, V. T. (1991). Two-dimensional NMR study of the conformation of (34–65) bacterioopsin polypeptide in SDS micelles. *J. Biomol. NMR.*, **1**, 313–322

Raleigh, D., Levitt, M. and Griffin, R. (1988). Rotational resonance in solid state NMR. *Chem. Phys. Lett.*, **146**, 71–76

Rees, D., Komiga, H., Yeates, T., Allen, J. and Feher, G. (1989). The photosynthetic reaction center as a model for membrane proteins. *Ann. Rev. Biochem.*, **58**, 607–633

Rice, D. M., Meinwald, Y., Scheraga, H. A., Das Gupta, S. K. and Griffin, R. G. (1987). Solid state NMR investigations of lipid bilayers, peptides, and proteins. In *Proc. SUNYA Convers. Disc. Biomol. Sterodyn. 2nd.*, Vol. 2, pp. 255–270

Russel, M. (1991). Filamentous phage assembly. *Mol. Microbiol.*, **5**, 1607–1613

Scheek, R. M., van Gusteren, W. F. and Kaptein, R. (1989). Molecular dynamics simulation techniques for determination of molecular structures from nuclear magnetic resonance data. *Meth. Enzymol.*, **177**, 204–218

Shinnar, A. E. and Kaiser, E. T. (1984). Physical and conformational properties of a synthetic leader peptide from M13 coat protein. *J. Am. Chem. Soc.*, **106**, 5006–5007

Shon, K. and Opella, S. J. (1989). Detection of ^{1}H homonuclear NOE between amide sites in proteins with ^{1}H/^{15}N heteronuclear correlation spectroscopy. *J. Magn. Reson.*, **82**, 193–197

Shon, K., Kim, Y., Colnago, L. and Opella, S. J. (1991a). NMR studies of the structure and dynamics of membrane bound bacteriophage Pf1 coat protein. *Science*, **252**, 1303–1305

Shon, K., Schrader, P., Kim, Y., Bechinger, B., Zasloff, M. and Opella, S. J. (1991b). NMR structural studies of membrane bound peptides and proteins. In *Biotechnology: Bridging Research and Applications.* Kamely, D., Chakrabarty, A. and Kornguth, S. (Eds), Kluwer, Dordrecht, pp. 109–124

Smith, S. O. and Griffin, R. G. (1988). High-resolution solid-state NMR of proteins. *Ann. Rev. Phys. Chem.*, **39**, 511–535

Smith, S. O., de Groot, H. J. M., Gebhard, R., Courgin, J. M., Lugtenburg, J., Herzfeld, J. and Griffin, R. G. (1989). Structure and protein environment of retinal chromophore in light- and dark-adapted bacteriorhodopsin studied by solid-state NMR. *Biochemistry*, **28**, 8897–8904

Spruijt, R. B., Wolfs, Cor. J. A. and Hemminga, M. (1989). Aggregation related conformational change of the membrane associated coat protein of bacteriophage M13. *Biochemistry*, **28**, 9158–9165

Tobias, D. J., Klein, M. L. and Opella, S. J. (1992). Molecular dynamics simulation of a membrane protein (submitted)

Weiss, M., Abele, U., Weckesser, J., Welte, W., Schiltz, E. and Schulz, G. (1991). Molecular architecture and electrostatic properties of a bacterial porin. *Science*, **254**, 1627–1630

Wennerberg, A. B. A., Cooke, R. M., Carlquist, M., Rigler, R. and Campbell, I. D. (1990). A ^{1}H NMR study of the solution conformation of the neuropeptide Galanin. *Biochem. Biophys. Res. Commun.*, **166**, 1102–1109

Williams, R. and Dunker, A. K. (1977). Circular dichroism studies of fd coat protein in membrane vesicles. *J. Biol. Chem.*, **252**, 6253–6255

Wüthrich, K. (1986). *NMR of Proteins and Nucleic Acids.* Wiley, New York

Yamamoto, Y., Ohkubo, T., Kohara, A., Tanaka, T., Tanaka, T. and Kikuchi, M. (1990). Conformational requirement of signal sequences functioning in yeast: circular dichroism and ^{1}H nuclear magnetic resonance studies of synthetic peptides. *Biochemistry*, **29**, 8998–9006

Zimmermann, R., Watts, C. and Wickner, W. (1982). The biosynthesis of membrane-bound M13 coat protein energetics and assembly intermediates. *J. Biol. Chem.*, **257**, 6529–6536

7

Heteronuclear NMR Studies of the Molecular Dynamics of Staphylococcal Nuclease

Dennis A. Torchia, Linda K. Nicholson, Holly B. R. Cole and Lewis E. Kay

1 Introduction

During the past decade tremendous progress has been made in developing NMR techniques to determine the structure of proteins in solution. For proteins containing less than c. 100 residues, essentially complete proton signal assignments can be obtained using two- and three-dimensional homonuclear (proton) pulse sequences (Wüthrich, 1986; Clore and Gronenborn, 1989). These assignments together with distance and dihedral angle constraints, derived from NOEs and coupling constants, respectively, provide sufficient information to determine the three-dimensional structure of the protein. Recently it has been shown that NMR methods can be applied to larger proteins, containing up to c. 200 residues, by combining heteronuclear labelling (with ^{13}C and/or ^{15}N) with sophisticated double- and triple-resonance multidimensional techniques (Bax et al., 1990; Fesik et al., 1990; Ikura et al., 1990; Kay et al., 1990a,b; Clore et al., 1991a,b; Pelton et al., 1991).

Because of the great interest in and importance of protein structure–function relationships, most NMR studies have been directed towards structure elucidation. However, it is widely recognized that proteins undergo significant internal motions on a wide range of time-scales (Karplus and McCammon, 1986). It will be necessary to delineate these motions in order to determine the range of structures available to proteins. In addition, protein dynamics have important functional roles in protein folding, enzyme action and molecular recognition. At a more fundamental level, measurement of quantities related to the rates and amplitudes of internal motions in proteins will provide data that can test the predictions of molecular dynamics calculations. The comparison of theory with experiment will result in improved potential functions that will enhance our understanding of the fundamental interactions between atoms in proteins.

While proton–proton NOEs and *J* couplings provide most of the information about protein structure, heteronuclear, rather than proton, relaxation data are better suited to yield information about motions of specific internuclear vectors in proteins. Although pioneering studies of natural abundance ^{13}C relaxation in proteins were carried out over ten years ago (Wittebort *et al.*, 1979, Richarz *et al.*, 1980; London, 1989), these studies were limited by lack of sequential assignments, and the low sensitivity and resolution afforded by one-dimensional ^{13}C spectra. Modern two-dimensional NMR experiments, in which heteronuclear relaxation is measured by proton detection, yield spectra having high sensitivity and resolution (Kay *et al.*, 1989; Nirmala and Wagner, 1989). In addition, double- and triple-resonance multidimensional NMR experiments provide essentially complete heteronuclear signal assignments (Clore *et al.*, 1990a; Ikura *et al.*, 1991; Pelton *et al.*, 1991).

Here we focus upon the application of these new NMR approaches to the study of the internal dynamics of recombinant staphylococcal nuclease, SNase, an enzyme containing 149 amino acid residues (Shortle, 1983). We begin with a brief summary of equations that relate the measured spin relaxation parameters to the model-free parameters (*S*, τ_m and τ_e) of Lipari and Szabo (1982a,b). We then discuss (a) the pulse sequences that are used to measure the relaxation parameters and (b) the method of determining the uncertainties in the relaxation parameters and the associated model-free parameters. Next we discuss how one derives information about molecular motion from the model-free parameters. We then review the application of heteronuclear relaxation measurements to study internal dynamics of SNase in solution and, for the purpose of comparison, in the crystalline state. We conclude with a discussion of the prospects for future NMR studies of protein dynamics.

2 Relaxation Theory

Nuclear Spin Relaxation

It has been recognized from the inception of the NMR technique (Bloembergen *et al.*, 1948) that nuclear spin relaxation rates are closely related to molecular motions. In the case of a spin $I = \frac{1}{2}$ heteronucleus (designated A) such as ^{13}C or ^{15}N, the three commonly measured relaxation parameters are the longitudinal relaxation time, T_1, the transverse relaxation time, T_2 and the A-{^{1}H} nuclear Overhauser enhancement (NOE) (Abragam, 1961). When the heteronucleus, A, is bonded to a single proton, X, the dominant relaxation mechanism is the AX dipolar interaction, while chemical shift anisotropy is a secondary but significant relaxation mechanism, particularly as field strength increases. Provided that the dipolar and CSA relaxation mechanisms are uncorrelated, the expressions for the relaxation parameters of a spin A bonded to a single proton are given by (Kay *et al.*, 1989)

$$1/T_1 = d^2[J(\omega_A - \omega_X) + 3J(\omega_A) + 6J(\omega_A + \omega_X)] + c^2J(\omega_A) \tag{7.1}$$

$$1/T_2 = 0.5d^2[4J(0) + J(\omega_A - \omega_X) + 3J(\omega_A) + 6J(\omega_X) +$$

$$6J(\omega_A + \omega_X)] + (1/6)c^2[4J(0) + 3J(\omega_A)] \tag{7.2}$$

$$\text{NOE} = 1 + \{(\gamma_X/\gamma_A)d^2[6J(\omega_A + \omega_X) - J(\omega_A - \omega_X)]T_1\} \tag{7.3}$$

In these Equations $d^2 = 0.1[\gamma_A\gamma_X h/(2\pi \langle 1/r_{AX}^3\rangle)]^2$ and $c^2 = (2/15)[\gamma_A B_0 (\sigma_\parallel - \sigma_\perp)]^2$, γ_i is the gyromagnetic ratio of spin i, h is Planck's constant, r_{AX} is the AX internuclear distance, B_0 is the magnetic field strength, $\sigma_\parallel$ and $\sigma_\perp$ are the parallel and perpendicular components of the axially symmetric A-spin CSA tensor and $J(\omega)$ is the spectral density function. For calculations described herein r_{AX} was set to 1.02 Å and $\Delta\sigma = \sigma_\parallel - \sigma_\perp$ was set to $- 160$ ppm in the case of A = ^{15}N. The quantities d and c are coupling constants that are proportional to the static dipolar and chemical shift interactions, respectively, and the spectral densities, $J(\omega)$, are proportional to the amplitudes of the fluctuations in the dipolar and CSA interactions at frequency ω.

In the absence of dipolar cross-correlations, Equations (7.1)–(7.3) apply to methyl carbon relaxation, provided that $1/T_i$ ($i = 1, 2$) is replaced by $(1/3T_i)$. In this case $r_{AX} = 1.09$ Å and the CSA is neglected. A rotating methyl typically has $\Delta\sigma \leqslant 25$ ppm, from which one calculates that relaxation due to CSA is insignificant (less than 2.5%) compared with relaxation due to the dipolar interaction.

It was tacitly assumed in the above discussion that the pulse sequences used to measure the relaxation parameters eliminate cross-correlation between the dipolar and CSA interactions. Although this can be readily accomplished, RF pulses cannot completely eliminate the effects of cross-correlations among the three carbon–proton dipolar couplings in a methyl spin system. The effect of dipolar cross-correlation upon methyl carbon relaxation has been thoroughly investigated (Werbelow and Grant, 1977). In general, neither longitudinal nor transverse methyl carbon magnetization relaxes in a single exponential manner, and the equations for the NMR relaxation parameters become quite formidable. However, a recent analysis of these equations has shown that dipolar cross-correlations make only small contributions to longitudinal relaxation rates and to NOEs for carbons in rapidly rotating methyl groups in proteins. This is the case for two reasons. First, using the three-site Woessner model (Woessner, 1962) to describe methyl group internal dynamics, it is found that cross-correlation spectral density terms nearly vanish when the overall and the internal correlation times are in the 5–20 ns and 15–65 ps ranges, respectively (Kay and Torchia, 1991). Second, in larger proteins, efficient cross-relaxation involving methyl protons and neighbouring protons attenuates the effects of cross-correlation on T_1 measurements. If cross-correlation effects are significant, one can still apply Equation (7.1) provided that one derives the relaxation rate from the initial decay of the longitudinal magnetization.

In contrast with longitudinal relaxation, transverse methyl carbon relaxation is strongly affected by dipolar cross-correlations. Therefore, one must derive the T_2 value from the initial decay of the transverse magnetization (Werbelow and Grant, 1977). Finally, we note that chemical exchange may also contribute to the transverse relaxation rate in proteins, a subject that we discuss later in this review.

Model-free Formulation of Relaxation Equations

A useful way of expressing $J(\omega)$ in terms of the minimum number of model independent motional parameters is provided by the formalism of Lipari–Szabo (1982a,b). According to this formalism

$$J(\omega) = S^2\tau_m/(1 + (\omega\tau_m)^2) + (1 - S^2)\tau/(1 + (\omega\tau)^2) \qquad (7.4)$$

where S is the generalized order parameter, τ_m is the overall correlation time and $1/\tau = 1/\tau_m + 1/\tau_e$, where τ_e is an effective correlation time characterizing the internal motions. Although Equation (7.4) contains a single effective internal correlation time, it is an exact expression for the spectral density function, even in the case of multiple internal motions, provided that all internal motions are in the extreme narrowing limit (Lipari and Szabo, 1982a), i.e. $(\omega\tau_i)^2 \ll 1$, where τ_i is the correlation time associated with an arbitrary internal motion.

Equation (7.4) applies to solids as well as to liquids. In solids, τ_m approaches infinity, and the first term in Equation (7.4) becomes vanishingly small. In this limit the expression for $J(\omega)$ becomes

$$J(\omega) = (1 - S^2)\tau_e/(1 + (\omega\tau_e)^2) \qquad (7.5)$$

Equation (7.4) can be generalized to apply to the case when the internal motions take place on two (or more) significantly different time-scales, at least one of which is outside the extreme narrowing limit. For example, if two internal correlation times, τ_f and τ_s, are required to define the internal correlation function, $J(\omega)$ becomes (Clore *et al.*, 1990b)

$$J(\omega) = S^2 \tau_m / (1 + (\omega \tau_m)^2) + (1 - S_f^2) \tau_f / (1 + (\omega \tau_f)^2)$$
$$+ S_f^2 (1 - S_s^2) \tau_s / (1 + (\omega \tau_s)^2) \qquad (7.6)$$

where S_f and S_s are the order parameters associated with the 'fast' and 'slow' internal motions and $S = S_f S_s$. Note that Equation (7.6) reduces to Equation (7.4) when $\tau_f = \tau_s$.

The physical interpretation of the model-free parameters is given in the next section, after discussion of how these parameters are derived from measured relaxation parameters.

3   Data Acquisition and Analysis

Pulse Sequences and Measurement Artifacts

In order to maximize resolution and sensitivity it is essential that heteronuclear (^{13}C or ^{15}N) relaxation parameters be measured by use of two-dimensional pulse schemes that employ proton detection (Kay *et al.*, 1989; Nirmala and Wagner, 1989). We discuss two classes of such experiments that are appropriate for AX and AX$_3$ spin systems, respectively.

Sequences ɟor AX Spin Systems

Pulse sequences designed to measure the A-spin T_1, T_2 and NOE values in AX spin systems (typically A = ^{13}C or ^{15}N; X = ^{1}H) are depicted schematically in Figure 7.1. The sequences used to measure T_1 and T_2 (Kay *et al.*, 1992a; Palmer *et al.*, 1992) are improved versions of sequences proposed originally (Kay *et al.*, 1989) in that 180° ^{1}H pulses are applied periodically during the relaxation delay period, T, in order to suppress cross-correlation of AX dipolar coupling with A-spin CSA (Goldman, 1984; Boyd *et al.*, 1990). If this procedure is not followed, the amide ^{15}N relaxation times measured for SNase are overestimated by 5–9% in the case of SNase (Kay *et al.*, 1992b). Note that the T_1 and T_2 pulse sequences yield signals that decay to zero as the relaxation delay increases. Therefore, the relaxation measurement is independent of the delay time between scans, permitting the decay curve to be fitted with two parameters.

In order to measure the NOE of spin A, it is necessary to record spectra (a) with and (b) without proton saturation. The sensitivity of the NOE experiment is significantly below that of the T_1 and T_2 experiments (particularly when A = ^{15}N) because the initial magnetization must be derived from the A spins rather than from the protons. Furthermore, in contrast with the T_1 and T_2 experiments, it is necessary

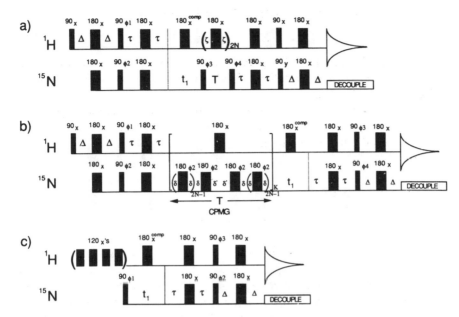

Figure 7.1 Pulse sequences used to measure (a) T_1, (b) T_2 and (c) the A-{^{1}H} NOE in AX spin systems. The sequences are discussed in detail in Kay *et al.* (1992a)

to wait a long period ($> 3T_{1L}$, where T_{1L} is the larger of the proton and A-spin T_1 values) in order to ensure that AX cross-relaxation does not prevent the A-spin magnetization from attaining its equilibrium value prior to the application of the initial A-spin pulse. When experiments are recorded in H$_2$O, water saturation must be achieved quickly in the experiment without NOE, in order to minimize the transfer of saturation by spin diffusion. An alternative to presaturation in this experiment (Barbato *et al.*, 1992) is the application of a short scrambling sequence (Messerle *et al.*, 1989).

Pulse Sequences for AX$_3$ Spin Systems

Pulse sequences for measuring methyl ^{13}C relaxation parameters differ in a number of important ways from sequences used in an AX spin system. Sequences designed to measure ^{13}C relaxation parameters in methyl groups are shown in Figure 7.2. Note that, in contrast to the T_1 and T_2 sequences depicted in Figure 7.1, magnetization is initially derived from the heteronuclei rather than the protons. In our experience this causes no loss in sensitivity because (a) the NOE of the methyl carbons is large (typically in the range 2–2.5) and (b) at the cost of

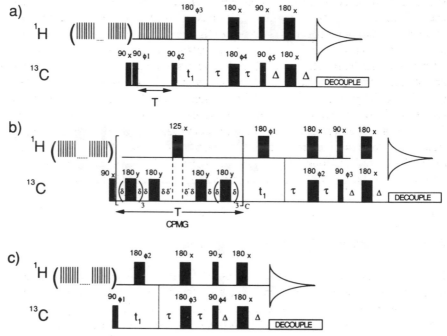

Figure 7.2 Pulse sequences used to measure (a) T_1, (b) T_2 and (c) the A-{^{1}H} NOE in AX$_3$ spin systems. The sequences are discussed in detail in Kay *et al.* (1992b)

reducing sensitivity, one must use magic angle pulses or delays in DEPT- and INEPT-type sequences, respectively, in order to suppress complicating effects that arise from dipolar cross-correlations (Palmer *et al.*, 1991b). Note that the delay time τ in the reverse INEPT portions of the T_1 and T_2 sequences in Figure 7.2(a,b) must satisfy the magic angle condition ($2\pi J\tau = 0.955 = 54.7°$) in order that magnetization be correctly transferred from ^{13}C transitions to proton transitions (Kay *et al.*, 1992a). Furthermore, the delay Δ is set to a short value, $1/(8J_{CH})$, so that rapid proton relaxation will not affect the measurement of the ^{13}C relaxation rates. These points and our reason for using the reverse INEPT rather than the reverse DEPT sequence are discussed in detail elsewhere (Kay *et al.*, 1992a). The sequences shown in Figure 7.2 have been tested (a) on [methyl-^{13}C]-leucine-labelled SNase and (b) on [methyl-^{13}C]-alanine, dissolved in deuterated glycerol, over a range of temperatures; they have been optimized to minimize artefacts when measuring methyl ^{13}C relaxation times in proteins by proton detection.

Although the pulse schemes in Figures 7.1 and 7.2 suppress cross-correlation of AX dipolar and A-spin CSA interactions, the pulse scheme in Figure 7.2(b) minimizes, but does not completely eliminate, the effects of dipolar cross-correlations, because RF pulses do not mix the 3/2 and 1/2 proton manifolds (Ernst, 1966). Hence, the decay of transverse magnetization is biexponential, and the initial slope of the decay curve must be determined in order to measure T_2 correctly (Werbelow and Grant, 1977). Finally, we note that accurate T_2 values are obtained by the CPMG method only when the condition $\delta \ll 1/(2J)$ is satisfied (Kay *et al.*, 1989, 1992b; Peng *et al.*, 1991a). Because small values of δ are employed in the T_2 sequence, chemical exchange affects T_2 in almost the same manner as $T_{1\rho}$ (particularly because spin locking fields for heteronuclei in liquids are limited to < 10 kHz). Hence, it is difficult to find evidence for chemical exchange by comparing measured values of T_2 with $T_{1\rho}$. For reasons discussed elsewhere (Kay *et al.*, 1992a), we prefer to measure T_2 values using a CPMG-type sequence rather than to measure $T_{1\rho}$ values using a spin lock (Peng *et al.*, 1991b).

Relaxation Parameters and Their Uncertainties

In one-dimensional NMR experiments that employ heteronuclear detection, low sensitivity is the primary factor that limits the accuracy of the relaxation measurements in proteins. In addition to significantly enhancing resolution, proton-detected two-dimensional techniques greatly improve sensitivity. In the case of 1.5 mM solutions of SNase, typical signal-to-noise ratios are in the range of 50–100 for well-labelled samples (incorporation levels greater than 50%). At these signal-to-noise levels, the main uncertainty in determining the signal intensity arises from baseline distortions and from overlap of the signals of interest with the tails of nearby strong signals. With careful data processing, signal intensities can be measured with an accuracy of a few per cent. Signal intensities can be derived from measurements of either peak heights or peak volumes, with the latter quantities obtained from surface-fitting routines. The relaxation parameters derived from these two types of measurements should usually agree to within 5% or better.

In order to extract values of T_1 and T_2 from the NMR data, we measure the signal intensities of the cross-peaks in the 2D spectra recorded as a function of the relaxation delay, T, and fit the measured volumes to an equation of the form $y = A \exp(-t/T_i)$, using conjugate gradient minimization techniques (Press *et al.*, 1988). The standard deviations in the parameters A and T_i are derived from the residuals of the fits to the data by using a Monte Carlo approach (Kamath and Shriver, 1989; Nicholson *et al.*, 1992).

In a similar manner, NOE values are determined by measuring the signal intensities in the presence (I), and absence (I') of the NOE. The NOE is equal to the ratio I/I', while the error in the NOE measurement is given by

$$E(\text{NOE}) = (I/I')[(\Delta I/I)^2 + (\Delta I'/I')^2]^{1/2} \tag{7.7}$$

where $\Delta I(\Delta I')$ is the error in the signal intensity measurement in the presence (absence) of the NOE. When signal intensity is obtained from peak height, the error is the standard deviation of the baseline noise. When signal intensity is obtained from a volume integral, the uncertainty is given by the standard deviation of the noise times the square root of the number of points in the area under integration.

Extracting Model-free Parameters from Relaxation Data

In analysing the relaxation data obtained on SNase, we assume that the overall (rigid body) reorientation of the molecule is isotropic and is characterized by a single correlation time, τ_m. This assumption is validated by (a) the fact that the principal components of the inertia tensor of the protein are calculated (using the crystal coordinates) to be in the ratio of 1.0:1.3:1.4, and (b) by the observation that the amide ^{15}N T_1/T_2 ratio is nearly the same (rms variation of 6%) for all residues in the protein.

In addition to their dependence on the global parameter τ_m, the relaxation parameters depend upon the residue-specific model-free parameters S and τ_e. A general approach to determine these three parameters consists of minimizing the error function (Dellwo and Wand, 1989)

$$E = \sum_i \{[(T_{1i}{}^{\text{obs}} - T_{1i}{}^{\text{cal}})/T_{1i}{}^{\text{obs}}]^2 + [(T_{2i}{}^{\text{obs}} - T_{2i}{}^{\text{cal}})/T_{2i}{}^{\text{obs}}]^2 +$$

$$[(\text{NOE}_i{}^{\text{obs}} - \text{NOE}_i{}^{\text{cal}})/\text{NOE}_i{}^{\text{obs}}]^2 \} \tag{7.8}$$

The minimization is carried out by assigning a value of τ_m to all residues in the protein, and then determining the set of $\{S_i, \tau_{ei}\}$ that minimize E, using conjugate gradient minimization. This process is repeated, stepping τ_m through a series of values until the minimum value of E is determined.

A Monte Carlo procedure (Kamath and Shriver, 1989) is used in order to ascertain the precision limits on the derived values of $\{S_i, \tau_{ei}\}$. Having determined the values and standard deviations of the NMR relaxation parameters (T_1, T_2 and the NOE) as described in the previous section, a Gaussian random number generator is used to generate synthetic values of these parameters, and the error function defined in Equation (7.8) is minimized with respect to S_i, and τ_{ei}. This procedure is repeated for several hundred synthetic data sets in order to determine the distribution of values of the model-free parameters. These distributions are used to derive uncertainties in the model-free parameters.

While the procedure outlined above for determining the values and precision limits of the model-free parameters is generally applicable, a few practical comments are perhaps in order. When measuring backbone relaxation parameters, the investigator should note whether the NOEs are close to their slow motion values. If this is the case, then the $\tau_e = 0$ approximation is valid (Kay *et al.*, 1989). This permits a simpler two-parameter, τ_m, S, fit of the data to be made. The investigator should also note whether a residue has an anomalously small backbone T_2 value, as this is strong evidence that chemical exchange makes a significant contribution to T_2; hence, the relaxation parameters for this residue should be excluded from Equation (7.8). At the other extreme, if a residue has an anomalously large T_2 value and an NOE value significantly different from the slow tumbling limiting value, it is possible that the residue exhibits a complex segmental motion that cannot be simply described by a single internal correlation time. For this reason the relaxation parameters for such a residue should be excluded from Equation (7.8). Once the optimum value of τ_m is determined, the relaxation parameters obtained for the flexible residue can be used to determine the values of τ_m, S and τ_e. If the optimum model-free parameters do not yield calculated relaxation parameters that agree with experiment, then a more complex analysis involving two internal order parameters and correlation times is warranted (Clore *et al.*, 1990b). One must be particularly careful about analysis of ^{15}N relaxation times of flexible residues, because the relaxation parameters of these residues may be affected by rapid exchange of the amide protons with solvent water molecules. Although we have assumed that the overall motion of the protein is isotropic (since this is appropriate in the case of SNase), this assumption is not a requirement of the model-free approach, and methods for dealing with anisotropic motion have been described (Lipari and Szabo, 1982a,b; Dellwo and Wand, 1989; Barbato *et al.*, 1992).

Deriving Motional Information from Model-free Parameters

The overall correlation time derived from the model-free approach, τ_m, can be compared with values calculated by hydrodynamic theory (Venable and Pastor, 1988). The correlation time of a sphere with volume V in a medium having viscosity η is given by

$$\tau_m = \eta \, V/kT \tag{7.9}$$

where k is Boltzmann's constant and T is the absolute temperature. It is necessary to use a 3 Å hydration sphere in order to bring the values of τ_m calculated from Equation (7.9) into agreement with values of τ_m measured by NMR techniques. The value of τ_m obtained from Equation (7.9) (including the 3 Å hydration layer) is sometimes less than the value of τ_m obtained from ^{15}N NMR relaxation measurements. At present the reason for this discrepancy is not clear, but may be a consequence of the high protein concentrations (*c*. 20–30 mg/ml) typically required for NMR studies.

The bulk of the information obtained from NMR experiments is contained in the residue-specific parameters S and τ_e. Although τ_e provides an estimate of the time-scale of the internal motion, it is difficult to interpret this parameter in a quantitative fashion, because it is a complex combination of geometric factors and internal correlation times (Lipari and Szabo, 1982a,b). In contrast, the generalized order parameter is precisely related to the equilibrium orientational distribution function associated with internal motions having correlation times that are smaller than the overall correlation time of the protein. When the NMR interaction responsible for spin relaxation is axially symmetric with the unique axis parallel to the AX bond vector (a condition that is rigorously satisfied by the AX dipolar interaction and is approximately true for the CSA interaction considered herein), the generalized order parameter is given by the expression (Lipari and Szabo, 1982a)

$$S^2 = \int d\Omega_1 \, p_{eq}(\Omega_1) \int d\Omega_2 \, p_{eq}(\Omega_2) \, P_2(\cos \theta_{12}) \qquad (7.10)$$

where $p_{eq}(\Omega)d\Omega$ is the normalized equilibrium probability that the vector $\boldsymbol{\mu}$, a unit vector along the AX bond axis, has orientation $d\Omega$, $\Omega = (\theta, \phi)$, in a coordinate system fixed in the protein, θ_{12} is the angle between two such vectors, $\hat{\boldsymbol{\mu}}_1$ and $\hat{\boldsymbol{\mu}}_2$, and $P_2(\cos x) = (3\cos^2 x - 1)/2$. Using Equation (7.10), one can calculate S for any model of AX bond reorientation. For instance, in the cone model, where $\hat{\boldsymbol{\mu}}$ diffuses freely in the angular region $0 < \theta < \theta_0$, $0 < \phi < 2\pi$, one finds that

$$S^2 = [(1/2) \cos \theta_0 (1 + \cos \theta_0)]^2 \qquad (7.11)$$

This expression reveals several general features about the relationship of the order parameter with the amplitude of the motion. The order parameter equals unity in the absence of motion, $\theta_0 = 0$, diminishes as θ_0 increases, vanishing when motion is isotropic, $\theta_0 = 180°$. Note, however, that S^2 is not a monotonic function of θ_0. It vanishes even when motion is not isotropic, i.e. $\theta_0 = 90°$, and attains a small local maximum at $\theta_0 = 120°$. In general, a small-amplitude motion implies that S^2 is large. However, a large-amplitude motion need not have a small order parameter. For example, if the AX bond reorients by jumps of 180°, $S^2 = 1$.

The order parameter is closely related to the orientation-dependent local magnetic fields (due to either dipolar coupling or CSA) that are responsible for relaxation. In the absence of reorientation, these local fields are the sources of the static powder line shapes observed in the spectra of unoriented solid samples (Haeberlen, 1976). The size of the local magnetic field parallel to the external field determines the width of the powder line shape, which is normally characterized by principal frequencies or equivalently by moments. In the presence of motion, the line shape narrows and the moments are reduced. The ratio between the motionally averaged second moment and the static second moment is proportional to S^2 (Torchia and Szabo, 1985). S^2 is therefore proportional to the value of the mean squared local field that remains after motional averaging.

In addition to the cone model, several other plausible models of internal motion have proved useful in analysing relaxation data. These models are now briefly described and expressions for their corresponding order parameters are given.

The Restricted Diffusion Model

If the unit vector $\hat{\mu}$, along the AX bond axis, makes an angle of 109.5° with a rotation axis and undergoes rotational diffusion (about the axis) through an angular range of $\pm \gamma_0$, S^2 is given by

$$S^2 = (1/9)[1 + 8 \sin^2 \gamma_0 (1 + 2 \cos^2 \gamma_0)/(3\gamma_0^2)] \tag{7.12}$$

The Two-site Jump Model

If $\hat{\mu}$ abruptly changes its orientation by 109.5°, S^2 is given by

$$S^2 = 1 - (8/3)[r/(1 + r)^2] \tag{7.13}$$

where $r = p_1/p_2$, and p_1 and p_2 are the relative populations of the two orientations of $\hat{\mu}$.

Threefold Rotation

If $\hat{\mu}$ reorients about a threefold axis, S^2 is given by

$$S^2 = [(3 \cos^2\theta - 1)/2]^2 \tag{7.14}$$

where θ is the angle between μ and the rotation axis.

4 Applications

SNase Backbone Dynamics in Solution

Using the pulse sequences described in Figure 7.1, backbone amide ^{15}N T_1, T_2 and NOE values were measured for *c.* 100 of the 149 amino acid residues in liganded SNase (complexed with pdTp and Ca^{2+}). The

T_1 and T_2 values are plotted as a function of residue number in Figure 7.3. The relaxation times of the SNase sample measured using the sequences in Figure 7.1 are 5–10% shorter than the relaxation times measured using sequences reported previously (Kay *et al.*, 1989), because the original sequences did not entirely eliminate the effects of cross-correlation of dipolar and CSA interactions. The first six residues and

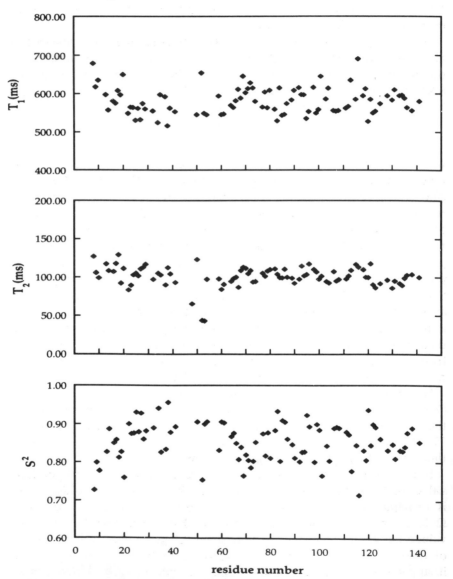

Figure 7.3 Amide ^{15}N T_1, T_2 and S^2 values obtained for the various residues of SNase

the last eight residues in SNase are not included in Figure 7.3, because the amide protons of these residues exchange rapidly with water. In addition, relaxation times were not obtained for a number of residues in or near the disordered Ω loop of SNase (Baldisseri *et al.*, 1991), because their signals were severely exchange broadened. The remaining residues not included in Figure 7.3 are either prolines or residues whose NMR signals are not well enough resolved to permit accurate measurements of their relaxation times.

As noted previously (Kay *et al.*, 1989), because the NOE values for all residues (except for the flexible residues in the N and C termini of the protein) are nearly equal to their slow limit $((\omega_A \tau_m)^2 \gg 1)$ value of 0.82, the $\tau_e = 0$ approximation is valid and Equations (7.1) and (7.2) can be replaced by the simple expressions

$$1/T_i = S^2(1/T_i)_{abs} \tag{7.15}$$

where $i = 1, 2$, and the subscript 'abs' indicates the value of the relaxation time in the absence of internal motion. Following Kay *et al.* (1989), we calculate τ_m on a residue-by-residue basis from the T_1/T_2 value for each residue, and then calculate the value of τ_m for SNase as the average of values obtained for all the residues. In this manner we obtain $\tau_m = 8.3 \pm 0.5$ ns for SNase, slightly less than the value of 9.1 ns reported previously, using relaxation times measured in the presence of dipolar/CSA cross-correlations. Three residues in the Ω loop region of SNase (K48, E52 and K53) were excluded from the calculation of the average τ_m value because they have anomalously small T_2 values, which indicates that chemical exchange makes a significant contribution to their T_2 values. In addition, the flexible terminal residues were excluded from the calculation of τ_m because their NOE values indicated that $\tau_e = 0$ was not a good approximation for these residues.

Having determined τ_m, S^2 is obtained for each residue, using Equation (7.1) or (7.2), respectively, with either the T_1 or T_2 value measured for the residue. The values of S^2 obtained in this manner, using the measured T_1 values (Figure 7.3), are in good agreement (rms difference 0.035) with those reported previously (Kay *et al.*, 1989).

It is evident from the figure that T_1 values and the corresponding values of S^2 do not show a large variation from residue to residue. This is also true of the T_2 values, except for those residues in the disordered Ω loop. As noted above, these residues have exceptionally small T_2 values but their T_1 and NOE values are close to the average values of these parameters observed for the other residues in SNase. Hence, it appears that relatively slow chemical exchange, $\tau_{exch} \approx 10^{-4}$–$10^{-6}$ s, modulation of amide chemical shifts is responsible for the small T_2 values of these residues.

The observation that the values of S^2 are rather uniform, $\langle S^2 \rangle = 0.85 \pm 0.05$, should not be taken to mean that the high-frequency back-bone motions are homogeneous throughout SNase. Using the cone model and Equation (7.11) to estimate the amplitudes of the internal motions, one finds that θ_0 increases from 10.5° to 22° as S^2 decreases from 0.95 to 0.80. Hence, a decrease in S^2 of 15% corresponds to a fourfold increase in the solid angle through which the NH bond reorients.

Unfortunately, one cannot conclude from these remarks that the observed variations in S^2 are a consequence of differences in the ampli-tudes of the internal motions of the NH bonds. In SNase where the $\tau_e = 0$ approximation is valid, it is easy to show that, to a good approximation, the relaxation rates are proportional to $S^2 \langle r_{NH}^{-3} \rangle^2$. Hence, a 13% variation in relaxation rate (and in S^2) would result from a 0.02 Å variation in NH bond length. For this reason, most of the observed variation in S^2 could result from small differences in the NH bond lengths.

Note that in the solid state, where the spectral density is totally determined by the internal motion (Equation 7.5), the relaxation rate in the extreme narrowing limit is proportional to $(1 - S^2)\tau_e \langle r_{NH}^{-3} \rangle^2$. One sees that the relaxation rate increases threefold as S^2 decreases from 0.95 to 0.8. Hence, these considerations suggest that one could detect heterogeneity in SNase backbone dynamics by measuring ^{15}N amide T_1 values in the crystalline protein.

SNase Backbone Dynamics in the Crystalline State

The cross-polarization magic angle sample spinning, CPMASS, spectrum of crystalline SNase labelled with [^{15}N]-valine (Figure 7.4), shows eight resolved signals, whose chemical shifts nearly coincide with the chemical shifts of sequentially assigned valine signals observed in solution. The close coincidence of chemical shifts observed in solution with shifts observed in the solid-state spectrum permits sequential assignment of signals in the solid-state spectrum of all residues except Val 51. This residue is in the disordered Ω loop of SNase, and its signal is not observed in the CPMASS spectrum, because it is presumably broadened by molecular motion. The spin–lattice relaxation rates of the eight observed valine signals were measured at 5.9 and 11.7T, and are com-pared with the corresponding solution state measurements in Figure 7.5. The figure shows that relaxation rates observed in solution are nearly 100 times larger than those observed in the crystalline state. This result is a consequence of the fact that, in solution, the overall motion of SNase is primarily responsible for relaxation of the backbone amide ^{15}N spins, whereas in the solid phase relaxation is due exclusively to small-amplitude internal motions. As anticipated above, Figure 7.5 also

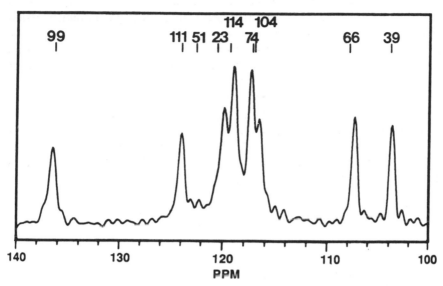

Figure 7.4 CPMASS spectrum of crystalline SNase labelled with [^{15}N]-valine. Resonance positions and signal assignments obtained in solution are indicated at the top of the figure

shows that the spin–lattice relaxation rates of the various valine residues show a much greater variation in the solid state than in solution.

Spin–lattice relaxation rates were also recorded at $5.9T$ for crystalline SNase labelled with ^{15}N His, and the relaxation rates of the His and Val residues are plotted against the X-ray temperature factors in Figure 7.6. While the plot shows that the relaxation rate increases as the order parameter increases, the correlation is by no means perfect. This is expected because B values contain contributions from static disorder, and from motions on a time-scale ranging from ps to the time required to collect the diffraction data. In addition, Equations (7.1) and (7.5) show that the relaxation rate depends upon a product, $(1 - S^2)\tau_e$, involving both the relaxation time and the internal correlation time. Hence, the measured relaxation time will reflect both the rate as well as the amplitude of the internal motion.

Although it is in principle possible to derive S^2 from measurements of spinning sidebands observed in CPMASS spectra (Cole and Torchia, 1991), signal-to-noise limitations as well as uncertainties in the static values of the ^{15}N CSA tensor elements limit the accuracy of such measurements to an uncertainty of 10%. It is also possible to determine τ_e from field-dependent T_1 measurements. However, the small field dependence that is observed implies that the derived τ_e value is extremely sensitive to uncertainties in the T_1 values (Cole and Torchia, 1991). The very large values of T_1 observed in the solid state limit the accuracy of the T_1 measurements to $c.$ 10%, thus precluding a precise determination of the internal correlation time.

While it is not possible to distinguish the separate contributions of the order parameter and the internal correlation time to the spin–lattice

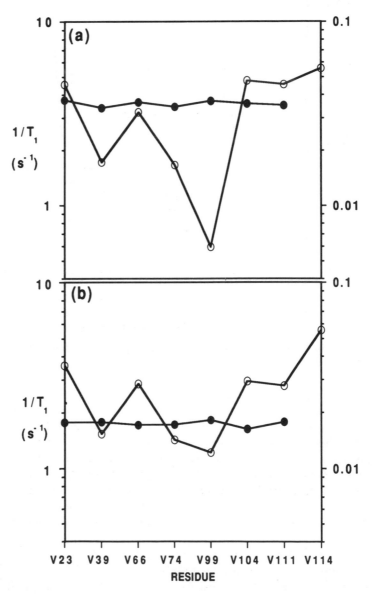

Figure 7.5 Comparison of SNase [^{15}N-Val spin–lattice relaxation rates measured in solution (●) and in the crystalline state (○) at (a) 6T and (b) 12T

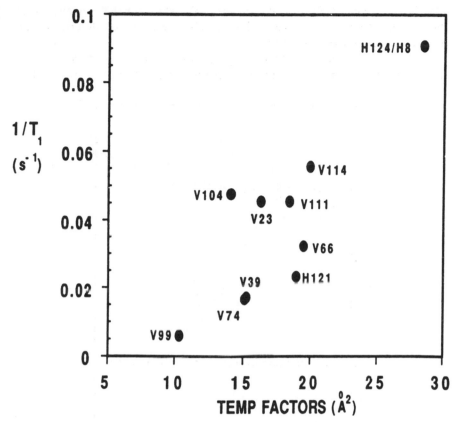

Figure 7.6 Comparison of X-ray B factors and spin–lattice relaxation rates of SNase measured in the crystalline state at $6T$

relaxation in the crystalline phase, it is clear that the variation in the measured T_1 must arise because of variations of one or both of these motional parameters from one residue to another. That is, the variations in the measured T_1 values are a consequence of heterogeneity in the internal motion of the protein backbone. It will thus be interesting to see whether these variations in internal dynamics of individual residues in SNase can be reproduced by molecular dynamics calculations.

Internal Dynamics of SNase Leucine Side-chains in Solution

We have seen that with the exception of a few amino acids in the N and C termini, and in the Ω loop of the protein, the internal motions of the backbone NH bond vectors are restricted to small-amplitude fluctuations. It is to be expected that internal motions of the amino acid side-chains, particularly those at or near the protein surface, will execute motions of considerably larger amplitude. Clear evidence for such motions are the narrow linewidths typically observed for the signals arising from the side-chains of Lys and Arg residues in proteins. Although one expects that motions of large hydrophobic residues will be considerably more restricted because they are usually buried within the protein, pioneering studies of methyl carbons in myoglobin (Wittebort *et al.*, 1979) and BPTI (Richarz *et al.*, 1980) suggested that a number of large hydrophobic side-chains in these proteins had significant internal motions in addition to rotation of the methyl group. Also, ^{2}H NMR studies of crystalline SNase showed that the side-chains of the Met and Phe residues experienced large-amplitude motions (Sparks *et al.*, 1989).

Until recently a major impediment to the study of side-chain dynamics in proteins has been the absence of sequential signal assignments. However, isotopic enrichment combined with multidimensional NMR spectroscopy now provides a method for obtaining nearly complete assignments of protons and heteronuclei of proteins in the 15–25 kDa range (Clore *et al.*, 1990a; Ikura *et al.*, 1991; Pelton *et al.*, 1991). In addition, this methodology provides spectra having high sensitivity and resolution. For example, the proton-detected HMQC spectrum of SNase/pdTp/Ca^{2+} labelled with [5,5'-^{13}C]-leucine (Figure 7.7) shows 22 resolved methyl signals corresponding to the 11 leucine residues in the protein. These signals were sequentially assigned, using a variety of two- and three-dimensional experiments (Nicholson *et al.*, 1992), as were the corresponding signals of SNase in the absence of ligands. Stereo-specific assignments were obtained by comparing NOE patterns predicted by the crystal structure of the protein (Loll and Lattman, 1989; Hynes and Fox, 1991) with those observed in NMR spectra. This procedure is justified, since all available evidence indicates that the crystal and solution structures of SNase backbone and buried side-chain atoms are essentially the same (Torchia *et al.*, 1989; Wang *et al.*, 1990a,b; Baldisseri *et al.*, 1991).

Although nearly all of the leucine methyl groups in SNase are completely buried, the previous evidence that buried hydrophobic groups in proteins exhibit internal flexibility encouraged us to measure ^{13}C relaxation parameters in order to investigate the dynamics of these methyl groups in SNase. The T_1, T_2 and NOE values of all 22 methyl carbons were measured, using the pulse sequences diagrammed in Figure 7.2 (Nicholson *et al.*, 1992).

Model-free parameters were extracted from the relaxation data (Nicholson *et al.*, 1992) in two different ways. First, the values of τ_m, S^2 and τ_e were derived from the relaxation data, using Equation (7.8). This approach yielded values of τ_m of 6.6 ns and 7.2 ns for liganded and unliganded SNase, respectively. These values are significantly smaller than the values of 8.3–8.5 ns obtained from measurements of backbone

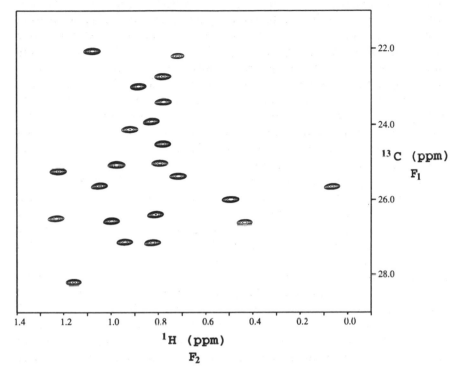

Figure 7.7 Two-dimensional shift correlation spectrum of SNase labelled with [5,5'-^{13}C$_2$]-leucine, showing the 22 resolved methyl signals of the 11 Leu residues

relaxation parameters. In addition, the NOE values predicted by the best-fit values of the model-free parameters were, in general, 10–15% larger than the experimental NOEs. These differences are well outside the 1–2% errors estimated for the NOE measurements.

We think that these inconsistencies arise because not all internal motions of the methyl group are in the extreme narrowing limit. Physical considerations suggest that in addition to rapid threefold rotation of the methyl group, a motion of the C^γ–C^δ bond axis on a slower time-scale also occurs. The order parameters and effective correlation times associated with the fast (methyl rotation) and slow (C^γ–C^δ bond axis reorientation) internal motions are designated S_f, S_s, τ_f and τ_s, respectively. The spectral density function, $J(\omega)$, is given in terms of these parameters and τ_m in Equation (7.6). In order to extract the model-free parameters from the relaxation data, we set τ_m equal to 8.5 ns and note that, according to Equation (7.14), $S_f^2 = 0.111$ for a methyl group having tetrahedral geometry and rotating about its threefold axis. With the values of τ_m and S_f^2 set equal to 8.5 ns and 0.111, respectively, the remaining model-free parameters, S_s, τ_s and τ_f, were determined for each leucine methyl carbon, from conjugate gradient minimization of Equation (7.8), and are listed in Table 7.1. For purposes of comparison, the values of the slow internal motion order parameters, S_{LZ}^2, are listed. These order parameters are obtained by dividing the squared order parameter derived using the original Lipari–Szabo formulation ($J(\omega)$ is given by Equation 7.5) by 0.111. Examination of the table shows that S_s^2 and S_{LZ}^2 typically differ by less than 5%.

Table 7.1 Effective correlation times[a] obtained for fast internal motions, τ_f, and the order parameters[b], S_s^2 and S_{LZ}^2, of leucine methyl carbons

Carbon		τ_f	S_s^2	Error	S_{LZ}^2	Error	τ_f	S_s^2	Error	S_{LZ}^2	Error
			Liganded SNase					*Unliganded SNase*			
L7	δ1	0.032	0.591	0.048	0.631	0.060	0.026	0.449	0.031	0.478	0.038
	δ2	0.034	0.466	0.022	0.479	0.023	0.047	0.518	0.040	0.524	0.031
L14	δ1	0.040	0.727	0.048	0.706	0.052	0.037	0.717	0.028	0.707	0.023
	δ2	0.053	0.890	0.028	0.885	0.034	0.058	0.887	0.036	0.897	0.043
L25	δ1	0.014	0.891	0.054	0.878	0.023	0.022	0.777	0.052	0.754	0.056
	δ2	0.029	0.815	0.028	0.906	0.025	0.035	0.910	0.039	0.898	0.041
L36	δ1	0.053	0.463	0.036	0.527	0.023	—	—		—	
	δ2	0.026	0.516	0.013	0.533	0.015	0.024	0.369	0.023	0.379	0.025
L37	δ1	0.083	0.958	0.030	0.955	0.028	0.049	0.640	0.088	0.725	0.028
	δ2	0.054	0.819	0.051	0.840	0.034	0.032	0.601	0.021	0.629	0.014
L38	δ1	0.014	0.697	0.061	0.755	0.029	0.023	0.502	0.037	0.506	0.042
	δ2	0.016	0.602	0.025	0.612	0.028	0.015	0.381	0.023	0.401	0.022
L89	δ1	0.034	0.840	0.031	0.881	0.024	0.029	0.577	0.085	0.667	0.021
	δ2	0.005	0.916	0.030	0.980	0.014	0.005	0.654	0.037	0.792	0.023
L103	δ1	0.040	0.881	0.038	0.878	0.029	—	—		—	
	δ2	0.060	0.931	0.039	0.913	0.044	0.065	0.688	0.037	0.806	0.050
L108	δ1	0.025	0.806	0.051	0.803	0.036	0.035	0.787	0.057	0.788	0.040
	δ2	0.028	0.884	0.036	0.855	0.027	0.027	0.757	0.046	0.840	0.033
L125	δ1	0.034	0.516	0.018	0.523	0.022	0.059	0.331	0.041	0.319	0.038
	δ2	0.014	0.441	0.018	0.451	0.022	0.038	0.373	0.012	0.364	0.014
L137	δ1	0.024	0.612	0.045	0.630	0.027	0.024	0.570	0.056	0.640	0.048
	δ2	0.025	0.522	0.011	0.544	0.011	0.023	0.443	0.050	0.491	0.045

[a]In ns.
[b]The values of S_f^2 are assumed to equal 0.111 and τ_m is 8.5 ns. The slow-motion order parameter obtained using the Lipari–Szabo analysis, Equation (7.3) divided by 0.111, S_{LZ}^2, is listed to allow comparison with S_s^2. The stereospecific assignments for L14, L103 and L137 are reversed from those published previously (Nicholson *et al.*, 1992). The previous assignments were based upon the crystal structure in an early stage of refinement.

Probably the major source of uncertainty in determining S_s^2 is the sensitivity of S_f^2 to the geometry of the methyl group. The methyl geometries determined from neutron diffraction studies of single crystals of L-alanine and L-valine are not ideal tetrahedral; rather they correspond to S_f^2 values of 0.104, 0.085 and 0.090. If the value of S_f^2 used to calculate the results in Table 7.1 had been 0.090 rather than 0.111, the S_s^2 values would be 20% larger, with several sites having physically unreasonable order parameters in excess of unity. In view of this result and the fact that the strong ionic forces present in the crystals of amino acids may slightly distort the geometry of the methyl group, we have chosen to analyse the dynamics of the Leu methyl groups assuming tetrahedral geometry, recognizing that the values of S_s^2 so obtained have uncertainties of 10–20% as a consequence of possible variations in methyl group geometry.

Examination of Table 7.1 reveals that the values of S_s^2 show significant variation from residue to residue. Figure 7.8 shows a plot of the values of S_s^2 against the X-ray temperature factors obtained for liganded (Loll and Lattman, 1989) and unliganded (Hynes and Fox, 1991) SNase. There is an evident correlation between the order parameters and B values, although, for the reasons mentioned when discussing Figure 7.6, it is imperfect. An interesting feature of the Leu temperature factors is that, in a given residue, they are not significantly larger for the side-chain methyl carbons than for the backbone atoms. A comparison of Figure 7.3 and Table 7.1 shows that this result is in accord with the observation that the order parameters of the amide groups of L14, L25, L37, L89, L103 and L108 in liganded SNase are approximately the same

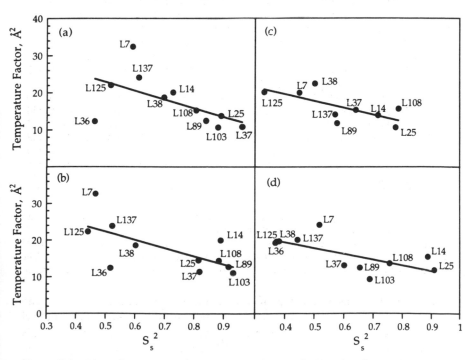

Figure 7.8 Plot of S_s^2 against the X-ray B factors for leucine δ_1 and δ_2 carbons of (a), (b) liganded and (c), (d) unliganded SNase

as the order parameters of the methyl carbons. In contrast, the order parameters of the methyl carbons of residues L36, L38, L125 and L137 are significantly smaller (S_s^2 is less than 0.71 for each residue) than the order parameters of their corresponding amide NH groups. This result is interesting, as it suggests that buried side-chains in liganded SNase undergo significant internal motions. We note that in the case of L7, the first residue in the ordered structure of SNase, the backbone amide S^2 has a low value of 0.54 (Kay *et al.*, 1989). However, the L7 side-chain is buried and has B factors (Figure 7.8) only slightly larger than those of the other Leu methyl carbons. We therefore also interpret the small order parameters observed in the case of the L7 methyl carbons in terms of internal motions of the buried L7 side-chain.

The values of the Leu methyl order parameters decrease further in the case of the unliganded protein (Table 7.1), with the result that, in addition to the five residues listed above, L37, L89 and L103 also have values of S_s^2 less than 0.7. Because the order parameters of the δ_1 and δ_2 carbons of each leucine side-chain are nearly the same (Table 7.1), the average order parameter of each pair is plotted in Figure 7.9. In order to compare the flexibility of the Leu side-chains in the liganded and unliganded states, we analyse the order parameters in terms of the following three specific models of internal motion of the Leu side-chains: (1) the cone model, in which the C^γ–C^δ bond axis diffuses in a cone of semiangle θ_0; (2) the restricted-diffusion model, in which the C^β–C^γ

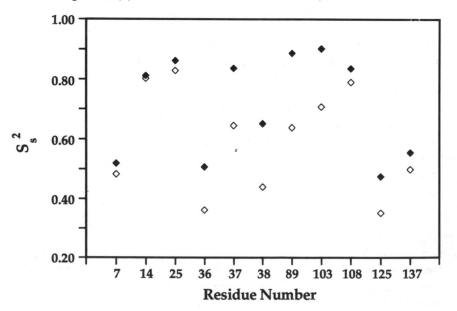

Figure 7.9 Plot of average S_s^2 for each pair of methyl groups in the presence (♦) and absence (◊) of ligands

bond axis diffuses through an angular range $\pm \gamma_0$; and (3) the two-site jump model, in which the $C^\gamma\text{-}C^\delta$ bond axis is found in one of two orientations that differ by 109.5°. In this latter model, the equilibrium probabilities of the two orientations are p_1 and p_2, with $r = p_1/p_2$, and the bond axis executes instantaneous jumps between the two orientations.

The values of θ_0, γ_0 and r derived from the measured ordered parameters and Equations (7.11)–(7.13) are listed in Table 7.2. Examination of the table shows that the flexible leucine side-chains, i.e. those for which $S^2 < 0.7$, undergo internal motions of significant angular amplitude in the case of either the cone or the restricted-diffusion model. Alternatively, in the case of the two-site jump model, where the angular amplitude is fixed to a large value, the minor conformation has a significant probability of occupation. Hence, about half of the buried leucine side-chains in SNase exhibit significant internal motions.

Table 7.2 Values of S_s^2 and corresponding values of θ_0, γ_0 and r derived from the cone, restricted diffusion and two-site jump models of leucine side-chain motion, respectively, in liganded (L) and unliganded (U) SNase

Residue		S_s^2	θ_0	γ_0	r
L7	(L)	0.53	36°	50°	0.30
	(U)	0.48	39°	54°	0.36
L36	(L)	0.49	38°	53°	0.35
	(U)	0.37	45°	65°	0.62
L37	(L)	0.89	16°	21°	0.05
	(U)	0.62	32°	43°	0.21
L38	(L)	0.65	30°	41°	0.18
	(U)	0.44	41°	58°	0.43
L89	(L)	0.88	17°	22°	0.05
	(U)	0.62	32°	43°	0.21
L103	(L)	0.91	14°	19°	0.04
	(U)	0.69	28°	38°	0.16
L125	(L)	0.48	39°	54°	0.36
	(U)	0.35	46°	67°	0.73
L137	(L)	0.57	34°	47°	0.25
	(U)	0.50	38°	52°	0.33

Another interesting feature of Table 7.2 is the evident difference in internal motion of the leucine side-chains in the liganded and unliganded proteins. Examination of Table 7.2 shows that the six leucine side-chains identified with an asterisk in Figure 7.10 (L36, L37, L38, L89, L103 and L125) show a significant decrease in internal flexibility in the presence of ligands. According to the crystal structure of liganded SNase (Loll and Lattman, 1989), the C^δ atoms of these six residues are within 10 Å

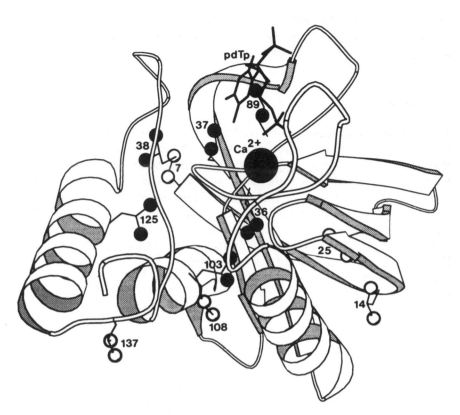

Figure 7.10 Structure of liganded SNase, showing the positions of the 11 pairs of leucine methyl carbons

of either Ca^{2+} or at least one of the heavy atoms of pdTp, while the C^δ atoms of the remaining five leucine residues are not.

The greater mobility of the leucine side-chains in unliganded SNase was not expected *a priori*. The comparison of the X-ray structures of liganded (Loll and Lattman, 1989) and unliganded (Hynes and Fox, 1991) SNase show relatively few differences in protein conformation, and these are confined to the protein surface near the ligand binding sites. It is true that the liganded protein undergoes thermal unfolding at a temperature 12–20° higher than unliganded SNase, and that the slowly exchanging amide protons have rates that are 2–3 orders of magnitude slower in the liganded protein (Baldisseri and Torchia, unpublished results). However, methyl groups of a number of residues in RCAM-BPTI and BPTI have essentially the same flexibility (Richarz *et al.*, 1980), in spite of the higher melting temperature and slower amide exchange rates of the latter protein. Evidently, in SNase, the increase in thermal stability and the decrease in hydrogen exchange rates that are caused by ligand binding are accompanied by a tightening of the internal structure of the protein in the neighbourhood of the ligand binding sites.

5 Conclusions and Perspective

We have reviewed pulsed NMR techniques for measuring A-spin relaxation, in AX and AX_3 spin systems, with high resolution and sensitivity. While we have focused upon studies of the dynamics of SNase, the backbone motions of several other ^{15}N-labelled proteins have been elucidated by use of similar techniques (Clore *et al.*, 1990b; Peng *et al.*, 1991b; Barbato *et al.*, 1992; Schneider *et al.*, 1992; Stone *et al.*, 1992; Takahashi *et al.*, 1992).

While [1]H-detected natural abundance ^{13}C relaxation studies are feasible at protein concentrations above 10 mM (Dellwo and Wand, 1989; Nirmala and Wagner, 1989; Palmer *et al.*, 1991a), solubility limitations will typically require enrichment levels of at least 15% to achieve the sensitivity required to measure relaxation parameters accurately. The approach taken herein, that of incorporating specific types of enriched amino acids into a protein, maximizes sensitivity and minimizes spectral overlap, but requires chemical synthesis of amino acids and multiple protein sample preparations, although the latter can be limited by incorporating several different types of amino acids into the protein.

It would be ideal if uniformly 98% ^{13}C-enriched protein samples could be used in ^{13}C relaxation studies, because such samples would yield the greatest sensitivity and, in any case, are needed to obtain side-chain carbon sequential assignments. The challenge to be overcome is the design of pulse sequences that can eliminate the effects of carbon–carbon couplings upon the relaxation measurements. In the event that this cannot be done, a sample that is 15–20% ^{13}C-enriched should provide satisfactory sensitivity and at the same time reduce the effect of homonuclear coupling to an acceptable level.

One anticipates that ^{13}C relaxation studies will provide a wealth of information about protein dynamics at a myriad of assigned sites throughout the protein, ranging from highly constrained backbone and side-chain atoms in the interior to flexible side-chain atoms near and at the protein surface. Of particular interest will be the information obtained about dynamics of protein surfaces involved in catalysis and in intermolecular interactions. Of course, in order to obtain such information one must measure thousands of signal intensities and then derive the relaxation parameters from these measurements. Newly developed processing software (Delaglio and Bax, private communication) greatly accelerates this task.

Interpreting the relaxation parameters will represent a final challenge. Although one will be able to obtain models of internal motion that are consistent with the values of S^2 and τ_e derived from the model-free analysis, a more fundamental approach would be to calculate spectral density functions and the associated relaxation parameters from molecular dynamics trajectories and compare these with the experimental results. Agreement between theory and experiment would establish the veracity of the potential functions used in the calculations, and confirm that the calculated trajectories are an accurate description of the internal motions of the protein. It seems reasonable to expect that advances in technology will make such calculations practical in the next few years.

References

Abragam, A. (1961). In Marshall, W. C. and Wilkinson, D. H. (Eds), *The Principles of Nuclear Magnetism*. Clarendon Press, Oxford, pp. 289–305

Baldisseri, D. M., Torchia, D. A., Poole, L. B. and Gerlt, J. A. (1991). Deletion of the Ω-loop in the active site of staphylococcal nuclease. 2. Effects on protein structure and dynamics. *Biochemistry*, **30**, 3628–3633

Barbato, G., Ikura, M., Kay, L. E., Pastor, R. W. and Bax, A. (1992). Backbone dynamics of calmodulin studied by ^{15}N relaxation using inverse detected two-dimensional NMR spectroscopy: the central helix is flexible. *Biochemistry*, **31**, 5269–5278

Bax, A., Clore, G. M. and Gronenborn, A. M. (1990a). ^{1}H–^{1}H correlation via isotropic mixing of ^{13}C magnetization, a new three-dimensional approach for assigning ^{1}H and ^{13}C spectra of ^{13}C-enriched proteins. *J. Magn. Reson.*, **88**, 425–431

Bloembergen, N., Purcell, E. M. and Pound, R. V. (1948). Relaxation effects in nuclear magnetic resonance absorption. *Phys. Rev.*, **73**, 679–712

Boyd, J., Hommel, U. and Campbell, I. D. (1990). Influence of cross-correlation between dipolar and anisotropic chemical shift relaxation mechanisms upon longitudinal relaxation rates of ^{15}N in macromolecules. *Chem. Phys. Lett.*, **175**, 477–482

Clore, G. M., Bax, A., Driscoll, P. C., Wingfield, P. T. and Gronenborn, A. M. (1990a). Assignment of the side-chain ^{1}H and ^{13}C resonances of interleukin-1β using double- and triple-resonance heteronuclear three-dimensional NMR spectroscopy. *Biochemistry*, **29**, 8172–8184

Clore, G. M. and Gronenborn, A. M. (1989). Determination of three dimensional structures of proteins and nucleic acids in solution by NMR spectroscopy. *CRC Crit. Rev. Biochem. Mol. Biol.*, **24**, 479–564

Clore, G. M., Kay, L. E., Bax, A. and Gronenborn, A. M. (1991a). Four-dimensional ^{13}C/^{13}C-edited nuclear overhauser enhancement spectroscopy of a protein in solution: Application to interleukin-1β. *Biochemistry*, **30**, 12–18

Clore, G. M., Szabo, A., Bax, A., Kay, L. E., Driscoll, P. C. and Gronenborn, A. M. (1990b). Deviation from the simple two-parameter model-free approach to the interpretation of nitrogen-15 nuclear magnetic relaxation of proteins. *J. Am. Chem. Soc.*, **112**, 4989–4991

Clore, G. M., Wingfield, P. T. and Gronenborn, A. M. (1991b). High resolution three-dimensional structure of interleukin-1β in solution by three- and four-dimensional nuclear magnetic resonance spectroscopy. *Biochemistry*, **30**, 2315–2330

Cole, H. B. R. and Torchia, D. A. (1991). An NMR study of the backbone dynamics of staphylococcal nuclease in the crystalline state. *Chem. Phys.*, **158**, 271–281

Dellwo, M. J. and Wand, A. J. (1989). Model-independent and model-dependent analysis of the global and internal dynamics of cyclosporin A. *J. Am. Chem. Soc.*, **111**, 4571–4578

Ernst, R. R. (1966). Nuclear magnetic double resonance with an incoherent radio-frequency field. *J. Chem. Phys.*, **45**, 3845–3854

Fesik, S. W., Eaton, H. L., Olejneczak, E. T., Zuiderweg, E. R. P., McIntosh, L. P. and Dahlquist, F. W. (1990). 2D and 3D NMR spectroscopy employing $^{13}C-^{13}C$ magnetization transfer by isotropic mixing. Spin system identification in large proteins. *J. Am. Chem. Soc.*, **112**, 886–888

Goldman, M. (1984). Interference effects in the relaxation of a pair of unlike spin-1/2 nuclei. *J. Magn. Reson.*, **60**, 437–452

Haeberlen, U. (1976). High resolution NMR in solids: Selective averaging. *Adv Magn. Reson., Suppl. 1*, 1–190

Hynes, T. R. and Fox, R. O. (1991). The crystal structure of staphylococcal nuclease refined at 1.7 Å resolution. *Protein Struct. Func.*, **10**, 92–105

Ikura, M., Kay, L. E. and Bax, A. (1990). A novel approach for sequential assignment of ^{1}H, ^{13}C, and ^{15}N spectra of larger proteins: Heteronuclear triple-resonance three-dimensional NMR spectrosocopy. Application to calmodulin. *Biochemistry*, **29**, 4659–4667

Ikura, M., Spera, S., Barbato, G., Kay, L. E., Krinks, M. and Bax, A. (1991). Secondary structure and side-chain ^{1}H and ^{13}C resonance assignments of calmodulin in solution by heteronuclear multidimensional NMR spectroscopy. *Biochemistry*, **30**, 9216–9228

Kamath, U. and Shriver, G. W. (1989). Characterization of thermotropic state changes in myosin subfragment-1 and heavy meromyosin by UV difference spectroscopy. *J. Biol. Chem.*, **264**, 5586 5592

Karplus, M. and McCammon, J. A. (1986). The dynamics of proteins. *Sci. Am.*, **254**, April, 42–51

Kay, L. E., Bull, T. E., Nicholson, L. K., Griesinger, C., Schwalbe, H., Bax, A. and Torchia, D. A. (1992a). On the measurement of heteronuclear transverse relaxation times in AX_3 spin systems via polarization transfer techniques. *J. Magn. Reson.*, **100**, 538–558

Kay, L. E., Clore, G. M., Bax, A. and Gronenborn, A. M. (1990a). Four-dimensional heteronuclear triple-resonance NMR spectroscopy of interleukin-1β in solution. *Science*, **249**, 411–414

Kay, L. E., Ikura, M. and Bax, A. (1990b). Proton–proton correlation via carbon–carbon couplings: A three dimensional NMR approach for the assignment of aliphatic resonances in proteins labeled with carbon-13. *J. Am. Chem. Soc.*, **112**, 888–889

Kay, L. E., Nicholson, L. K., Delaglio, F., Bax, A. and Torchia, D. A. (1992b). Pulse sequences for removal of the effects of cross correlation between dipolar and chemical-shift anisotropy relaxation mechanisms on the measurement of heteronuclear T_1 and T_2 values in proteins. *J. Magn. Reson.*, **97**, 359–375

Kay, L. E. and Torchia, D. A. (1991). The effects of dipolar cross correlation on ^{13}C methyl-carbon T_1, T_2 and NOE measurements in macromolecules. *J. Magn. Reson.*, **95**, 536–547

Kay, L. E., Torchia, D. A. and Bax, A. (1989). Backbone dynamics of proteins as studied by ^{15}N inverse detected heteronuclear NMR spectroscopy: Application to staphylococcal nuclease. *Biochemistry*, **28**, 8972–8979

Lipari, G. and Szabo, A. (1982a). Model-free approach to the interpretation of nuclear magnetic resonance relaxation in macromolecules. 1. Theory and range of validity. *J. Am. Chem. Soc.*, **104**, 4546–4559

Lipari, G. and Szabo, A. (1982b). Model-free approach to the interpretation of nuclear magnetic resonance relaxation in macromolecules. 2. Analysis of experimental results. *J. Am. Chem. Soc.*, **104**, 4559–4570

Loll, P. J. and Lattman, E. E. (1989). The crystal structure of the ternary complex of staphylococcal nuclease, Ca^{2+}, and the inhibitor pdTp, refined at 1.65 Å. *Proteins: Struct. Func. Genet.*, **5**, 183–201

London, R. E. (1989). Interpreting protein dynamics with NMR relaxation experiments. *Meth. Enzymol.*, **176**, 358–375

Messerle, B. A., Weder, G., Otting, G., Weber, C. and Wüthrich, K. (1989). Solvent suppression using a spin lock in 2D and 3D NMR spectroscopy with H_2O solutions. *J. Magn. Reson.*, **85**, 608–613

Nicholson, L. K., Kay, L. E., Baldisseri, D. M., Arango, J., Young, P. E., Bax, A. and Torchia, D. A. (1992). Dynamics of methyl groups in proteins as studied by proton detected ^{13}C NMR spectroscopy. Application to the leucine residues of staphylococcal nuclease. *Biochemistry*, **31**, 5253–5263

Nirmala, N. R. and Wagner, G. (1989). Measurement of ^{13}C spin–spin relaxation times by two-dimensional heteronuclear ^{1}H–^{13}C correlation spectroscopy. *J. Magn. Reson.*, **82**, 659–661

Palmer, A. G., III, Rance, M. and Wright, P. E. (1991a). Intramolecular motions of a zinc finger DNA-binding domain from Xfin characterized by proton-detected natural abundance ^{13}C heteronuclear NMR spectroscopy. *J. Am. Chem. Soc.*, **113**, 4371–4380

Palmer, A. G., III, Skelton, N. J., Chazin, W. J., Wright, P. E. and Rance, M. (1992). Suppression of the effects of cross-correlation between dipolar and anisotropic chemical shift relaxation mechanisms in the measurement of spin–spin relaxation rates. *Mol. Phys.*, **75**, 699–711

Palmer, A. G., III, Wright, P. E. and Rance, M. (1991b). Measurement of relaxation time constants for methyl groups by proton-detected heteronuclear NMR spectroscopy. *Chem. Phys. Lett.*, **185**, 41–46

Pelton, J. G., Torchia, D. A., Meadow, N. D., Wong, C. Y. and Roseman, S. (1991). ^{1}H, ^{15}N, and ^{13}C NMR signal assignments of III^{Glc}, a signal-transducing protein of *Escherichia coli*, using three-dimensional triple-resonance techniques. *Biochemistry*, **30**, 10043–10057

Peng, J. W., Thenabal, V. and Wagner, G. (1991a). Improved accuracy of heteronuclear transverse relaxation time measurements in macromolecules. Elimination of antiphase contributions. *J. Magn. Reson.*, **95**, 421–427

Peng, J. W., Thenabal, V. and Wagner, G. (1991b). 2D heteronuclear NMR measurements of spin–lattice relaxation times on the rotating frame of X nuclei in heteronuclear HX spin systems. *J. Magn. Reson.*, **94**, 82–100

Press, W. H., Flannery, B. P., Teukolsky, S. A. and Vetterling, W. T. (1988). In *Numerical Recipes in C*. Cambridge University Press, Cambridge, pp. 317–323

Richarz, R., Nagayama, K. and Wüthrich, K. (1980). Carbon-13 nuclear magnetic resonance relaxation studies of internal mobility of the polypeptide chain in basic pancreatic trypsin inhibitor and a selectively reduced analogue. *Biochemistry*, **19**, 5189–5196

Schneider, D. M., Dellwo, M. J. and Wand, A. J. (1992). Fast internal mainchain dynamics of ubiquitin. *Biochemistry*, **31**, 3645–3652

Shortle, D. (1983). A genetic system for analysis of staphylococcal nuclease. *Gene*, **22**, 181–189

Sparks, S. W., Cole, H. B. R., Torchia, D. A. and Young, P. E. (1989). Molecular dynamics and structure of staphylococcal nuclease in the crystalline state and in solution. *Chem. Scripta*, **29A**, 31–38

Stone, M. J., Fairbrother, W. J., Palmer, A. G., III, Reizer, J., Saier, M. H., Jr., and Wright, P. E. (1992). The backbone dynamics of the *Bacillus subtilis* glucose permease IIA domain determined from ^{15}N nmr relaxation measurements. *Biochemistry*, **31**, 4394–4406

Takahashi, H., Suzuke, E., Shimada, I. and Arata, Y. (1992). Dynamical structure of the antibody combining site as studied by ^{1}H–^{15}N shift correlation NMR spectroscopy. *Biochemistry*, **31**, 2464–2468

Torchia, D. A., Sparks, S. W. and Bax, A. (1989). Staphylococcal nuclease: Sequential assignments and solution structure. *Biochemistry*, **28**, 5509–5524

Torchia, D. A. and Szabo, A. (1985). The information content of powder lineshapes in the fast motion limit. *J. Magn. Reson.*, **64**, 135–141

Venable, R. M. and Pastor, R. W. (1988). Frictional models for stochastic simulations of proteins. *Biopolymers*, **27**, 1001–1014

Wang, J., Hinck, A. P., Loh, S. N. and Markley, J. M. (1990a). Two dimensional studies of staphylococcal nuclease: Evidence for conformational heterogeneity from hydrogen-1, carbon-13, and nitrogen-15 spin system assignments of the aromatic amino acids in the nuclease H124L–thymidine 3′-5′-bisphosphate–Ca^{2+} ternary complex. *Biochemistry*, **29**, 4242–4253

Wang, J., LeMaster, D. M. and Markley, J. M. (1990b). Two-dimensional NMR studies of staphylococcal nuclease. 1. Sequence-specific assignments of hydrogen-1 signals and solution structure of the nuclease H124L–thymidine 3′-5′-bisphosphate–Ca^{2+} ternary complex. *Biochemistry*, **29**, 88–101

Werbelow, L. G. and Grant, D. M. (1977). Intramolecular dipolar relaxation in multispin systems. *Adv. Magn. Reson.*, **9**, 189

Wittebort, R. J., Rothgeb, T. M., Szabo, A. and Gurd, F. R. N. (1979). Aliphatic groups of sperm whale myoglobin: ^{13}C NMR study. *Proc. Natl Acad. Sci. USA*, **76**, 1059–1063

Woessner, D. E. (1962). Spin relaxation processes in a two-proton system undergoing anisotropic reorientation. *J. Chem. Phys.*, **36**, 1–8

Wüthrich, K. (1986). *NMR of Proteins and Nucleic Acids*. Wiley, New York

8

Study of Protein Dynamics by NMR

Gerhard Wagner, Sven Hyberts and Jeffrey W. Peng

1 Introduction

The three-dimensional structure of globular proteins is responsible for their biological activity, since it arranges crucial side-chains, prosthetic groups or cofactors in a way to make the protein functional. Owing to the fact that most proteins work at ambient temperature, they have some internal mobility. It is not immediately obvious whether this interferes with or enhances protein function. Some internal mobility may be important for protein function; other kinds of mobility may be irrelevant. Generally, most parts of protein structures are not directly relevant for protein function, since only small surface patches or enzyme cavities constitute the active sites. Most of the regular secondary structures seem to be relatively rigid. They provide the scaffold to present the functional groups to target molecules. In many cases, the functional sites are located at regions of irregular secondary structure, on external loops, or even on disordered sites. Therefore, there seems to be a general tendency that functional sites have higher mobility than the scaffold of the protein. Moreover, many proteins undergo significant structural changes when they do their work. This is the case for many enzymes when they interact with substrates or inhibitors, or for nucleotide-binding proteins when they dock to DNA or RNA. Thus, mobility may be necessary for proteins to be adaptable to target molecules. Protein mobility may be even more important for a large group of proteins that escape our attention because they cannot be crystallized or show poor NMR spectra, so that structural analysis is too difficult or impossible. A main reason for this unfortunate behaviour may be extensive internal mobility. A large number of proteins are entirely flexible and fold up only when they interact with their target molecules, such as receptors.

Protein mobility has been discussed extensively in the literature (see, for example, Hirs and Timasheff, 1986). Techniques that can provide information about internal mobility of proteins include fluorescence spectroscopy, hydrogen exchange measurements by infrared or tracer techniques, EPR spin label studies, etc. NMR is a particularly useful method for studies of protein mobility, and the purpose of this chapter is to give an assessment of the method to study this aspect of proteins. In this review different methods for characterizing internal mobility in proteins are discussed. The main focus, however, is on application of relaxation time measurements for studies of protein mobility. A new strategy is discussed for mapping of spectral density functions for the motions of N–H bond vectors. This is essentially a characterization of the frequency distribution of rotational diffusive motions relative to the external magnetic field. It is shown that measurements of the conventional parameters T_1, T_2 and the heteronuclear NOE are not sufficient for this characterization. Additional types of auto- and cross-relaxation rates have to be measured. The formalism for deriving the relations between these rates and the spectral density functions, methods for measuring relaxation rates and experimental results are discussed.

2 Manifestations of Internal Mobility in NMR Spectra of Proteins

Most of our studies of protein mobility were performed with the proteinase inhibitor eglin c. The structure of this protein was solved by X-ray crystallography in complexes with several proteinases (Bode *et al.*, 1987; McPhalen and James, 1988; Gros *et al.*, 1989; Heinz *et al.*, 1991) and by NMR spectroscopy (Hyberts *et al.*, 1992a). Figure 8.1 shows a stereo representation of an ensemble of structures determined on the basis of NMR data. Many aspects of what follows will be discussed on this protein.

Indirect Manifestations in RMSDs of Structure Calculations

Determination of protein structures in solution by NMR relies on identification of distance and dihedral angle constraints. Distances and dihedral angles cannot be determined precisely, and in some regions of the proteins there are fewer constraints than in others. Moreover, the intensities of cross-peaks in NOESY spectra depend on the interproton distances and the mobility of the interproton connecting vector (variation of length and orientation relative to the external field). This mobility aspect is generally ignored in structure calculations, since there is no good method yet available to handle mobility in structure calculations. The crude way of coping with imprecise distance measurements

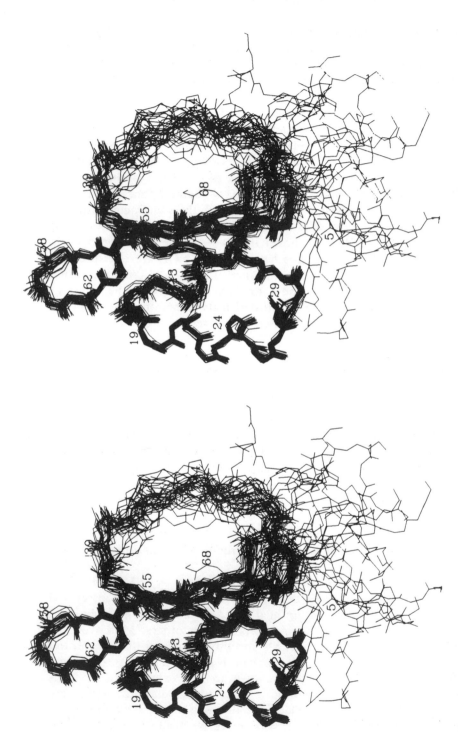

Figure 8.1 Stereo diagram of an ensemble of structures of eglin c calculated with the distance geometry program DG-II (Hyberts et al., 1992)

and internal mobility is to identify a particular measurement of an NOE or dihedral angle with a range of distances or dihedral angles, respectively. The width of this range is chosen entirely arbitrarily. There is no experimental method to obtain information on the width of this range. On the other hand, the widths of the ranges given (next to the number of constraints identified) are directly related to the sizes of the average root mean square distances (rmsds) between the structures obtained. In the best structural studies performed to date, up to 15 or 18 meaningful NOE distance constraints per residue have been derived in the average. However, these constraints are not evenly distributed. There may be 30–40 constraints per residue for the interior of the protein, but only 3 or 4 constraints for surface residues. The methods for structure calculations generally used result in ensembles of structures that are all compatible with the experimental data. As an example, Figure 8.2(a) shows a plot of the average rmsds versus sequence for the elastase inhibitor eglin c. It is a measure of the variability seen in Figure 8.1. There is a high ⟨rmsd⟩ for the seven N-terminal residues, for the residues of the proteinase binding loop (Gly40–Leu47) and for the hairpin loop of an antiparallel β-sheet (Pro58–Thr60). The number of constraints per residue are shown in Figure 8.2(b). There is a clear anticorrelation between the number of constraints and the ⟨rmsd⟩. Thus, a high rmsd is not direct evidence for internal mobility. Independent evidence has to be found from other data. It will be shown below that the profile of the ⟨rmsd⟩s is indeed misleading for some parts of the sequence. Direct characterization of protein mobility via relaxation parameters shows that there is less variation of mobility and there are different patterns from what the ⟨rmsd⟩s are suggesting. Thus, independent studies are important for characterization of protein mobility.

Comparison of NMR ⟨rmsd⟩ Values with Crystallographic B-factors

The B factors of protein crystal structures are related to the deviations of the coordinates from their mean positions, and to crystal disorder. The B factors and the mean square deviation $\langle u^2 \rangle$ are related by (Debye, 1914):

$$B = \frac{8\pi^2}{3} \langle u^2 \rangle \tag{8.1}$$

One is tempted to compare this with an 'NMR B factor' obtained as:

$$B_{\text{NMR}} = \frac{8\pi^2}{3} \langle \text{rmsd} \rangle^2 \tag{8.2}$$

Such comparisons have been made recently (Billeter *et al.*, 1989; Clore *et al.*, 1991), while the factor $\frac{1}{3}$ in Equation (8.2) was neglected. Whether

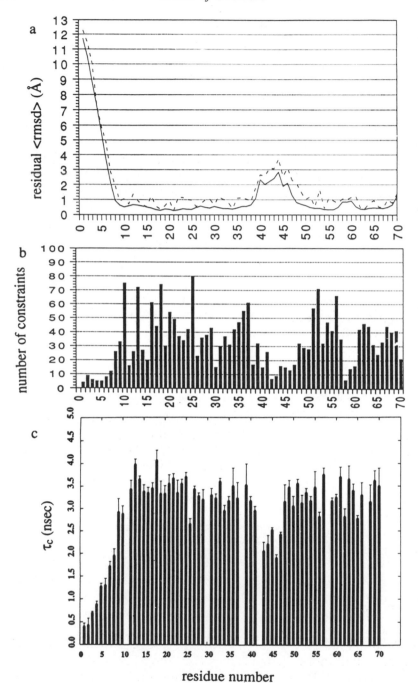

Figure 8.2 (a) Plot of the average rmsd of eglin c against the amino acid sequence. (b) Number of constraints per residue. (c) Generalized coherence time τ_c plotted against the amino acid sequence calculated as $\frac{5}{2} J(0)$

such a comparison is justified has been tested recently (Hyberts *et al.*, 1993). A set of 100 NMR structures of the protein eglin c were calculated from NMR data. From these 100 structures an average electron density and average structure factor were calculated. Standard crystallographic methods of model building and refinement were then used to fit a single 'best' structure to this average structure factor. This resulted in a protein model with an *R* factor of 17% and *B* factors for all atoms. For the well-defined part of the protein, the *B* factors obtained in this way were indeed very similar to but slightly larger than those obtained with Equation (8.2). For the flexible parts of the protein, Equation (8.2) yields much higher values than the method described by Hyberts *et al.* (1993). This also holds for most of the side-chains. On the other hand, the ⟨rmsd⟩s and the NMR *B* factors suggest for some parts of the protein higher mobility. This is not supported by measurements of relaxation parameters, as discussed below. Crystallographic *B* factors for the well-defined core of the same protein eglin c (Bode *et al.*, 1987; McPhalen and James, 1988; Gros *et al.*, 1989; Heinz *et al.*, 1991) are at least three times higher than the values obtained with Equation (8.2) or with the method described by Hyberts *et al.* (1993).

Exchange Effects—A Lower Limit for Amplitudes of Fluctuations in Protein Cores

If motions in a protein occur at the time-scale of chemical shift differences, this may be manifested in line broadening and in the symmetry of certain groups of resonances. A prominent example is the flipping of aromatic side-chains in proteins. The H^δ and H^ϵ protons of phenylalanines or tyrosines give rise to four resonances if the side-chain is immobilized in the interior of a protein. If the side-chain is rotating rapidly, two resonances of intensity 2 are observed—one for the H^δ, one for the H^ϵ (Wüthrich and Wagner, 1978). If the exchange is intermediate, the lines are broadened, and the exchange rates can be estimated from line shape analysis. Transitions from slow to fast rotations with increasing temperature have first been observed for the basic pancreatic trypsin inhibitor (Wüthrich and Wagner, 1975) and in ferrocytochrome c (Campbell *et al.*, 1976). Such effects have since been seen in a number of proteins. However, in most proteins that have been studied in detail by NMR, all or most aromatic side-chains rotate rapidly about the C^β–C^γ axis. Rotation that is slow on the NMR time-scale is the exception. This has some implications for the estimate of lower limits of amplitudes of internal motions. In all protein structures the aromatic side-chains appear tightly packed. An aromatic side-chain can be approximated as an oblate ellipsoid with half-axes of 1.7 Å and 3.5 Å (Richards, 1974). Thus, if the ring undergoes a 180° flip, the environment has to go apart by at least 1.8 Å on either side. This has to be considered when ⟨rmsd⟩ values for ensembles of NMR structures are interpreted. In the most precise NMR structures, the ⟨rmsd⟩s are well below 1.8 Å for interior side-chains. Thus motions with larger amplitudes occur occasionally that are not covered by the spread of the ensembles of NMR structures obtained from structure calculations.

Side-chain Mobility from Coupling Constants

Measurements of vicinal coupling constants can also provide information on internal mobility of proteins. This method is particularly suitable for assessing the mobility around the χ_1 angle. It seems that side-chains of amino acid residues are generally locked in one of the three rotamer states with $\chi_1 = +60°$, $-60°$ or $180°$ (Ponder and Richards, 1987). Alternatively, they may be hopping between the three rotamer states. This is well manifested in the values of the vicinal $^1H^\alpha$–$^1H^\beta$ and ^{15}N–$^1H^\beta$ coupling constants (Bystrov, 1976; Wagner, 1990). The homonuclear vicinal coupling constants are either small (3 Hz) or large (13 Hz) if the $^1H^\alpha$ and $^1H^\beta$ are *gauche* or *trans*, respectively. If the side-chain is hopping between the three rotamer states, this coupling is averaged and has values between these numbers. For an equal population of the three rotamer states, this value is *c*. 7 Hz. The same holds for the heteronuclear vicinal couplings between the peptide nitrogen and the β protons. For a side-chain locked in one of the three rotamer states, these heteronuclear coupling constants are around 0.5 Hz or around 5.5 Hz. If the side-chain is rotating, these coupling constants are around 2.5 Hz. Methods for measuring these coupling constants have been found recently (Montelione *et al.*, 1989) so that this aspect of side-chain mobility can be studied conveniently. In theory, measurements of the four homonuclear and heteronuclear coupling constants would allow one to determine the populations of the three rotamer states (Bystrov, 1976; Wagner, 1990). However, it remains to be seen whether the precision of the measurements and the calibration of the relations between the coupling constants and the dihedral angles are accurate enough to obtain relevant results.

Mobility around the χ_1 angle via measurements of coupling constants has not yet been studied extensively. The few cases include the basic pancreatic trypsin inhibitor (Nagayama and Wüthrich, 1981; Hoch *et al.*, 1985; Hyberts *et al.*, 1992a). One might expect that rotational mobility of the side-chains around the χ_1 angle is higher for the side-chains on the protein surface than for the protein interior. The data available so far show that this is not generally true. There are a number of cases for internal side-chains (for example, Tyr 21 in BPTI or Val 13 in eglin c) where the side-chain is highly mobile around the C^α–C^β bond, whereas most side-chains on protein surfaces seem to be locked in one of the three rotamer states.

Hydrogen Exchange Studies

Isotope exchange of amide protons of the protein interior was recognized long ago as a manifestation of protein mobility (Linderstrøm-Lang and Schellman, 1959). Despite a large number of hydrogen exchange studies on proteins, mechanistic models for the internal motions that promote hydrogen exchange have not conclusively been derived. Measurements of the activation energies for the exchange rates have shown that exchange of the innermost amide protons of stable proteins requires enthalpy changes comparable to those for protein unfolding (Woodward and Hilton, 1980; Wagner, 1983). In the basic pancreatic trypsin inhibitor, for example, there is such evidence for occasional complete unfolding 60° below the melting temperature. This is evidence for occasional large-amplitude fluctuations in proteins. This has to be kept in mind when considering the average rmsds obtained in structure calculations. They cannot be considered to be representative of all internal motions in a protein. Certainly, this type of motion is also not manifested in relaxation time measurements.

3 Protein Dynamics from Measurements of Relaxation Parameters

Relaxation time measurements provide a rich source of information on internal protein motions. There are numerous excellent treatments of the principles of NMR relaxation (see, for example, Solomon, 1955; Abragam, 1961; Werbelow and Grant, 1977; Vold and Vold, 1978). Relaxation occurs after the equilibrium of a system of nuclear spins is perturbed—for example, by applying radiofrequency pulses. After the perturbation, the system of spins will go back to equilibrium with a characteristic time profile. This is essentially due to a coupling of the nuclear energies of the spin system with motional energies of the molecule. From the time profile of the relaxation we may learn about motions of the proteins. The goal of measuring relaxation parameters is to obtain this information.

We are here mainly concerned with relaxation of ^{15}N or ^{13}C nuclei in proteins that are bound to a single hydrogen. This simplifies the interpretation of the relaxation data (Allerhand *et al.*, 1971). These nuclei predominantly relax by dipole–dipole interaction with the coupled protons. Dipole–dipole relaxation is governed by the time fluctuation of the N–H or C–H vector in length and orientation relative to the external magnetic field. Because of the chemical bond, the length of the vector can be considered constant, and only rotational motions of the bond vector have to be considered. The ^{15}N nuclei in peptide groups have a significant chemical shift anisotropy (CSA), so that relaxation of ^{15}N via CSA has also to be considered. Cross-correlation between dipole–dipole and CSA is also important for relaxation of ^{15}N spins in proteins.

General Aspects of Relaxation

Relaxation is due to a coupling between nuclear spins in the magnetic field of the spectrometer and molecular motion. If the equilibrium of a system is perturbed by application of radiofrequency pulses, it will return to equilibrium by exchange of nuclear energy with kinetic energy of molecular motion (longitudinal relaxation) or by dephasing of coherence by fluctuating fields arising from the reorientation of the spin environment in the framework of the randomly tumbling molecule.

The principles of nuclear relaxation have been treated in numerous articles and reviews. An well-known early treatment was given by Solomon (1955). More complex relaxation phenomena can be handled with Redfield's theory (Redfield, 1965), which uses essentially time-dependent perturbation theory to second order. First, we discuss dipole–dipole relaxation on the basis of Solomon's treatment, since it provides a good insight into the relations between mobility and relaxation, before going on to a more general treatment.

Dipole–Dipole Relaxation

Dipole relaxation is due to a coupling between nuclear energies of the spins in the magnetic field of the spectrometer and kinetic energies of molecular motion. Figure 8.3 shows a simple picture of a protein containing a pair of spins located in the interior of a protein, such as a ^{1}H and a ^{15}N. The dipolar Hamiltonian giving rise to relaxation is:

$$\mathcal{H}_{d\bar{d}}'(t) = [H_z N_z - \tfrac{1}{4}(H_+ H_- + H_- N_+)]F_0(t)$$

$$+ [H_+ N_z + H_z N_+]F_1(t) + [H_- N_z + H_z N_-]F_1^*(t)$$

$$+ H_+ N_+ F_2(t) + H_- N_- F_2^*(t)$$

$$= \sum_q F_q(t)A_q \tag{8.3}$$

Here H and N are used to denote the operators for proton and nitrogen spins, respectively, and the functions $F_i(t)$ describe the orientation of the N–H vector relative to the external magnetic field, with

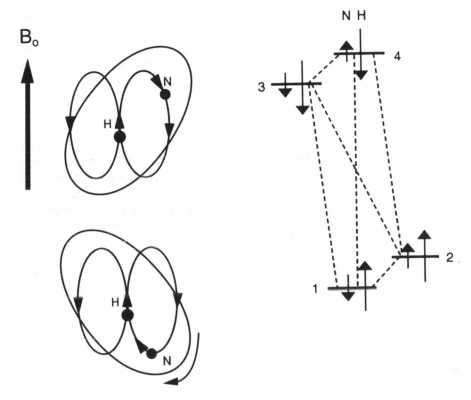

Figure 8.3 Illustration of the variation due to molecular motion of the dipole field, at the position of the nitrogen generated by a proton in an external magnetic field. The energy diagram for an N–H spin pair is also given. To induce transitions between the levels m and n, molecular motions on the time-scales $\tau = \dfrac{2\pi}{\omega_{mn}}$ are required

$$F_0(t) = k[1 - 3\cos^2\theta(t)]$$

$$F_1(t) = -\tfrac{3}{2}k\sin\theta(t)\cos\theta(t)\,e^{i\phi(t)}$$

$$F_2(t) = -\tfrac{3}{4}k\sin^2\theta(t)\,e^{i2\phi(t)}$$

$$k = \frac{\hbar^2\gamma_H\gamma_N}{r_{HN}^3} \tag{8.4}$$

The proton produces a dipole field that is experienced by the nitrogen nucleus (Figure 8.3). With rotational diffusion of the protein or internal motions of the protein, the dipole field vector changes orientation and size with the frequency of the protein rotational motion. According to time-dependent perturbation theory, fluctuation of the dipole–dipole energy can induce transitions between two energy levels if the modulation frequency corresponds to the energy difference. The oscillating dipole–dipole energy has an effect analogous to irradiation of a radio-frequency field inducing transitions. The transition probabilities per unit time are obtained with time-dependent second-order perturbation theory:

$$\overline{w_{mn}}^{NH} = \frac{1}{t\hbar^2} \overline{\left| \int_0^t \langle n|\mathcal{H}'(t')|m\rangle e^{-i\omega_{mn}t'} \, dt' \right|^2} \tag{8.5}$$

or

$$\overline{w_{mn}}^{NH} = \frac{1}{t\hbar^2} \int_0^t \int_0^t \overline{\langle n|\mathcal{H}'(t')|m\rangle\langle m|\mathcal{H}'^*(t'')|n\rangle e^{-i\omega_{mn}(t'-t'')}} \, dt' \, dt''$$

$$(t \rightarrow \infty) \tag{8.6}$$

$$= \frac{1}{\hbar^2} \int_0^\infty \overline{\langle n|\mathcal{H}'(0)|m\rangle\langle m|\mathcal{H}'^*(\tau)|n\rangle e^{-i\omega_{mn}\tau}} \, d\tau$$

$$= \sum_{qq'} \langle n|A_q|m\rangle\langle m|A_{q'}|n\rangle \frac{1}{\hbar^2} \int_0^\infty \overline{F_q(0)F_{q'}^*(\tau)} \, e^{i-\omega_{mn}\tau} \, d\tau \tag{8.7}$$

$$= \sum_{q} \langle n|A_q|m\rangle\langle m|A_q|n\rangle \frac{1}{\hbar^2} \int_0^\infty \overline{F_q(0)F_q^*(\tau)} \, e^{-i\omega_{mn}\tau} \, d\tau \tag{8.8}$$

The bar indicates the ensemble average, which is usually calculated as an integral over all spatial orientations. The sum over q' drops out because the F_q are spherical harmonics that are orthogonal. The letters m and n refer to the levels in Figure 8.3. The frequencies ω_{mn} correspond to the energy differences between the levels m and n. It is obvious that the transition probability is only non-zero if the dipolar Hamiltonian, $\mathcal{H}'_{dd}(t')$, or rather the autocorrelation function of the dipolar Hamiltonian, oscillates with the frequency ω_{mn}. This means the orientation of the NH vector relative to the external field has to oscillate with the frequencies corresponding to the energy differences of Figure 8.3 in order to induce the corresponding transitions. In the case of the N–H system these frequencies are ω_N, ω_H, $\omega_H + \omega_N$ and $\omega_H - \omega_N$; they are indicated in the figure. It is obvious that the term $-\frac{1}{4}(H_+N_- + H_-N_+)F_0(t)$ is responsible for transitions between levels 1 and 4 which require protein motions around $\omega_H - \omega_N$. The term $H_-N_+F_1(t) + H_-N_-F_1^*(t)$ is responsible for (1,2) and (3,4) transitions which require protein motions around ω_N. $H_+N_z F_1(t) + H_-N_z F_1^*(t)$ is responsible for (1,3) and (2,4) transitions which require protein motions around ω_H, and the term $H_+N_+F_2(t) + H_-N_-F_2^*(t)$ is responsible for (2,3) transitions and requires motions around $\omega_H - \omega_N$. The term $H_zN_z F_0(t)$ has no influence on longitudinal relaxation. The term describes a fluctuation of the energy levels. As such, it randomly modulates the precession frequencies with the reorientation of the molecule and contributes to transverse relaxation. If $F_0(t)$ varies slowly, this interaction is very efficient, and coherence dephases rapidly, leading to broad lines.

Spectral Density Functions

Equations (8.6)–(8.8) show that relaxation is related to molecular motion. This is contained in the function

$$\int_0^\infty F_q(0) \, F_q^*(\tau) \, e^{-i\omega\tau} \, d\tau \tag{8.9}$$

which is the Fourier transform of the product of spherical harmonics describing the motion. It gives essentially the frequency spectrum of the rotational motions of the N–H vector in a single molecule relative to the external field. For a single molecule in a single historic event, this may be a spectrum of a limited number of frequencies. However, we can only observe the ensemble average. Thus, the whole distribution of frequencies comes to play a role. It does not matter whether the average is taken over the whole integral or only the product of spherical harmonics, although the latter was instrumental in establishing $q' = q$ in Equation (8.7). It is more convenient to do the latter. The quantity obtained is the so-called spectral density function:

$$J_q(\omega) = \int_0^\infty \overline{F_q(0)F_q^*(\tau)} \, e^{-i\omega\tau} \, d\tau \tag{8.10}$$

where the quantity

$$G_q(\tau) = \overline{F_q(0)F_q^*(\tau)} \tag{8.11}$$

is called the autocorrelation function, and the spectral density function is the Fourier transform of the autocorrelation function. It has been shown that, for isotropic motion, the spectral density function and the autocorrelation function are independent of q (Vold and Vold, 1978; Lipari and Szabo, 1982a).

$$J(\omega) = \int_0^\infty \overline{F_0(0)F_0^*(\tau)}e^{-i\omega\tau}\,d\tau \tag{8.12}$$

$$G(\tau) = \overline{F_0(0)F_0^*(\tau)} \tag{8.13}$$

In this way, different relaxation parameters measured can be related to a single spectral density function. If more than one interaction contributes to relaxation, or if one spin interacts with more than one other spin, then products of matrix elements of different interactions appear in Equation (8.6), so that cross-correlation has to be considered:

$$J_q^{ij}(\omega) = \int_0^\infty \overline{F_q^i(0)F_q^{j*}(\tau)}\,e^{-i\omega\tau}\,d\tau \tag{8.14}$$

and

$$G_q^{ij}(\tau) = \overline{F_q^i(0)F_q^{j*}(\tau)} \tag{8.15}$$

where i and j stand for different interactions. For the sake of a simple interpretation, it may be advantageous to eliminate cross-correlation effects by a suitable design of the experiments. However, at a different level, interesting dynamic information may come out of studies of cross-correlation effects.

The spectral densities represent raw data about protein internal motions. In the two-spin system of a ^{15}N bound to a ^{1}H, transitions at the frequencies ω_N, ω_H, $\omega_H + \omega_N$ and $\omega_H - \omega_N$ are relevant. Motions of the molecule as a whole as well as internal motions can cause such transitions. The motions have to have rotational characteristics, since translational motions would not modulate the dipolar interaction in the N–H system. Thus, if a non-equilibrium state is produced for a nuclear system (for example, by application of radiofrequency pulses), re-equilibration occurs by energy exchange via the channels of motions at four discrete frequencies. To obtain access to the spectral density functions, one has to deconvolute the data obtained from relaxation experiments. The simplest unambiguous relations between spectral density functions and relaxation parameters have to be established. It is important to design clean experiments where the recovery of the system after a perturbation follows an interpretable, in the best case a monoexponential, recovery curve.

A General Treatment of Relaxation in Proteins

A more formal treatment of relaxation is given in the following. The perturbing interactions relevant for relaxation can be written as a sum of interactions:

$$\mathcal{H}'(t) = \sum_k \mathcal{H}'k(t) \tag{8.16}$$

where each $H'(t)$ can be expressed in the interaction frame as

$$\mathcal{H}'_k(t) = -\sqrt{\frac{10}{3}}\, A_k \sum_{q=-2}^{2} \sum_{\mu=-1}^{1} (-)q\, \mathcal{D}_{q0}^{(2)}\, (\Omega_{\text{lab}}^k(t)) \begin{pmatrix} 1 & 1 & 2 \\ \mu & q-\mu & -q \end{pmatrix}$$

$$\exp[i(\mu\omega_S + (q-\mu)\omega_T)\, t]\, S_\mu^k\, T_{q-\mu}^k \tag{8.17}$$

The perturbing Hamiltonian is written in the spherical basis using the Wigner 3-J symbols (Zare, 1988). The index k in Equation (8.17) sums over the possible interactions causing relaxation of the spin system. These may include the dipole–dipole and chemical shift anisotropy (CSA) interactions. It is assumed that dipole–dipole and CSA interaction have axially symmetric interaction tensors. To illustrate, k would run from 1 to 3 in a system of three non-equivalent spins to account for the three distinct dipole–dipole interactions. If one of these spins had significant anisotropy in its chemical shift shielding tensor, the upper limit of k would then extend to 4. For an isolated N–H two-spin system, $k = 1$ represents the dipole interaction between the two spins, and $k = 2$ represents the interaction of the nitrogen with the CSA tensor. A_k is a physical constant which depends on the specific nature of the interaction k. For the dipolar interaction between the N and H^N spins this constant is $\dfrac{3\gamma_H\gamma_N\hbar^2}{r_{NH^N}^3}$, where r_{NH^N} is the internuclear distance between N and H^N. In the cases treated here, the internuclear distance is taken to be fixed. In the case of proton relaxation, A_k for dipole–dipole interaction would be $3\gamma_H^2\hbar^2$, and $\mathcal{D}_{q0}^{(2)}(\Omega_{\text{lab}}^k(t))$ should be replaced with $\dfrac{\mathcal{D}_{q0}^{(2)}(\Omega_{\text{lab}}^k(t))}{r_{HH}^3(t)}$. For the interaction between the CSA tensor and the nitrogen, the constant A_k is $\gamma_N\hbar\Delta$, where Δ is the chemical shift anisotropy. A value of -160 ppm is generally used for the peptide NH group (Hiyama *et al.*, 1988).

The operators S_μ^k and $T_{q-\mu}^k$ represent the dependence on the spin states of the interaction and are related to the spin operators in Equation (8.3). They are expressed as the spherical components of the two interacting vector operators specified by the kth relaxation mechanism. For the dipole–dipole interaction, S_μ^k and $T_{q-\mu}^k$ are spin operators of the interacting nuclei. For the CSA interaction, we can take $T_{q-\mu}^k$ to be a spin operator of the nucleus with anisotropic shielding. S_μ^k then represents the components of the static external field, in which case q runs only from -1 to 1, since μ is restricted to 0. The laboratory spatial degrees of freedom are contained in the $\mathscr{D}^{(2)}_{q0}(\Omega^k_{lab}(t))$ terms, which are elements of the Wigner rotation matrices and are proportional to the second-order spherical harmonics (Zare, 1988). These are essentially the functions shown in Equations (8.4). The $\Omega^k_{lab}(t)$ symbol denotes the polar angles $\theta_k(t)$, $\phi_k(t)$ of the symmetry axis belonging to the kth interaction tensor. Thus, the $\mathscr{D}^{(2)}_{q0}(\Omega^k_{lab}(t))$ terms are simply trigonometric functions which describe the axis' orientation with respect to the laboratory magnetic field (Wittebort and Szabo, 1978). For the dipole–dipole interaction, this axis is the vector connecting the two nuclei in the interacting spin pair. For the CSA interaction it is the symmetry axis of the shielding tensor.

The time dependence of the density matrix in the interaction frame is obtained from

$$\frac{d\sigma}{dt} = -\frac{1}{\hbar^2}\int_0^\infty \langle\langle[\mathscr{H}'(t),[\mathscr{H}'(t-\tau),\sigma(t)-\sigma_{eq}]]\rangle\rangle\,d\tau \tag{8.18}$$

The relaxation of spin order $\langle Q\rangle$, associated with the spin operator Q, is described by a first-order differential equation (Abragam, 1961; Ernst *et al.*, 1987; Goldman, 1988). In this context, 'spin order' includes longitudinal and transverse magnetization (e.g. N_z, $N_{x,y}$), as well as antiphase coherences, longitudinal multi-spin orders (e.g. $2H_z^N N_z$), and more general p-quantum coherences ($p \neq 1$). When using the term 'spin order', we follow the convention used by Ernst *et al.* (1987). If we use the form of the perturbing Hamiltonian given in Equation (8.17), we obtain for $\dfrac{d\langle Q\rangle}{dt}$

$$\frac{d\langle Q\rangle}{dt} = -\frac{5}{3}\sum_{(k,k')}\frac{A_k A_{k'}}{\hbar^2}\sum_{\substack{(q=2,\mu=1)\\(q=-2,\mu=-1)}}(-)^q J_{kk'}(\mu\omega_S+(q-\mu)\omega_T)\begin{pmatrix}1&1&2\\\mu&q-\mu&-q\end{pmatrix}^2$$

$$\times \operatorname{Tr}\{[S^{k'}_{-\mu}\,T^{k'}_{\mu-q},[S^k_\mu\,T^k_{q-\mu},Q]]\,(\sigma-\sigma_{eq})\} \tag{8.19}$$

The traces of the products of the double commutator $[S^{k'}_{-\mu}\,T^{k'}_{\mu-q}, [S^k_\mu\,T^k_{q-\mu},\,Q]]$ with the spin density operator, σ, represent macroscopic spin orders. These will consist of $\langle Q\rangle$ itself, and possibly other distinct spin orders. The relaxation rate of $\langle Q\rangle$ is given by its own net coefficient; the coefficients of any other spin orders are then cross-relaxation rates. Again, for the CSA interaction, the index μ is restricted to 0, and therefore, q only runs from -1 to 1.

The relaxation rate of $\langle Q \rangle$ depends on the generalized spectral density functions, $J_{kk'}(\omega)$ at particular frequencies that are selected by the rules of Equation (8.19). $J_{kk'}(\omega)$ is the cosine Fourier transform (Abragam, 1961)

$$J_{kk'}(\omega) = 2 \int_0^\infty \cos(\omega\tau)\, G_{kk'}(\tau)\, d\tau \qquad (8.20)$$

where $G_{kk'}(\tau)$ is a time correlation function given by

$$G_{kk'}(\tau) = \langle \mathscr{D}_{q0}^{(2)*} (\Omega_{\mathrm{lab}}^{k'}(t + \tau))\, \mathscr{D}_{q0}^{(2)} (\Omega_{\mathrm{lab}}^{k}(t)) \rangle \qquad (8.21)$$

When $k = k'$, $G_{kk'}(\tau)$ becomes an autocorrelation function and describes the rotational diffusion of axis k alone. This case is appropriate for studying the motion of an NH^N bond, when the NH^N dipole–dipole interactions (and possibly the chemical shift anisotropy of the ^{15}N) are considered dominant. In the case that $k \neq k'$, $G_{kk'}(\tau)$ is a cross-correlation function describing the decay of correlations between the k and k' tensor axes during a time τ, resulting from their respective rotational diffusions. Examples include correlations between distinct dipole–dipole vectors, and between a dipole–dipole vector and a CSA symmetry axis. In both cases, one spin must be shared between the two vectors or axes (Canet, 1989). This general formalism allows one to calculate autorelaxation and cross-relaxation rates (Peng and Wagner, 1992a,b).

Relaxation Equations for an N–H Spin System

Equation (8.19) can be considered a system of coupled linear differential equations:

$$\frac{d\langle Q \rangle}{dt} = - R \langle Q \rangle \qquad (8.22)$$

The 15 elements of the vector $\langle Q \rangle$ are $\langle N_z \rangle$, $\langle H_z \rangle$, $\langle N_x \rangle$, $\langle N_y \rangle$, $\langle H_x \rangle$, $\langle H_y \rangle$, $\langle 2N_xH_z \rangle$, $\langle 2N_yH_z \rangle$, $\langle 2N_zH_x \rangle$, $\langle 2N_zH_y \rangle$, $\langle 2N_xH_y \rangle$, $\langle 2N_xH_x \rangle$, $\langle 2N_yH_x \rangle$, $\langle 2N_yH_y \rangle$, $\langle 2N_zH_z \rangle$. Note that $\langle Q \rangle$ means the deviation of the expectation value of the operator Q from the equilibrium value. R is the relaxation superoperator. The matrix elements of R are obtained from evaluation of Equation (8.19). Fortunately, many of the matrix elements are zero, for symmetry reasons, and only a subset of spin orders can interact with each other. Ideally one would like R to be diagonal so that one could perturb a single spin order and analyse its recovery unperturbed by interaction with other spin orders. Unfortunately, this is not the case. As a consequence, all recovery of spin order after a perturbation is multiexponential. In special cases, all other spin orders that interact with the perturbed spin order can be forced to be constant during recovery—for example, by radiofrequency irradiation. This is the strategy for measuring longitudinal relaxation rates of heteronuclei while saturating the protons. It should be noted that Equation (8.22) can be diagonalized. One then obtains relaxation modes, and each of them relaxes with a single exponential (Peng and Wagner, unpublished). To our knowledge, this has never been made use of experimentally. So far, one has tried to design pulse sequences that force the system to be pseudodiagonal. Rather complicated pulse sequences are necessary to achieve this approximately (see below).

Spectral Density Functions for an N–H Spin System

In an isolated N–H spin system the important interactions are the dipole–dipole interaction between the two spins ($k = d$, for dipole) and CSA of the nitrogen ($k = c$, for CSA). Evaluation of Equation (8.19) shows that we may have terms containing $J_{dd}(0)$, $J_{dd}(\omega_N)$, $J_{dd}(\omega_H + \omega_N)$, $J_{dd}(\omega_H)$, $J_{dd}(\omega_H - \omega_N)$, $J_{cc}(0)$, $J_{cc}(\omega_N)$, $J_{cd}(0)$ and $J_{cd}(\omega_N)$. To simplify this, we consider the correlation functions. It has been shown that the autocorrelation function for dipole–dipole interaction can be written as

$$G_{dd}(\tau) = \frac{1}{5} \langle P_2(r_{NH}(0) \cdot r_{NH}(\tau)) \rangle \tag{8.23}$$

where $P_2(x)$ is the second Legendre polynomial, and the argument is the scalar product of the N–H vector at time zero and the same vector at time τ in the laboratory frame (Lipari and Szabo, 1982a,b):

$$P_2(x) = \frac{1}{2}(3x^2 - 1) \tag{8.24}$$

For the autocorrelation of CSA and the cross-correlation between dipole–dipole and CSA we have:

$$G_{cc}(\tau) = \frac{1}{5} \langle P_2(r_{CSA}(0) \cdot r_{CSA}(\tau)) \rangle \tag{8.25}$$

and

$$G_{dc}(\tau) = \frac{1}{5} \langle P_2(r_{NH}(0) \cdot r_{CSA}(\tau)) \rangle \tag{8.26}$$

Since the orientation of the principal axis of the CSA tensor is fixed relative to the N–H bond vector, and if $\| r_{NH} \|$ is constant, it is a good approximation to assume that $J_{dd}(\omega) = J_{cc}(\omega)$. For the same reason, the spectral density functions for the cross-correlation are the same, normalized with $\langle P_2(\cos\beta) \rangle$, where β is the angle between the r_{NH} vector and the principle CSA axis, r_{CSA} (Palmer *et al.*, 1992). A value of *c.* 20° has been given in the literature for this angle (Hiyama *et al.*, 1988). We therefore can drop the subscripts of the spectral density functions and are left with the set of five values of the same spectral density function $\{J(0), J(\omega_N), J(\omega_H + \omega_N), J(\omega_N), J(\omega_H - \omega_N)\}$ which we like to determine.

At this point it may be useful to obtain a pictorial understanding of what the spectral density function means. The energy of the dipole–dipole interaction between N and H in a single molecule (and the CSA energy) has a certain value depending on the orientation of the N–H vector, relative to the external field. This energy fluctuates with the random rotational motion of the vector. The spectral density function $J(\omega)$ describes how this energy is distributed over different frequencies. This means that in the ensemble of independent molecules, the mean value of the dipole–dipole energy is constant. Therefore, the integral over the spectral density function is also constant. Of particular interest is $J(0)$. From Equation (8.20) it can be seen that it is simply the integral over the autocorrelation function. Thus, $J(0)$ is proportional to the average time it takes until the autocorrelation function decays. In this sense it is proportional to a generalized correlation time or coherence time τ_c (Peng and Wagner, 1992a):

$$J(0) = 2 \int_0^\infty G(\tau) \, d\tau = 2 \, G(0) \, \tau_c = \frac{2}{5} \tau_c \tag{8.27}$$

$J(0)$ is particularly important for transverse relaxation. It describes the time the orientation of the r_{N-H} vector relative to the external field does not change in the average. So the precession frequencies of the nitrogens are different in the different molecules, owing to the different orientations of the vectors leading to dephasing of the signal and line broadening. The term of the dipole–dipole Hamiltonian important for this effect is:

$$H_z N_z \, F_0(t) = H_z N_z k [1 - 3 \cos^2 \theta(t)] \tag{8.28}$$

with

$$k = \frac{\hbar^2 \gamma_H \gamma_N}{r_{HN}^3} \tag{8.29}$$

Obviously, the longer this correlation time, the more efficient is the dephasing effect and thus the transverse relaxation. A plot of $J(\omega)$ against ω gives the relative probability that the orientation of the r_{NH} vector changes on the time scale $\frac{2\pi}{\omega}$.

Mapping of Spectral Density Functions

As we have seen above, it will require special experiments to extract well-defined autorelaxation or cross-relaxation rates from the decay of signals after perturbation of a spin system. The decay may be mono-exponential or multiexponential. If we are successful in achieving this, each rate will depend on values of the spectral density function at up to five frequencies. Thus, if we want to determine the spectral density function at these frequencies, we have to measure at least five relaxation rates. Usually only three parameters are measured, such as T_1, T_2 and the heteronuclear NOE. Since this is not sufficient to map $J(\omega)$, models had to be made about the motion of the N–H vector, such as the 'wobbling in a cone' model (Woessner, 1962; Kinoshita *et al.*, 1977; Richarz *et al.*, 1980), and parameters defined by this model were fitted to the experimental parameters. Alternatively, a 'model-free approach' was developed (King and Jardetzky, 1978; King *et al.*, 1978; Lipari and Szabo, 1982a,b) which does not invoke a mechanistic model of the motion but assumes that the autocorrelation function is a sum of exponentials. As a consequence, the spectral density function is a sum of Lorentzians centred at $\omega = 0$.

Recently, a strategy was developed to map spectral density functions independent of mechanistic models or models of the time dependence of the autocorrelation function (Peng *et al.*, 1991a; Peng and Wagner, 1992a,b). This requires measurements of six independent relaxation parameters. It remains to be seen whether this approach has the sensitivity to provide additional insight into aspects of internal protein motions.

Dependence of Relaxation Parameters on Spectral Density Functions

Evaluation of Equation (8.19) shows that measurement of the following parameters could give a complete set of data to map the spectral density function: the longitudinal ^{15}N relaxation rate,

$$R_N(N_z) = \frac{\gamma_H^{2N}\gamma_N^2\hbar^2}{4r_{NH}^{6N}} \{J(\omega_{H^N} - \omega_N) + 3J(\omega_N) + 6J(\omega_{H^N} + \omega_N)\}$$

$$+ \frac{\Delta^2\omega_N^2}{3} J(\omega_N) \qquad (8.30)$$

the transverse relaxation rate of in-phase ^{15}N coherence:

$$R_N(N_{x,y}) = \frac{\gamma_H^{2N}\gamma_N^2\hbar^2}{8r_{NH}^{6N}} \{4J(0) + J(\omega_{H^N} - \omega_N) + 3J(\omega_N) + 6J(\omega_{H^N})$$

$$+ 6J(\omega_{H^N} + \omega_N)\} + \frac{\Delta^2\omega_N^2}{3} \left\{\frac{2}{3} J(0) + \frac{1}{2} J(\omega_N)\right\} \qquad (8.31)$$

the ^{1}H–^{15}N cross-relaxation rate:

$$R_N(H^{N_z} \rightarrow N_z) = \frac{\gamma_H^{2N}\gamma_N^2\hbar^2}{4r_{NH}^{6N}} \{6J(\omega_{H^N} + \omega_N) - J(\omega_{H^N} - \omega_N)\} \qquad (8.32)$$

Note that the cross-relaxation rate and the longitudinal relaxation rate combine to give the steady-state heteronuclear NOE (η) defined by

$$\eta = \frac{I_{sat} - I_{eq}}{I_{eq}} = \frac{\gamma_H}{\gamma_N} \frac{R_N(H^{N_z} \to N_z)}{R_N(N_z)} \tag{8.33}$$

This definition of η as the steady-state NOE is that of Noggle and Schirmer (1971). I_{sat} is the cross-peak intensity with proton saturation, and I_{eq} is the equilibrium Zeeman intensity. NOE values consistent with definitions given in previous studies of ^{15}N protein relaxation are obtained by using $(1 + \eta)$ (Kay *et al.*, 1989; Clore *et al.*, 1990a). Furthermore, we measure the two-spin relaxation rates of longitudinal two-spin order and of anti-phase ^{15}N coherence (Goldman, 1988; Bax *et al.*, 1990; Boyd *et al.*, 1990; Peng *et al.*, 1991a,b; Peng and Wagner, 1992a,b)

$$R_{NH}(2H_z^N N_z) = \frac{\gamma_H^{2}{}^{N_1}\gamma_N^{2}\hbar^2}{4r_{NH^N}^{6}} \{3J(\omega_N) + 3J(\omega_{H^N})\}$$

$$+ \frac{\Delta^2 \omega_N^2}{3} J(\omega_N) + \rho_{H^N H^i} \tag{8.34}$$

$$R_{NH}(2H_z^N N_{x,y}) = \frac{\gamma_H^{2}{}^N \gamma_N^2 \hbar^2}{8r_{NH^N}^{6}} \{4J(0) + J(\omega_{H^N} - \omega_N) + 3J(\omega_N)$$

$$+ 6J(\omega_{H^N} + \omega_N)\} + \frac{\Delta^2 \omega_N^2}{3} \left\{ \frac{2}{3} J(0) + \frac{1}{2} J(\omega_N) \right\} + \rho_{H^N H^i} \tag{8.35}$$

Unfortunately, the latter two parameters depend on the relaxation of the amide proton which interacts with other protons such that the term $\rho_{H^N H^i}$ has to be included.

$$\rho_{H^N H^i} = \sum_i \frac{\gamma_H^{2}{}^N \gamma_H^{2}{}^i \hbar^2}{4r_{H^N H^i}^{6}} \{J_{H^N H^i}(\omega_{H^N} - \omega_{H^i}) + 3J_{H^N H^i}(\omega_{H^N})$$

$$+ 6J_{H^N H^i}(\omega_{H^N} + \omega_{H^i})\} \tag{8.36}$$

The constants γ_N and γ_{H^N} are the gyromagnetic ratios for the ^{15}N and amide proton nuclei. r_{NH^N} is the length of the NH bond vector. The $r_{H^N H^i}$ are the lengths of interspin vectors connecting a given amide H^N proton with other distinct protons H^i. ω_N is the Larmor frequency of the ^{15}N nucleus in rad/s, and Δ is the chemical shift anisotropy value for an axially symmetric shift tensor (Wittebort and Szabo, 1978). The ^{15}N shift tensor is reported to be approximately symmetric with a symmetry axis nearly coincident with the NH bond vector (Hiyama *et al.*, 1988). The relaxation rates for the two-spin orders $R_{NH}(2H_z^N N_z)$ and $R_{NH}(2H_z^N N_{x,y})$ are complicated by a significant dependence on the spin–lattice relaxation rate of the attached amide proton via the $\rho_{H^N H^i}$ term.

To solve the equations, the longitudinal relaxation of the amide proton, H^N, has also to be measured:

$$R_H(N_z^N) = \frac{\gamma_H^{2N}\gamma_N^2\hbar^2}{4r_{NH}^{6N}} \{J(\omega_{H^N} - \omega_N) + 3J(\omega_{H^N}) + 6J(\omega_{H^N} + \omega_N)\}$$

$$+ \rho_{H^NH}i \qquad (8.37)$$

It should be mentioned that cross-correlation between dipole–dipole interaction of the amide proton with other protons or heteronuclei and its own CSA exists. However, it is expected to be small for proteins, since the amide proton CSA is small, and the effect depends only on $J(\omega_H)$ of the proton CSA tensor, which is small in proteins in high-field spectrometers.

The system of linear equations (8.30)–(8.32), (8.34)–(8.37) can be solved to obtain the spectral density values immediately from the experimental rates. These formulae are given as (Peng and Wagner, 1992b)

$$J(0) = \frac{3}{4}\frac{1}{3d+c}\left\{-\frac{1}{2}R_N(N_z) + R_N(N_{x,y}) + R_{NH}(2H_z^NN_{x,y})\right.$$

$$\left. -\frac{1}{2}R_{NH}(2H_z^NN_z) - \frac{1}{2}R_H(H_z^N)\right\} \qquad (8.38)$$

$$J(\omega_{H^N} - \omega_N) = \frac{1}{4}\frac{1}{d}\{R_N(N_z) - R_{NH}(2H_z^NN_z) + R_H(H_z^N)$$

$$- 2R_N(H_z^N \rightarrow N_z)\} \quad (8.39)$$

$$J(\omega_N) = \frac{1}{2}\frac{1}{3d+c}\{R_N(N_z) + R_{NH}(2H_z^NN_z) - R_H(H_z^N)\} \qquad (8.40)$$

$$J(\omega_{H^N}) = \frac{1}{12}\frac{1}{d}\{-R_N(N_z) + 2R_N(N_{x,y}) - 2R_{NH}(2H_z^NN_{x,y})$$

$$+ R_{NH}(2H_z^NN_z) + R_H(H_z^N)\} \qquad (8.41)$$

$$J(\omega_{H^N} + \omega_N) = \frac{1}{24}\frac{1}{d}\{R_N(N_z) - R_{NH}(2H_z^NN_z) + R_N(H_z^N)$$

$$+ 2R_N(H_z^N \rightarrow N_z)\} \quad (8.42)$$

$$\rho_{H^NH}i = \left\{-\frac{1}{4}R_N(N_z) - \frac{1}{2}R_N(N_{x,y}) + \frac{1}{2}R_{NH}(2H_z^NN_{x,y})\right.$$

$$\left. + \frac{1}{4}R_{NH}(2H_z^NN_z) + \frac{1}{4}R_H(H_z^N)\right\} \qquad (8.43)$$

The latter quantity, $\rho_{H^NH}i$, is the contribution of proton–proton relaxation to the amide proton longitudinal relaxation. The constants are defined as:

$$d = \frac{\gamma_{HN}^2 \gamma_N^2 \hbar^2}{4 r_{NHN}^6} \quad \text{and} \quad c = \frac{\Delta^2 \omega_N^2}{3}$$

For a CSA value of $\Delta = -160$ ppm (Hiyama *et al.*, 1988) and an internuclear $^{15}N-^1H$ bond distance of $r_{NH^N} = 1.02$ Å (Keiter, 1986), d and c become $\approx 1.3 \times 10^9$ (rad/s)^2 and 0.9×10^9 (rad/s)^2, respectively.

Experimental Techniques for Measuring Relaxation Rates

The relaxation rates can best be measured in series of heteronuclear $^1H-^{15}N$ correlated spectra where the relaxation rates are obtained from a fit of the peak intensities vs. a relaxation delay, as first used for measuring ^{13}C relaxation data (Nirmala and Wagner, 1988). In most experiments an INEPT transfer (Morris and Freeman, 1979) is used at the beginning and a reverse INEPT at the end of the sequence to enhance sensitivity. Pulse sequences have been described by Peng and Wagner (1992a,b). Here they are briefly reviewed under the aspect of cross-relaxation with other spin orders.

The Longitudinal Autorelaxation Rate $R_N(N_z)$

Pulse sequences for measuring longitudinal relaxation rates of heteronuclei in 2D experiments have been used in a number of laboratories (Nirmala and Wagner, 1988; Kay *et al.*, 1989). The pulse sequence we have been using is shown in Figure 8.4(a). It is an inversion recovery sequence. The $90°(^{15}N)$ pulse after the initial refocused INEPT sequence turns the enhanced ^{15}N coherence into the negative z axis. After the recovery delay we have a trivial read-out sequence, consisting of a $90°(^{15}N)$ pulse followed by frequency labelling and reverse INEPT. When we perturb $\langle N_z \rangle$, we have to consider cross-relaxation with other spin orders. Analysis of Equation (8.19) shows that this can be $\langle H_z \rangle$ and $\langle H_z N_z \rangle$. Thus, there is a system of three coupled differential equations:

$$\frac{d\langle N_z \rangle}{dt} = -R_N(N_z)\langle N_z \rangle - R_N(H_z \to N_z)\langle H_z \rangle$$
$$- R_N(2H_z N_z \to N_z)\langle 2H_z N_z \rangle \tag{8.44}$$

$$\frac{d\langle H_z \rangle}{dt} = -R_N(N_z \to H_z)\langle N_z \rangle - R_H(H_z)\langle H_z \rangle \tag{8.45}$$

$$\frac{d\langle 2H_z N_z \rangle}{dt} = -R_N(N_z)\langle N_z \rangle - R_N(2H_z N_z \to N_z)\langle 2H_z N_z \rangle \tag{8.46}$$

Communication of the amide proton with other protons is neglected at this level. The cross-relaxation between $\langle H_z N_z \rangle$ and $\langle N_z \rangle$ is only possible

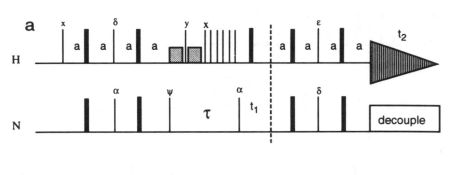

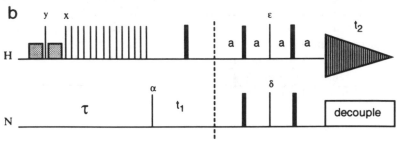

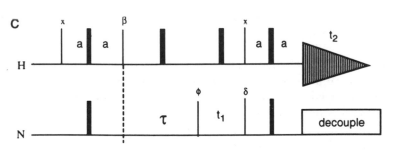

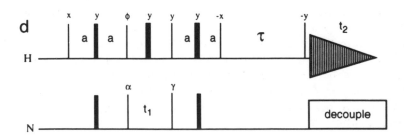

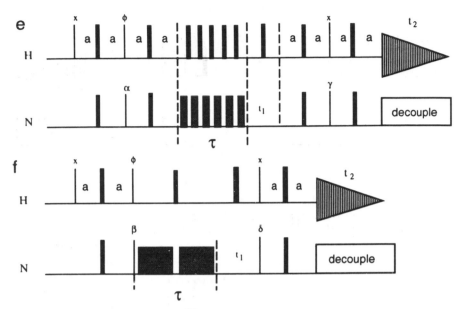

Figure 8.4 Pulse sequences for ^{15}N measuring relaxation parameters in proteins: (a) $R_N(N_z)$; (b) $R_N(H^{N_z} \rightarrow N_z)$; (c) $R_{NH}(2H_z^N N_z)$; (d) $R_H(H_z^N)$; (e) $R_N(N_{x,y})$; (f) $R_{NH}(2H_z^N N_{x,y})$. Thin bars indicate 90° pulses; thick bars are 180° pulses. Shaded blocks indicate spin lock pulses

because of cross-correlation between dipole–dipole and CSA. This system of differential equations has three eigenvalues. If we would just invert $\langle N_z \rangle$ and let it recover unperturbed, we would get a recovery profile consisting of a sum of three exponentials. The only way to obtain monoexponential decay is to force the magnetization of the interacting spins constant. The initial INEPT transfer saturates the amide proton polarization, and if we pulse the amide protons during the recovery delay, $\langle H_z \rangle$ is kept constant, and the proton–nitrogen cross-relaxation rate does not show up in the recovery curve. However, $\langle N_z \rangle$ recovers to the steady state NOE value rather than to the equilibrium value. Thus, in such an experiment the longitudinal relaxation rate and the steady state NOE can be measured. Boyd *et al.* (1990) have also shown that proton irradiation during the recovery delay eliminates the cross-relaxation with $\langle 2H_z N_z \rangle$. Thus, the sequence shown in Figure 8.4(a) yields monoexponential recovery curves.

The Longitudinal ^{1}H–^{15}N Cross-relaxation Rate of $R_N(H_z \rightarrow N_z)$

The pulse sequence of Figure 8.4(b) measures cross-relaxation rates. Saturation of the proton magnetization leads to heteronuclear NOE build-up at the nitrogen. The read-out sequence consists of a frequency labelling period followed by an inverse INEPT step. Because of the proton irradiation, we can neglect the effect of cross-relaxation with $\langle H_z N_z \rangle$. Interaction of the nitrogen with other protons can be neglected. Thus the build-up is a single exponential, ending at the steady-state NOE value.

The Longitudinal Autorelaxation Rate of $\langle 2H_z N_z \rangle$

A pulse sequence for measuring the autorelaxation of $\langle 2H_z N_z \rangle$ has been shown by Boyd *et al.* (1990). A slightly different sequence was used by Kay *et al.* (1992). Figure 8.4(c) shows the pulse sequence we are using (Peng and Wagner, 1992a). The pulse sequences differ by placing the t_1 period before or after the relaxation period. According to Equation (8.46), cross-relaxation between $\langle 2H_z N_z \rangle$ and $\langle N_z \rangle$ could occur. To reduce this effect we have applied a 180° proton pulse in the middle of the relaxation delay. Application of trains of 180° proton pulses resulted in qualitatively unsatisfactory spectra. Kay *et al.* (1992) have claimed that cross-relaxation due to dipolar–CSA cross-correlation is small in larger proteins, while Boyd *et al.* (1990) have shown that it is significant for smaller proteins. In this experiment also cross-relaxation with other protons has to be considered. The pulse sequences shown here and the analysis of the data are not yet fully satisfactory.

Longitudinal Relaxation of $\langle H_z \rangle$

A pulse sequence for measuring longitudinal proton relaxation rates is shown in Figure 8.4(d). The sequence is essentially a 2D HSQC-NOESY, with ^{15}N shifts along ω_1 and ^1H shifts along ω_2. The decay of the direct N–H cross-peaks with τ is used to obtain proton longitudinal relaxation times. Usually, proton inversion recovery exhibits multiexponential behaviour. One reason for this is that other protons are also inverted and create cross-relaxation effects. Here this is no problem, since the other protons would be labelled with another frequency along ω_1 (if they are bound to a ^{15}N), or eliminated by the phase cycle. Furthermore, the phase cycle is adjusted so that all protons except the H^N are turned back to the equilibrium orientation (along $+z$) by the 90_{-x} (^1H) pulse preceding the mixing time τ. Conceptually, it makes sense to saturate ^{15}N during the mixing time τ, although this may have little influence on the results. The more serious problem is that the other protons interacting with the H^N lead to multiexponential decay, irrespective of the initial conditions. This is analogous to the classical transient NOE (Solomon, 1955). Selective saturation of all protons except the amides seems to be impracticable. Only complete deuteration of the protein could solve this problem. However, when we analyse the recovery with a Taylor series, the linear coefficient gives the autorelaxation rate of the amide protons. In our experience, Taylor expansions and exponential fits using only short mixing times yield very similar results.

Transverse Relaxation Rates of $\langle 2N_{xy} \rangle$

When measuring transverse relaxation rates of ^{15}N, we have to consider cross-relaxation with $\langle 2H_zN_{xy} \rangle$ mediated via cross-correlation with CSA (Palmer *et al.*, 1992; Kay *et al.*, 1992).

$$\frac{d\langle N_{xy} \rangle}{dt} = -R_N(N_{xy})\langle N_{xy} \rangle - R_N(2H_zN_{xy} \rightarrow N_{xy})\langle 2H_zN_{xy} \rangle \tag{8.47}$$

$$\frac{d\langle 2H_zN_{xy} \rangle}{dt} = -R_N(2H_zN_{xy} \rightarrow N_z)\langle N_{xy} \rangle - R_N(2H_zN_{xy})\langle 2H_zN_{xy} \rangle \tag{8.48}$$

Cross relaxation from $\langle H_z \rangle$ to $\langle 2H_zN_{xy} \rangle$ can be neglected. However, $R_N(2H_zN_{xy})$ contains contributions of relaxation due to the interaction of the amide proton with other protons (see above).

The pulse sequence shown in Figure 8.4(c) differs from the earlier concepts of measuring transverse relaxation times using Carr–Purcell (1954) or Meiboom–Gill (1958) sequences. There are a number of complications that have been discovered only recently. The most straightforward way of designing a ^{15}N or ^{13}C T_2 experiment would be to place a $180°(^{15}N)$ pulse in the centre of the τ period of Figure 8.4(e) (Carr and Purcell, 1954; Nirmala and Wagner, 1989). Later it was found that trains of $180°$ ^{15}N pulses yield longer T_2 values (Peng *et al.*, 1991a,b). The relaxation times are longest for a continuous spin lock, and we measure essentially $T_{1\rho}$. In the protein eglin c, $T_{1\rho}$ is typically twice as long as T_2 measured with a simple Carr–Purcell sequence (Peng *et al.*, 1991a). This observation was qualitatively true for all nitrogens, those on-resonance with the ^{15}N spin lock field and at the end of the ^{15}N spectral width. This effect can be explained with the fact that in-phase coherence evolves into antiphase coherence during the delay τ. It is refocused to in-phase coherence by the end of τ. However, the relaxation properties represent an average between in-phase and antiphase relaxation (Peng *et al.*, 1991b). This effect can be eliminated by application of closely spaced $180°$ pulses or continuous spin locks.

A second complication arises due to cross-relaxation of $\langle 2H_zN_{xy} \rangle$ with $\langle N_{xy} \rangle$ caused by cross-correlation of dipole–dipole interaction with CSA (Palmer *et al.*, 1992; Kay *et al.*, 1992). This effect causes one component of the $^{15}N-^{1}H$ doublet to relax faster than the other. This is equivalent to the creation of antiphase coherence. This effect can be eliminated by applying 180°(^{1}H) pulses during the delay τ. This effectively exchanges the two doublet components, and the effect of CSA–dipole cross-correlation is averaged out. This decoupling effect is perturbed by the application of the 180° pulses on ^{15}N. If they are applied synchronously with the proton pulses, the decoupling effect is quenched. Thus they need to be applied interleaved, and the ^{1}H pulses should be applied only after a multiple of a 360° ^{15}N rotation (Palmer *et al.*, 1992; Kay *et al.*, 1992; Peng and Wagner, 1992a,b). In the pulse sequence of Figure 8.4(e), the shaded blocks in the τ period on the nitrogen channel mean even numbers of 180° pulses.

Transverse Relaxation Rates of $\langle 2H_zN_{x,\,y}\rangle$

Figure 8.4(d) shows a sequence for measuring the transverse relaxation rate of antiphase coherence, $R_N(2H_zN_{xy})$ (Peng and Wagner, 1992a). The shaded blocks in the nitrogen channel mean spin locks or trains of 180° pulses. Their purpose is to prevent oscillation between antiphase and in-phase coherence. The single 180° (^{1}H) pulse in the centre of the τ period is to reduce the effect of cross-relaxation with $\langle N_{xy}\rangle$ due to dipole–CSA cross-correlation. In this experiment, cross-relaxation is also a consideration.

The six experiments of Figure 8.4 provide a large enough set of data to calculate the spectral densities at five frequencies. Experiments at a different field strength can add spectral densities at an additional four frequencies. For proteins it is advantageous to perform these experiments at lower field strength, because the most dramatic changes in the spectral density function seem to appear in the range below 450 MHz. In addition to the experiments described in Figure 8.4, other parameters could be measured, such as the relaxation of proton–nitrogen double-quantum or zero-quantum coherence, or the cross-relaxation rates between $\langle 2H_zN_{xy}\rangle$ and $\langle N_{xy}\rangle$, or between $\langle 2H_zN_z\rangle$ and $\langle N_z\rangle$. This would yield additional (redundant) information on the spectral density function. In fact, Brüschweiler and Ernst (1992) have investigated the use of cross-relaxation rates arising from cross-correlation effects for the purpose of monitoring molecular dynamics.

Experimental Results on the Elastase Inhibitor Eglin c

A complete set of the experiments outlined in Figure 8.4 has been recorded for the protein eglin c at 11.74 T (Peng and Wagner, 1992b), and spectral densities $J(0)$, $J(50)$, $J(450)$, $J(500)$ and $J(550)$ have been calculated for nearly all NH nitrogens in the protein. Figure 8.1 shows an ensemble of the structures as obtained from distance geometry calculations. Inspection of these structures and the plot of the $\langle rmsd\rangle$s of Figure 8.2 suggests that there is higher mobility for the eight N-terminal residues, the proteinase binding loop (residues 38–49) and the hairpin loop of the antiparallel β sheet (residues 57–60). Figure 8.2(c) is a plot of the generalized correlation time τ_c obtained from $J(0)$ according to Equation (8.27). Table 8.1 lists some of these results for representative residues. These are Thr 1 at acetylated N terminus, Val 13 in the 3_{10} helix, Ala 21 in the α helix, Tyr 35 from a parallel β sheet, Asp 46, the P1′ residue of the binding loop, Arg 53 from an antiparallel β sheet and Thr 60 from the hairpin loop. It can be seen from Table 8.1 that $J(0)$ and $J(50)$ vary significantly with the amino acid sequence. The spectral density functions at higher frequencies are, however, nearly independent of the sequence. This means that all conformation specific differences in mobility are in the frequency range below 450 MHz. Thus, experiments at different field strengths are important to cover the frequency range between 50 and 450 MHz.

Table 8.1 Parameters of mobility for some representative residues of eglin c, obtained from spectral density mapping[a]

Residue	Secondary structure	J(0)	τ_c	J(50)	J(450)	J(500)	J(550)
Thr 1	—	0.16	0.40	0.08	0.036	0.041	0.054
Val 13	3_{10} helix	1.59	3.98	0.47	0.028	0.040	0.095
Ala 21	α-helix	1.42	3.55	0.49	0.024	0.078	0.082
Tyr 35	β-parallel	1.27	3.18	0.45	0.028	0.104	0.097
Asp 46	binding loop	0.76	1.90	0.30	0.026	0.064	0.073
Arg 53	β-antiparallel	1.34	3.35	0.48	0.015	0.081	0.040
Thr 60	turn	1.30	3.25	0.44	0.013	0.071	0.039

[a]Spectral densities and τ_c are in ns.

There is no significant difference between the mobility of the hairpin loop of the antiparallel β sheet (residues 57–60) and the core of the protein. Thus, the apparently higher variability in the structures as seen in distance geometry structures of Figure 8.1 and the ⟨rmsd⟩ of Figure 8.2(a) is definitely due to lack of data rather than higher mobility (Peng and Wagner, 1992b). Considering the estimated errors in the data analysis, there was no significant correlation between spectral density functions and involvement in hydrogen bonds of either the NH or the CO of the peptide group.

Another interesting observation is that the spectral density seems to increase from 450 MHz ($\omega_H + \omega_N$) to 550 MHz ($\omega_H - \omega_N$). This increase is small but significant. It is observed for almost all residues (Peng and Wagner, 1992b). This phenomenon is not yet understood. Systematic errors cannot be ruled out completely. Possible sources of such errors have been discussed (Peng and Wagner, 1992b). If the increase of the spectral density function at high frequencies were true, it would contradict the generally made assumption that the autocorrelation function decays as a sum of exponentials. It would indicate that there is an oscillatory behaviour in the autocorrelation function. Experiments at higher field strength (14.09 T) are in progress to further investigate this effect. The experimental basis for the increase in the spectral density function at high frequencies is that the size of the heteronuclear cross-relaxation rate (or the size of the heteronuclear NOE) is smaller than expected for a decreasing spectral density function (compare Equations 8.32, 8.39 and 8.42).

Comparison with Other Relaxation Studies

Early relaxation studies on proteins were focused on the relaxation of ^{13}C resonances—in particular, methyl resonances, since these could be observed even at natural abundance in 1D spectra. Most of these early studies used the 'wobbling in a cone' model (Woessner, 1962; Kinoshita, 1977; Richarz et al., 1980). After methods for measuring relaxation times in 2D heteronuclear correlated spectra were introduced (Nirmala and Wagner, 1988; Kay et al., 1989), and ^{15}N-enriched proteins have become available, numerous relaxation studies in proteins have been published. All of these later studies used the formalism of the 'model-free approach' (Lipari and Szabo, 1982a,b) to interpret the data.

The Approach of King and Jardetzky (1978)

A general model for the shape of spectral density functions was proposed. It was assumed that an arbitrary number of motions of arbitrary nature contribute to the relaxation in a protein. It was further assumed that a probability can be assigned to each step in the motion and that the system has no memory (conditions for a Markov process). Under these conditions, it was shown that the autocorrelation function behaves as a sum of decaying exponentials, and the spectral density function is a sum of Lorentzian lines centred at $\omega = 0$. This formalism was used to fit the sparse experimental relaxation data available at that time (Ribeiro et al., 1980). These were data on the basic pancreatic trypsin inhibitor. No a priori assumptions about a model were made, and it was attempted to determine frequency and amplitude of intramolecular motions. Finally, mechanistic proposals were made to assign these modes with motions of particular protein moieties.

The Model-free Approach of Lipari and Szabo (1982a,b)

This approach makes a minimum of assumptions about the NH vector dynamics, and introduces a generalized order parameter that retains its physical interpretation for a variety of more specific models of motion. The model of $J(\omega)$ is based on a very simple form for the autocorrelation function:

$$G(\tau) = \frac{1}{5} G_0(\tau) G_i(\tau) \tag{8.49}$$

with

$$G_0(\tau) = \exp(-\tau/\tau_m) \tag{8.50}$$

and

$$G_i(\tau) = S^2 + (1 - S^2) \exp(-\tau/\tau_i) \tag{8.51}$$

After Fourier transformation as indicated in Equation (8.20), the resulting spectral density is

$$J(\omega) = \frac{2}{5} \left\{ \frac{S^2 \tau_m}{1 + (\omega\tau_m)^2} + \frac{(1 - S^2)\tau_i}{1 + (\omega\tau_i)^2} \right\} \tag{8.52}$$

Furthermore, an internal correlation time τ_e is used. τ_i, τ_m and τ_e are related as

$$\frac{1}{\tau_i} = \frac{1}{\tau_m} + \frac{1}{\tau_e} \tag{8.53}$$

Equations (8.49)–(8.53) assume isotropic tumbling of the protein, and that internal fluctuations of the NH vectors are independent of those due to molecular tumbling. $G_0(\tau)$ is the familiar monoexponentially decaying autocorrelation function for overall molecular tumbling (Abragam, 1961). τ_m is the overall rotational correlation time for molecular tumbling, and therefore pertains to all residues. $G_i(\tau)$ is the autocorrelation function for the internal motions of a given NH vector with respect to a molecular-fixed frame. The parameters S^2 and τ_e are adjusted along with τ_m to fit the observed relaxation data.

S^2 is the generalized order parameter and is the limiting value of $G_i(\tau)$ as τ goes to ∞. S^2 is interpreted as describing the amount of spatial freedom the NH vector has as a result of internal motion. If there are no restrictions on the internal motion, then S^2 is 0, and if there is no internal motion, S^2 is 1.0. In the limit that S^2 is unity, $J(\omega)$ reduces to the familiar Lorentzian distribution for the rotational diffusion of a rigid, spherical tumbling molecule (Abragam, 1961). The τ_e values cannot be directly equated to microscopic time constants for the internal motion, since they generally depend on both the rate and amplitude of the internal motion (Lipari and Szabo, 1982a,b). Therefore, a physical interpretation of the τ_e values requires a more specific model of motion, such as the 'wobbling in a cone' approach (Woessner, 1962; Kinoshita *et al.*, 1977; Richarz *et al.*, 1980). In this model, the NH vector is assumed to diffuse in a cone of semiangle α, which is related to $\sqrt{S^2}$ through the relation (Lipari and Szabo, 1982a,b)

$$S = \frac{1}{2} (\cos \alpha)(1 + \cos \alpha) \qquad (8.54)$$

Dellwo and Wand (1989) applied this model-free approach to studies of motions in cyclosporin A in chloroform solution. T_1 and heteronuclear NOEs were measured for most of the carbon resonances at two field strengths. For the C^α carbons, a variation in S^2 between 0.91 and 0.24 was found. The data were also interpreted in the framework of the diffusion in a cone model. The same authors have also carried out a detailed study of the mobility of ^{15}N-enriched ubiquitin (Schneider *et al.*, 1992). They find no correlation of order parameters with involvement in regular secondary structures. On the other hand, a correlation of high order parameters with involvement in hydrogen bonds of the NH or the CO of the peptide group is claimed. Furthermore, it is claimed that molecular packing is an important factor determining the internal mobility as monitored by these relaxation studies.

Extension of the Model-free Approach

Studies of relaxation parameters in the proteins staphylococcal nuclease and interleukin-1β (Clore *et al.*, 1990a,b) have indicated that for some residues the fit of the model-free parameters is very poor. This was found in the following way: the model-free parameters were determined from measurements of T_1 and T_2 at different field strengths. The model-free parameters obtained predict a much larger size of the heteronuclear NOE (larger reduction of the peak intensities) than observed experimentally. Thus, an extension of the model-free approach was made so that the spectral density function can be written as a sum of three Lorentzian lines. With this extension all experimental data could be fitted. It should be mentioned that the experimental observation that the heteronuclear NOEs are smaller than expected on the basis of the simple model-free approach is related to the effect discussed above for the protein eglin c which gives rise to an increase in the spectral density function at high frequencies. Here the number of the parameters in the model were increased; in the method of spectral density mapping, it appears as an increase in the spectral density function.

Glucose Permease IIA

Extensive studies of the relaxation properties were carried out for the *Bacillus subtilis* glucose permease IIA domain (Stone *et al.*, 1992). Quite a significant variation of mobility was found in this protein as expressed in S^2 and τ_e. All regions of higher mobility were located at the termini, surface loops or regions of irregular secondary structure.

Calbindin D_{9k}

Similar studies were carried out on the calcium-loaded protein calbindin D_{9k} (Kördel *et al.*, 1992). Here the extended model-free approach was used to analyse the data. Here also the crystallographic B factors and the ⟨rmsd⟩ of structure calculations were compared with S^2. It was found that the termini, the link between the two domains and the two Ca binding sites have higher NMR ⟨rmsd⟩ values. The relaxation studies showed, however, that there is no higher mobility in the metal binding sites, and the higher ⟨rmsd⟩ values are a consequence of fewer distance constraints in these regions.

Calmodulin

Recently, the backbone dynamics of *Drosophila* calmodulin have been studied with ^{15}N relaxation time measurements (Barbato *et al.*, 1992). The protein consists of two globular calcium-binding domains that are connected by a long α helix. The relaxation time experiments have shown that this connecting helix is mobile. The correlation times obtained for the protein are shorter than expected for a protein of this size (6.3 ns and 7.1 ns for the two domains). This indicates that both domains move relatively independently, and the relaxation properties of this protein are similar to those of significantly smaller globular proteins. This fortuitous result resolves questions about the enigmatic performance of triple-resonance experiments developed on this protein as a test case for a large protein. The results and the sensitivity obtained for this protein cannot be generalized for larger proteins without such module mobility.

^1H–^1H Cross-relaxation in Antamanide

An extensive study of proton relaxation was performed recently for the peptide antamanide (Brüschweiler *et al.*, 1992). Experimental results were compared with predictions from molecular dynamics simulations. In contrast to the studies described above, here the variation of the length and the orientation of the ^1H–^1H connecting vector has to be considered. It was found that the effects of radial and angular motions have generally opposite effects on the relaxation and can be separated to a good approximation. The benefit of measuring NOE and ROE effects was also discussed.

Studies of Slow Events

The relaxation studies described so far can characterize essentially internal motions that are faster than the overall tumbling of the molecule. An interesting approach to investigate slow events has been proposed by Deverell *et al.* (1970). This technique measures essentially $T_{1\rho}$, and the strength of the spin lock field is varied. If the frequency of the spin lock field coincides with a slow intramolecular process, a resonance effect can be observed. This has been applied to the study of the chair-to-chair conversion in cyclohexane. The method should be readily applicable to study slow internal motions in proteins. To our knowledge, no application for a protein has been published.

4 Conclusion

Characterization of internal motions in proteins has recently focused mainly on measurements of relaxation parameters. Since ^{15}N labelling of proteins has become commonly available for proteins with efficient expression systems, studies of mobility based on the model-free approach by Lipari and Szabo (1982a,b) are becoming routine. The information that can be obtained is complementary to structural data from both NMR and X-ray. In this way, motions of domains relative to each other can be analysed, and high variability of certain protein regions as obtained from structure calculations can be identified as being due to mobility or lack of sufficient constraints. The main work horse of relaxation analysis in proteins is the ^{15}N nucleus. ^{13}C relaxation studies have had less impact so far. On the one hand, this is due to the higher cost of carbon labelling (experiments at natural abundance of ^{13}C are too time consuming). On the other hand, relaxation in ^{13}C-enriched proteins is complicated by $^{13}C-^{13}C$ interaction. Selective labelling would be the method of choice to study carbon relaxation. To reach beyond the analysis on the basis of the model-free approach, the method of spectral density mapping has been developed (Peng and Wagner, 1992a,b). It will provide a direct access to spectral density functions, unbiased by mechanic models of the motions or models about the functional dependence of the autocorrelation functions. Another avenue to expand our information about relaxation properties will come from increased studies of cross-correlation effects. Overall, studies of internal mobility of proteins by relaxation time studies are gaining momentum, and it is expected that they will provide new insight into the properties of globular proteins. This will be achieved by combination of the experimental data with results of molecular mechanics simulations.

Acknowledgements

This work was supported by NSF (Grant DMB-9007878) and NIH (Grant GM38608).

References

Abragam, A. (1961). *The Principles of Nuclear Magnetism*. Clarendon Press, Oxford

Allerhand, A., Doddrell, D., Glushko, V., Cochran, D. W., Wenkert, E., Lawson, P. J. and Gurd, F. R. N. (1971). Conformation and segmental motion of native and denatured ribonuclease A in solution. Application of natural-abundance carbon-13 partially relaxed Fourier transform nuclear magnetic resonance. *J. Am. Chem. Soc.*, **93**, 544–546

Barbato, G., Ikura, M., Kay, L. E., Pastor, R. W. and Bax, A. (1992). Backbone dynamics of calmodulin studied by ^{15}N relaxation using inverse detected two-dimensional NMR spectroscopy: The central helix is flexible. *Biochemistry*, **31**, 5269–5278

Bax, A., Ikura, M., Kay, L. E., Torchia, D. A. and Tschudin, R. (1990). Comparison of different modes of two-dimensional reverse-correlation NMR for the study of proteins. *J. Magn. Reson.*, **86**, 304–318

Billeter, M., Kline, A., Braun, W., Huber, R. and Wüthrich, K. (1989). Comparison of the high-resolution structure of the α-amylase inhibitor tendamistat determined by nuclear magnetic resonance in solution and by X-ray diffraction in single crystals. *J. Mol. Biol.*, **206**, 677–687

Bode, W., Papamokos, E. and Musil, D. (1987). The high-resolution X-ray crystal structure of the complex formed between subtilisin Carlsberg and eglin c, and elastase inhibitor from the leech *Hirudo medicinalis*. *Eur. J. Biochem.*, **166**, 673–692

Boyd, J., Hommel, U. and Campbell, I. D. (1990). Influence of cross-correlation between dipolar and anisotropic chemical shift relaxation mechanisms upon longitudinal relaxation rates of ^{15}N in macromolecules. *Chem. Phys. Lett.*, **175**, 477–482

Brüschweiler, R. and Ernst, R. R. (1992). Molecular dynamics monitored by cross-correlated cross relaxation of spins quantized along orthogonal axes. *J. Chem. Phys.*, **96**, 1758–1766

Brüschweiler, R., Roux, B., Blackledge, M., Griesinger, C., Karplus, M. and Ernst, R. R. (1992). Influence of rapid intramolecular motion on NMR cross-relaxation rates. A molecular dynamics study of antamanide in solution. *J. Am. Chem. Soc.*, **114**, 2289–2302

Bystrov, V. F. (1976). Spin–spin coupling and the conformational states of peptide systems. *Prog. NMR Spectrosc.*, **10**, 41–81

Campbell, I. D., Dobson, C. M., Moore, G. R., Perkins, S. J. and Williams, R. J. P. (1976). Temperature dependent molecular motion of a tyrosine residue of ferrocytochrome C. *FEBS Lett.*, **70**, 96–100

Canet, D. (1989). Construction, evolution and detection of magnetization modes designed for treating longitudinal relaxation of weakly coupled spin 1/2 systems with magnetic equivalence. *Prog. NMR Spectrosc.*, **21**, 237–291

Carr, H. Y. and Purcell, E. M. (1954). Effects of diffusion on free precession in nuclear magnetic resonance experiments. *Phys. Rev.*, **94**, 630–638

254 *NMR of Proteins*

Clore, G. M., Driscoll, P. C., Wingfield, P. T. and Gronenborn, A. (1990a). Analysis of the backbone dynamics of interleukin-1β using two-dimensional inverse detected heteronuclear ^{15}N–^{1}H NMR spectroscopy. *Biochemistry*, **29**, 7387–7401

Clore, G. M. and Gronenborn, A. M. (1991). Comparison of the solution nuclear magnetic resonance and crystal structure of interleukin-8. Possible implications for the mechanism of receptor binding. *J. Mol. Biol.*, **217**, 611–620

Clore, G. M., Szabo, A., Bax, A., Kay, L. E., Driscoll, P. C. and Gronenborn, A. (1990b). Deviations from the simple two-parameter model-free approach to the interpretation of nitrogen-15 nuclear magnetic relaxation of proteins. *J. Am. Chem. Soc.*, **112**, 4989–4991

Debye, P. (1914). Interferenz von Röntgenstrahlen und Wärmebewegung. *Ann. Phys.*, **43**, 49–95

Dellwo, M. J. and Wand, A. J. (1989). Model-independent and model-dependent analysis of the global and internal dynamics of cyclosporin A. *J. Am. Chem. Soc.*, **111**, 4571–4578

Deverell, C., Morgan, R. E. and Strange, J. H. (1970). Studies of chemical exchange by nuclear magnetic relaxation in the rotating frame. *Mol. Phys.*, **18**, 553–559

Ernst, R. R., Bodenhausen, G. and Wokaun, A. (1987). *Principles of Nuclear Magnetic Resonance in One and Two Dimensions.* Oxford University Press, Oxford

Goldman, M. (1984). Interference effects in the relaxation of a pair of unlike spin-1/2 nuclei. *J. Magn. Reson.*, **60**, 437–452

Goldman, M. (1988). *Quantum Description of High-resolution NMR in Liquids.* Clarendon Press, Oxford

Gros, P., Betzel, C., Dauter, Z., Wilson, K. S. and Hol, W. G. J. (1989). Molecular dynamics refinement of a thermistase–eglin-c complex at 1.98 Å resolution and comparison of two crystal forms that differ in calcium content. *J. Mol. Biol.*, **210**, 347–367

Havel, T. F. (1991). An evaluation of computational strategies for use in the determination of protein structure from distance constraints obtained by nuclear magnetic resonance. *Prog. Biophys. Mol. Biol.*, **56**, 43–78

Heinz, D. W., Priestle, J. P., Rahuel, J., Wilson, K. S. and Grütter, M. G. (1991). Changing the inhibitory specificity and function of the proteinase inhibitor eglin c by site-directed mutagenesis: Functional and structural investigation. *J. Mol. Biol.*, **217**, 353–371

Hirs, C. H. W. and Timasheff, S. N. (1986). Structural dynamics and mobility of proteins. *Meth. Enzymol.*, **131**, 283–607

Hiyama, Y., Niu, C., Silverton, J. V., Bavoso, A. and Torchia, D. A. (1988). Determination of ^{13}C chemical shift tensor via ^{15}N–^{2}H dipolar coupling in Boc-glycylglycyl[^{15}N]glycine benzyl ester. *J. Am. Chem. Soc.*, **110**, 2378–2383

Hoch, J. C., Dobson, C. M. and Karplus, M. (1985). Vicinal coupling constants and protein dynamics. *Biochemistry*, **24**, 3831–3841

Hyberts, S. G., Goldberg, M. S., Havel, T. F. and Wagner, G. (1992). The solution structure of eglin c based on measurements of many NOEs and coupling constants and its comparison with X-ray structures. *Protein Sci.* (in press)

Hyberts, S. G., Peng, J. W. and Wagner, G. (1993). Comparison of structure variations and mobility in the NMR solution structure of eglin c with the crystal structure based on coordinates and B-factors. *J. Mol. Biol.*, (submitted)

Kay, L. E., Nicholson, L. K., Delaglio, F., Bax, A. and Torchia, D. A. (1992). The effects of cross-correlation between dipolar and chemical shift anisotropy relaxation mechanisms on the measurement of heteronuclear T_1 and T_2 values in proteins: Pulse sequences for the removal of such effects. *J. Magn. Reson.*, **97**, 359–375

Kay, L. E., Torchia, D. A. and Bax, A. (1989). Backbone dynamics of proteins as studied by ^{15}N inverse detected heteronuclear NMR spectroscopy: Application to staphylococcal nuclease. *Biochemistry*, **28**, 8972–8979

Keiter, E. A. (1986). PhD Thesis, University of Illinois

King, R. and Jardetzky, O. (1978). A general formalism for the analysis of NMR relaxation measurements on systems with multiple degress of freedom. *J. Chem. Phys. Lett.*, **55**, 15–18

King, R., Maas, R., Gassner, M., Nanda, R. K., Conover, W. W. and Jardetzky, O. (1978). Magnetic relaxation analysis of dynamics processes in macromolecules in the pico- to microsecond range. *Biophys. J.*, **6**, 103–117

Kinoshita, K., Kawato, W., Jr. and Ikegami, A. (1977). A theory of fluorescence polarization decay in membranes. *Biophys. J.*, **20**, 289

Kördel, J., Skelton, N. J., Akke, M., Palmer, A. G. and Chazin, W. J. (1992). Backbone dynamics of calcium-loaded calbindin D_{9k} studied by two-dimensional proton-detected ^{15}N NMR spectroscopy. *Biochemistry*, **31**, 4856–4866

Linderstrøm-Lang, K. and Schellman, J. A. (1959). Protein structure and enzyme activity. In Boyer, P. D. (Ed.), *The Enzymes*, Vol. 1. New York, Academic Press, 443–510

Lipari, G. and Szabo A. (1982a). Model-free approach to the interpretation of nuclear magnetic resonance relaxation in macromolecules. 1. Theory and range of validity. *J. Am. Chem. Soc.*, **104**, 4546–4559

Lipari, G. and Szabo A. (1982b). Model-free approach to the interpretation of nuclear magnetic resonance relaxation in macromolecules. 2. Analysis of experimental results. *J. Am. Chem. Soc.*, **104**, 4559–4570

McPhalen, C. A. and James, M. N. C. (1988). Structural comparison of two serine proteinase–protein inhibitor complexes: eglin-c–subtilisin Carlsberg and CI-2–subtilisin Novo. *Biochemistry*, **27**, 6582–6598

Meiboom, S. and Gill, D. (1958). Modified spin-echo method for measuring nuclear relaxation times. *Rev. Sci. Instrum.*, **29**, 688–691

Montelione, G. T., Winkler, M. E., Rauenbuehler, P. and Wagner, G. (1989). Accurate measurements of long-range heteronuclear coupling constants from homonuclear 2D NMR spectra of isotope-enriched proteins. *J. Magn. Reson.*, **82**, 198–204

Morris, G. A. and Freeman, R. (1979). Enhancement of nuclear magnetic resonance signals by polarization transfer. *J. Am. Chem. Soc.*, **101**, 760–762

Nagayama, K. and Wüthrich, K. (1981). Structural interpretation of vicinal proton–proton coupling constants $^{3}J_{H\alpha H\beta}$ in the basic pancreatic trypsin inhibitor measured by two-dimensional *J*-resolved NMR spectroscopy. *Eur. J. Biochem.*, **115**, 653–657

Nirmala, N. R. and Wagner, G. (1988). Measurements of ^{13}C relaxation times in proteins by two-dimensional heteronuclear [1]H–[13]C correlation spectroscopy. *J. Am. Chem. Soc.*, **110**, 7557–7558

Nirmala, N. R. and Wagner, G. (1989). Measurement of ^{13}C spin–spin relaxation times by two-dimensional heteronuclear [1]H–[13]C correlation spectroscopy. *J. Magn. Reson.*, **82**, 659–661

Noggle, J. H. and Schirmer, R. E. (1971). *The Nuclear Overhauser Effect.* Academic Press, New York

Palmer, A. G., III, Skelton, N. J., Chazin, W. J., Wright, P. E. and Rance, M. (1992). Suppression of the effects of cross-correlation between dipolar and anisotropic chemical shift relaxation mechanisms in the measurement of spin–spin relaxation rates. *Mol. Phys.*, **75**, 699–711

Peng, J. W., Thanabal, V. and Wagner, G. (1991a). 2D heteronuclear NMR measurements of spin–lattice relaxation times in the rotating frame of X nuclei in heteronuclear HX spin systems. *J. Magn. Reson.*, **94**, 82–100

Peng, J. W., Thanabal, V. and Wagner, G. (1991b). Improved accuracy of heteronuclear transverse relaxation time measurements in macromolecules: Elimination of antiphase contributions. *J. Magn. Reson.*, **95**, 421–427

Peng, J. W. and Wagner, G. (1992a). Mapping of spectral density functions using heteronuclear NMR relaxation measurements. *J. Magn. Reson.*, **98**, 308–332

Peng, J. W. and Wagner, G. (1992b). Mapping of spectral densities of N–H bond motions in eglin c using heteronuclear relaxation experiments. *Biochemistry* (in press)

Ponder, J. and Richards, F. M. (1987). Tertiary templates for proteins: Use of packing criteria in the enumeration of allowed sequences for different structural classes. *J. Mol. Biol.*, **193**, 775–791

Redfield, A. G. (1965). The theory of relaxation processes. *Adv. Magn. Reson.*, **1**, 1

Ribeiro, A. A., King, R., Restivo, C. and Jardetzky, O. (1980). An approach to the mapping of internal motions in proteins. Analysis of ^{13}C NMR relaxation in the bovine pancreatic trypsin inhibitor. *J. Am. Chem. Soc.*, **102**, 4040–4051

Richards, F. M. (1974). The interpretation of protein structures: Total volume, group volume distributions and packing density. *J. Mol. Biol.*, **82**, 1–14

Richarz, R., Nagayama, K. and Wüthrich, K. (1980). Carbon-13 nuclear magnetic resonance relaxation studies of internal mobility of the polypeptide chain in basic pancreatic trypsin inhibitor and a selectively reduced analogue. *Biochemistry*, **19**, 5189–5196

Schneider, D. M., Dellwo, M. and Wand, A. J. (1992). Fast internal main-chain dynamics of human ubiquitin. *Biochemistry*, **31**, 3645–3652

Solomon, I. (1955). Relaxation processes in a system of two spins. *Phys. Rev.*, **99**, 559–565

Stone, M. J., Fairbrother, W. J., Palmer, A. G., Reizer, J., Saier, M. H. and Wright, P. E. (1992). Backbone dynamics of the *Bacillus subtilis* glucose permease IIA domain determined from ^{15}N relaxation measurements. *Biochemistry*, **31**, 4394–4406

Vold, R. L. and Vold, R. R. (1978). Nuclear magnetic relaxation in coupled spin systems. *Prog. NMR Spectrosc.*, **12**, 79–133

Wagner, G. (1983). Characterization of the distribution of internal motions in the basic pancreatic trypsin inhibitor using a large number of internal NMR probes. *Quart. Rev. Biophys.*, **16**, 1–57

Wagner, G. (1990). NMR investigations of protein structure. *Prog. NMR Spectrosc.*, **22**, 101–139

Wagner, G., Hyberts, S. G. and Havel, T. F. (1992). NMR structure determination in solution: a critique and comparison with X-ray crystallography. *Ann. Rev. Biophys. Biomol. Struct.*, **21**, 167–198

Werbelow, L. G. and Grant, D. M. (1977). Intramolecular dipolar relaxation in multispin systems. In Waugh, J. S. (Ed.), *Advances in Magnetic Resonance*, Vol. 9. Academic Press, San Diego, pp. 189–299

Wittebort, R. J. and Szabo A. (1978). Theory of NMR relaxation in macromolecules: Restricted diffusion and jump models for multiple internal rotations in amino acid side chains. *J. Chem. Phys.*, **69**, 1723–1736

Woessner, D. E. (1962). Spin relaxation processes in a two-proton system undergoing anisotropic reorientation. *J. Chem. Phys.*, **36**, 1–4

Woodward, C. K. and Hilton, B. D. (1980). Hydrogen isotope exchange kinetics of single protons in bovine pancreatic trypsin inhibitor. *Biophys. J.*, **32**, 561–575

Wüthrich, K. (1986). *NMR of Proteins and Nucleic Acids*. Wiley, New York

Wüthrich, K. and Wagner, G. (1975). NMR investigations of the dynamics of the aromatic amino acid residues in the basic pancreatic trypsin inhibitor. *FEBS Lett.*, **50**, 265–268

Wüthrich, K. and Wagner, G. (1978). Internal motions in globular proteins. *Trends Biochem. Sci.*, **3**, 227–230

Zare, R. N. (1988). *Angular Momentum*. Wiley, New York

9

The Folding, Stability and Dynamics of T4 Lysozyme: A Perspective Using Nuclear Magnetic Resonance

D. Eric Anderson, Jirong Lu, Lawrence McIntosh and
Frederick W. Dahlquist

1 Introduction

In general, the primary amino acid sequence of a polypeptide chain is thought to encode all the information necessary to determine the final folded structure of a protein. However, the contributions made by individual amino acid residues to the overall structure, dynamics and stability of the folded protein as well as the role(s) of individual residues in the kinetic pathways between the unfolded and folded states remains less clear. In recent years, there has been a substantial amount of progress in addressing these questions (Alber, 1989; Goldenberg *et al.*, 1989; Lim and Sauer, 1989; Bowie *et al.*, 1990; Dill, 1990; Kim and Baldwin, 1990; Pace *et al.*, 1990; Matthews, 1991). The lysozyme produced by bacteriophage T4 offers an excellent experimental system to address these important questions. In addition, the collaborative research efforts of the Matthews (crystallography, structural bases of protein stability), Schellman (thermodynamics), Hudson (nanosecond dynamics) and our laboratories at the University of Oregon offer a nearly unique environment for the investigation of these issues.

The lysozyme from bacteriophage T4 is a 164 residue (18 700 dalton), single-chain polypeptide. The wild-type protein has two free sulphydryl groups and no disulphide bridges. As shown schematically in Figure 9.1(a), the protein is folded into a bilobed structure centred on a 19 residue α helix (Matthews and Remington, 1974). The protein is primarily composed of helical secondary structure but also contains a small antiparallel β sheet in the N-terminal lobe. Figure 9.1(b) shows the

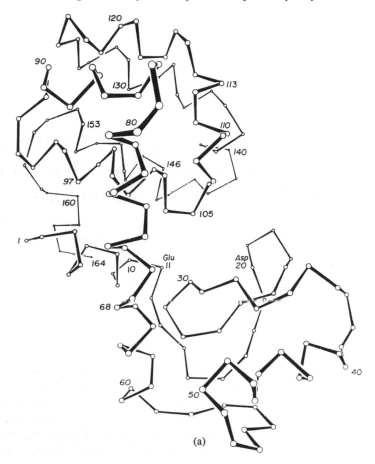

(a)

1   MNIFEMLRIDEGLRLKIYKDTEGYYTIGIGHLLTKSPSLNAAKSELDKAI
          Helix A   β-sheet and turns   Helix B

(b)

5 1   GRNCNGVITKDEAEKLFNEDVDAAVRGILRNAKLKPVYDSLDAVRRCALI
                    Helix C   Helix D   Helix

101   NMVFQMGETGVAGFTNSLRMLQQKRWDEAAVNLAKSRWYNQTPNRAKVI
          E   Helix F   Helix G   Helix H

1 5 1   TTFRTGTWDAYKNL

Figure 9.1 (a) The α carbon backbone of T4 lysozyme as deduced from the refined model derived from X-ray data by Brian Matthews (with permission). (b) The amino acid sequence of wild-type T4 lysozyme. The α-helical regions are indicated with the letters A–H. The region H consists of two α helices separated by Thr141 and Pro142. In addition, there are two distorted α-helical segments corresponding to residues 107–113 and the C terminus

amino acid sequence of the protein with the α helices indicated with the letters A–H. The C-terminal domain contains the first helix (helix A) of the N terminus of the protein.

A wide variety of mutations have been introduced into the protein. These include the replacement of the two cysteine residues to produce a thiol-free version of the wild-type protein with nearly identical thermodynamic stability as the original (Matsumura and Matthews, 1989), introduction of new disulphides at several positions which usually increase the thermodynamic stability of the protein (Matsumura *et al.*, 1989), and a host of systematic changes introduced at many positions to investigate hydrophobic (Matsumura *et al.*, 1988), electrostatic (Nicholson *et al.*, 1988), core packing (Karpusas *et al.*, 1989) and hydrogen bonding interactions (Alber *et al.*, 1987). The great majority of these mutant proteins have been crystallized and their crystal structures have been determined in the Matthews laboratory. To date, the structures of about 150 T4 lysozyme variants have been determined. Some of these involve different crystal packing arrangements (Faber and Matthews, 1990; Dixon *et al.*, 1992) which have allowed seven distinct views of the protein to be obtained. In addition, Tony Poteete's laboratory has recently developed a genetic system in which T4 lysozyme is used to complement a lysozyme-deficient P22 bacteriophage (Rennell *et al.*, 1991). Using this system, Poteete and co-workers have described the complementation properties of a huge collection of lysozyme variants in which residues 2–164 have each been replaced by 13 different residues using nonsense suppressors. This provides a qualitative indication of the activity and stability of these variants which can be very useful in the design of new lysozyme variants. Finally, we developed a very efficient bacterial expression system which allows the isolation of about 50 mg of pure protein from a litre of minimal medium (Muchmore *et al.*, 1990). To date, this expression system has produced about 300 lysozyme variants in quantities suitable for physical studies. Thus, the T4 lysozyme system offers an excellent opportunity to examine the interrelationships of amino acid sequence, structure, stability and dynamics.

We have recently published the assignments of the proton and ^{15}N magnetic resonance spectra of the backbone and about 25% of the side-chain protons of wild-type T4 lysozyme (McIntosh *et al.*, 1990) and these are shown in Figure 9.2. In this case the protein was expressed in bacterial cells grown with $^{15}N(99\%)$ ammonium sulphate as the sole source of nitrogen. The figure shows an HSMQC spectrum (Bax *et al.*, 1983; Zuiderweg, 1990) in which each cross-peak represents an $^{15}N–H$ directly bonded pair in the protein. The ^{15}N and H coordinates of each cross-peak correspond to the chemical shifts of the ^{15}N and proton environments of the pair. The spectrum was assigned in two steps. First,

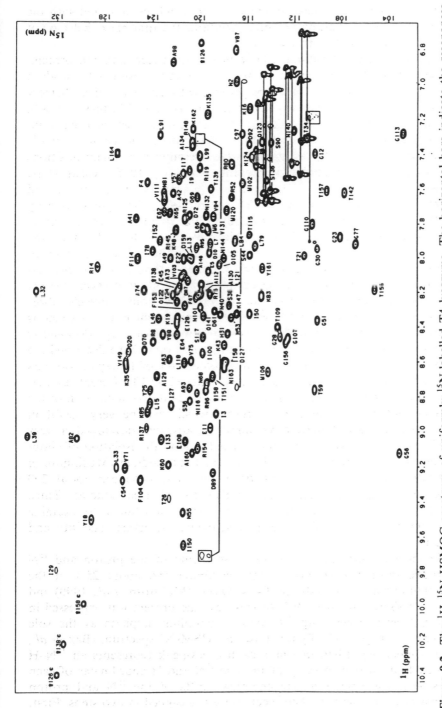

Figure 9.2 The ^{1}H–^{15}N HSMQC spectrum of uniformly ^{15}N-labelled T4 lysozyme. The horizontal bars indicate the resonances of the primary amides of asparagine and glutamine residues. From McIntosh *et al.* (1990)

the amino acid type of each cross-peak was determined by comparison of the uniformly labelled spectrum shown in Figure 9.2 to the spectra obtained when individual ^{15}N-labelled amino acids were biosynthetically incorporated into the protein. The residues were then assigned to particular amino acid residues by establishing the neighbouring amino acid type using a combination of isotope edited NOESY experiments (McIntosh et al., 1990) and combined ^{15}N and ^{13}C labelling experiments (Griffey et al., 1986) designed to label specific peptide bonds along the backbone. In a few cases, mutational substitutions were used to resolve particularly vexing ambiguities in the assignments (see below). These assignments of the backbone resonances serve as the basis for much of the work discussed below.

This chapter describes some of the recent work from our laboratory which is directed towards understanding the dynamics of the folded state, and the dynamics of the path used by the protein while it folds, and the roles of electrostatic interactions in stabilizing the folded state. Much of this work takes advantage of some of the recent advances in high-resolution nuclear magnetic resonance as applied to biological systems. When combined with site-directed mutation, thermodynamic analysis and rapid X-ray diffraction analysis of mutant structures, questions in these areas can often be addressed with exquisite sensitivity at atomic resolution.

2 Dynamics of the Folded State

Backbone Amide Hydrogen Exchange

T4 lysozyme contains 160 exchangeable backbone amide protons. To measure the hydrogen exchange rates of these amides, we utilized the two-dimensional ^{1}H–^{15}N correlation experiment shown in Figure 9.2 to study samples of uniformly ^{15}N-enriched T4 lysozyme. This experiment was useful for three reasons: (1) the ^{1}H–^{15}N spectrum of ^{15}N-labelled T4 lysozyme is a complete 'fingerprint' of the backbone of the protein, with most amides yielding a resolved crosspeak; (2) the experiment provides a direct measure of the relative proton occupancy of each amide residue and thus a site-specific measure of the proton-to-deuterium exchange reactions of T4 lysozyme; (3) the experiment is highly sensitive, such that complete two-dimensional spectra of uniformly ^{15}N-enriched T4 lysozyme (1–4 mM) can be recorded in 28 min or less. The hydrogen exchange kinetics of T4 lysozyme were measured by sequentially recording two-dimensional ^{1}H–^{15}N spectra of the protein after transfer into D$_2$O buffer. The exchange rate of each individual amide was determined from the time dependence of the decrease in the volume of the cross-peak corresponding to that amide ^1H–^{15}N pair.

A disadvantage of this 'real-time' approach is that exchange rates can only be conveniently determined for amides with exchange half-lives ranging from about an hour to a few days or weeks. To expand this range, we have measured the exchange rates of the individual amides in T4 lysozyme at five pH* (meter reading) values, ranging from 3.53 to 7.76 (where pH* is the uncorrected pH meter reading in D_2O). The exchange of most backbone amides is base-catalysed in this pH range, so that about a 10^4-fold range of base-catalysed rate constants can be monitored in the same time of exposure to the exchange conditions. The protein is in a stable, folded conformation under these conditions. In order to compare the exchange data measured at these five different pH*s, we have normalized the observed exchange rates for each amide to the expected rates for model peptides at each pH*. These protection factors, calculated for each amide using the calibrated exchange kinetics of poly-D,L-alanine with correction for intrinsic sequence effects (Molday *et al.*, 1972), is shown in Figure 9.3, using the geometric mean of the protection factors measured at each pH. If the exchange rates of each amide were dependent upon the hydronium and hydroxide concentration in a strictly first-order fashion and if the pH^*_{min} for exchange is similar to that of the reference peptide, then the protection factors measured for a given amide at several different pH*s should be constant, within the error of the measurements. However, the amide hydrogen exchange kinetics of proteins in general show a complex dependence upon pH. The protection factors of the amides in T4 lysozyme do vary somewhat with the sample pH*. The data presented in Figure 9.3 should be considered as an average representation of the amide exchange kinetics of T4 lysozyme at 20 °C over a range of pH* conditions. The observable protection factors span from a lower limit of approximately 10^2 to an upper limit of approximately 10^8. Outside of this range, the exchange rates are either too fast or too slow, respectively, to be measured by this experimental approach.

Some of the most protected amides correspond to helix E which includes residues 93–106. The most slowly exchanging of these correspond to residues (95–106) contained in the most buried helix in the structure. These residues form the hydrophobic core of the protein. We have not been able to quantify the exchange rates for these residues. The minimum protection factor of 10^8 is only about 10^2–10^4 times less than that expected from the estimated stability of the folded state under these conditions. It is possible that exchange of these residues proceeds via the global unfolding of the entire protein. This view is supported by the observation that the half-time for exchange of the amides is about 5 h at pH 11. This high pH value results in a

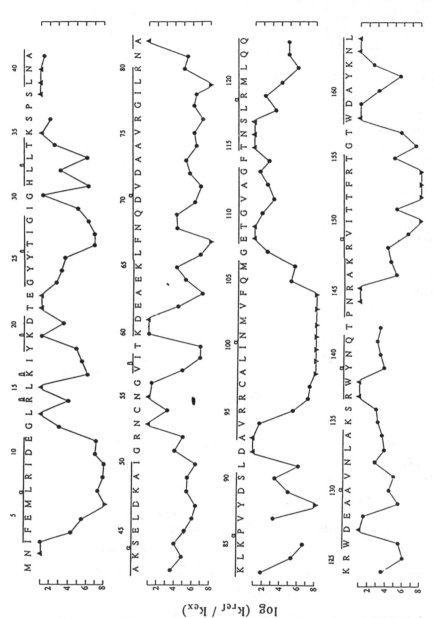

Figure 9.3 The hydrogen exchange properties of the backbone amides of T4 lysozyme determined by real-time exchange at several pH values expressed as the logarithm of the bimolecular base-catalysed rate constant k_{ex} normalized to the calculated rate constant for an exposed residue k_{ref} using the method of Molday *et al.* (1972). The ▼ symbol indicates amides whose exchange rates were too fast to measure by this method. The ▲ symbol indicates amides whose exchange rates were too fast to measure by this method

destabilization of the folded state by 100–1000-fold as compared with values near neutrality.

The bulk of the amides in T4 lysozyme have measurable exchange rates (as indicated by filled circles in Figure 9.3). Since these residues exchange more quickly than the slowest of helix E, the exchange involves something other than global unfolding of the protein.

Hydrogen Exchange of Mutant Proteins

We have used the hydrogen exchange properties of these amides as a qualitative indicator of the effect of mutation on the structure and dynamics of the protein. We chose a series of four mutations which each lower the thermodynamic stability of the folded state by 3–5 kcal/mol. These include a surface mutation which breaks a critical hydrogen bond, T157I, two mutations which change the packing of the hydrophobic core of the C-terminal domain by the introduction of a bulky side-chain, A98V and A146T, and one mutation which changes the hydrophobic core of the N-terminal domain, Y25G. In each case,

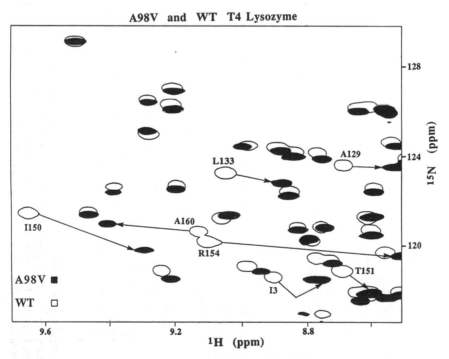

Figure 9.4 A comparison of a portion of the ^{1}H–^{15}N HSMQC spectra of the wild type and the mutant A98V at pH 5.6 and 20 °C. Arrows indicate the largest shifts observed. Smaller changes in the spectra are indicated as overlapping or partially overlapping cross-peaks

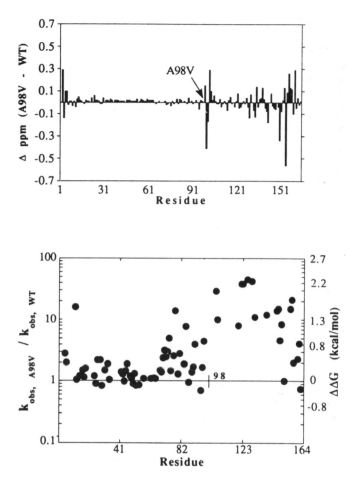

Figure 9.5(a) A comparison of the changes in proton chemical shift (above) and ratios of base-catalysed hydrogen exchange (below) for the backbone amides for the mutant A98V compared with wild type

the ^1H–^{15}N spectra showed substantial shifts in cross-peak resonance positions. The spectra of the mutants were reassigned by a combination of ^{15}N edited NOESY and COSY spectra, selective labelling, and HSMQC-NOESY spectra in two dimensions. Following assignment of the spectra, the hydrogen exchange properties of each mutant were determined by the methods outlined above. The results of these experiments are summarized in Figures 9.4–9.6. Figure 9.4 shows a comparison of a region of the ^1H–^{15}N spectra of the wild type and the mutant A98V in which alanine at position 98 has been replaced by valine. As can be seen, there are extensive changes in the positions of several resonances. This mutation introduces a bulkier group into the hydrophobic core of

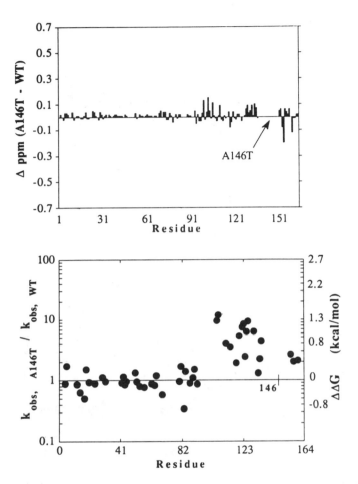

Figure 9.5(b)   A comparison of the changes in proton chemical shift (above) and ratios of base-catalysed hydrogen exchange (below) for the backbone amides for the mutant A146T compared with wild type

the C-terminal domain and there are fairly extensive changes in packing to accommodate this change.

   An overview of these effects is shown in Figure 9.5(a and b), which presents the magnitude of the changes in proton chemical shifts and the ratios of the observed hydrogen exchange rates for each measured residue of the mutants A98V and A146T as compared with wild-type. These figures show two important effects. Both chemical shift and hydrogen exchange are very sensitive to these changes. Second, it appears that the effects of mutation can propagate far from the site of mutation. However, it is also clear that these effects seem to be contained to the domain containing the mutation. Note that the residues

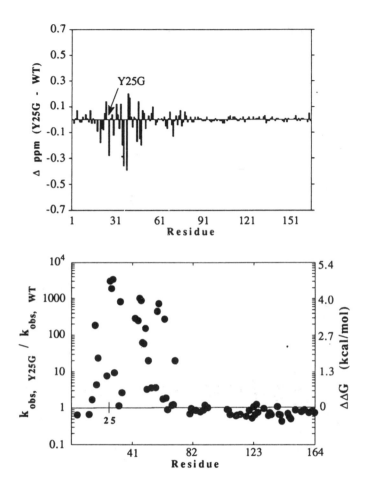

Figure 9.6(a) A comparison of the changes in proton chemical shift (above) and ratios of base-catalysed hydrogen exchange (below) for the backbone amides for the mutant Y25G compared with wild type

of helix A are perturbed along with the other members of the C-terminal domain. Thus, the effects propagate through space as opposed to along the sequence. This conclusion is confirmed by the results shown in Figure 9.6(a and b). This shows the magnitude of the changes in proton chemical shifts and the ratios of the observed hydrogen exchange rates for each measured residue of the mutants Y25G and T157I as compared with wild-type. In the case of Y25G, there are extensive changes in chemical shift and hydrogen exchange as the result of complete removal of the aromatic side-chain from the hydrophobic core of the N-terminal domain but these changes do not appear to propagate beyond the N-terminal domain. The surface mutation T157I serves as a control for

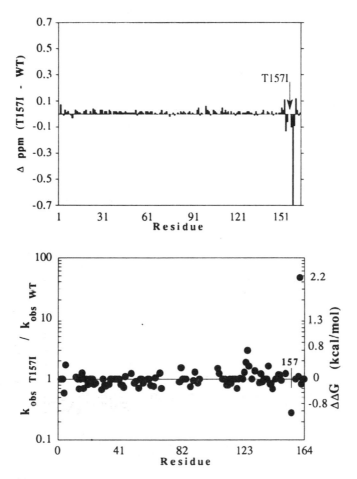

Figure 9.6(b) A comparison of the changes in proton chemical shift (above) and ratios of base-catalysed hydrogen exchange (below) for the backbone amides for the mutant T157I compared with wild type. The upfield shift of the amide proton resonance of Asp159 upon loss of the wild-type hydrogen bond from the side-chain hydroxyl of Thr157 is 0.74 ppm. This shift is truncated in the figure, owing to the scale chosen

these studies. X-ray crystallography of this mutant has shown that structural changes are very localized to the site of mutation. Both chemical shift and hydrogen exchange show very localized changes near the site of mutation.

Our analyses of these results are not complete, but the striking similarity in the changes observed in shift and hydrogen exchange reinforce the view of proteins as plastic entities which are able to accommodate amino acid changes by propagating structural changes.

These experiments also reinforce the view that T4 lysozyme is organized in two domains in both a structural and a dynamic sense.

Core Packing Mutants and the Dynamic Access to the Hydrophobic Core

Recently, the Matthews laboratory has discovered that several mutations, in which bulky side-chains in the hydrophobic core are replaced by smaller side-chains, can create or expand apparent cavities in the core (Karpusas *et al.*, 1989; Eriksson *et al.*, 1992a). In the case of the mutation L99A, replacement of the leucine side-chain by a methyl group expands an existing cavity in the core of the C-terminal domain. The mutant protein is substantially destabilized when compared with its wild type. They found that some of these mutant proteins can bind hydrophobic compounds such as benzene in the apparent cavities (Eriksson *et al.*, 1992b). This binding has been detected as a benzene-dependent stabilization of the mutant proteins towards reversible unfolding and by the observation of bound benzene molecules in the crystal structures of these mutant proteins in the presence of benzene.

We have examined the effects of these apparent cavity mutants on the dynamic properties of T4 lysozyme in solution. Enoch Baldwin, working in the Matthews laboratory, has constructed a series of alanine substitution mutants for residues 94–106. These mutant proteins are especially interesting because they involve changes at some of the same residues which are protected early in the folding pathway and may help in defining the roles of these residues in the folding pathway (see below). The mutants that bind benzene in the core include L99A, F104A, M102A and two substitutions outside the E helix, L133G and F153A. The mutations L99A, M102A and N101A all create apparent cavities that have no static route allowing access to the cavity by a molecule of the bulk solvent.

Are the cavities empty in the absence of small hydrophobic molecules? There are few, if any, hydrogen-bonding partners in the cavity to stabilize water in the cavity. The X-ray crystallographic analysis of most of these mutants does not show the presence of water in these cavities, but it is possible that the water is disordered and is not detected for that reason. The presence or absence of solvent in the cavities gives rise to different views of energetic changes associated with these types of mutations. In one view, the empty cavity results in decreased van der Waals interactions and that accounts for larger than expected destabilization of the folded state. In another view, the presence of solvent in the cavity causes a larger than anticipated destabilization due to unfavourable contacts between the remaining hydrophobic groups of the core and solvent molecules.

We were able to monitor [^{13}C]-benzene binding at sub-stoichiometric levels to the mutant protein L99A, using ^{13}C editing of the protein–ligand complex. The proton resonance of protein bound benzene is shifted by about 0.2 ppm upfield and broadened by about 10 Hz as compared with benzene in aqueous solution. As more benzene is added, the resonance shifts towards the position of free benzene and sharpens. The benzene concentration dependence of the shift suggests a dissociation constant of about 1mM for benzene to L99A. Such effects were not observed when benzene was added to the wild-type protein. These observations indicate that benzene exchanges in and out of its binding site in the core of L99A at a rate of ~ 10^3 s^{-1} or faster.

Examination of the crystal structure of L99A and the benzene–L99A complex shows that there is 7 Å or more between the surface and the bound benzene (Eriksson *et al.*, 1992a,b). There must be a fluctuation in the structure of about 7 Å (also the van der Waals thickness of benzene) to allow the benzene to gain access to the interior of the protein. The apparent dissociation constant of 1 mM for benzene binding combined with a 1 millisecond exchange lifetime places the bimolecular rate constant for benzene binding to be of the order of 10^6 M^{-1} s^{-1} or about two orders of magnitude slower than diffusion-limited. This suggests that benzene can gain access to the core rather readily despite the apparent lack of a sterically allowed path.

What is the nature of the fluctuations that allow such access? We hope to examine the mechanism by which a molecule such as benzene gains access to the core. There are a few protein resonances which appear to show broadenings in the presence of subsaturating concentrations of benzene. It is likely that exchange broadening of these resonances may allow us to determine the binding kinetics for benzene to the core under a variety of conditions and in a variety of mutant proteins. Coupled with specific structural studies these approaches may allow the mechanistic details of the binding event to be determined.

Interestingly, the hydrogen exchange properties of the residues that line the cavity remain quite slow. The amides of helix E are still the most protected from exchange. Their rate of exchange increases about 1000 fold in L99A as compared with wild-type. This increase in exchange rate agrees well with the decrease in stability caused by the L99A substitution and is consistent with the view that exchange of the amides in helix E may be limited by the global unfolding of the molecule. These results suggest that benzene gains access to the cavity in L99A by a qualitatively different route from the water and/or hydroxide needed for base-catalysed amide exchange.

These observations raise a number of interesting questions that we are now attempting to answer. What is the pathway of benzene access to the core? Is disordered water present in the cavity? In addition to replacements of large hydrophobic sidechains in the core that generate apparent cavities, there are mutants in which the surrounding sidechains collapse to fill in the space that had been previously occupied by the original side-chain. Do these mutations have qualitatively different dynamics than those that appear to create cavities?

3 Folding Pathway

We were particularly interested in investigating the folding kinetics of T4 lysozyme under conditions which strongly favour the native state (Lu and Dahlquist, 1992). Under these conditions the folding reaction is fast and there is an enhanced probability of accumulating folding intermediates. For these studies, the two cysteine residues of T4 lysozyme (Cys54/Cys97) were replaced by threonine and alanine respectively to increase the reversibility of folding and unfolding at high pH values. Both the stability and structure of this protein, referred to as WT*, are similar to those of wild-type T4 lysozyme.

Proline Isomerization

Refolding of WT* T4 lysozyme was first monitored by the change in the fluorescence of its three tryptophan residues (Trp126, Trp138 and Trp158), using stopped-flow methods. The protein was denatured in 3 M urea at pH 2.0 and refolding was initiated by a combined pH and urea concentration jump to a final concentration of 1.5 M urea at pH 6.8. As shown in Figure 9.7, the fluorescence-detected refolding time-course consists of two kinetic phases. The fast phase has a relaxation time of 120 ms, accounting for 90% of the total amplitude, and the slow phase has a relaxation time of 650 ms, accounting for 10% of the total amplitude. The low amplitude of the slower phase was suggestive of a potential role for proline isomerization in this phase. T4 lysozyme contains three proline residues, all with a *trans* conformation in the native crystal structure. These three proline residues of WT* were replaced by alanine by site-specific mutation by Hale Nicholson. The resulting protein (P37A, P86A, P143A) has similar thermodynamic stability to the original and is fully active. The fluorescence-detected refolding time-course of this protein lacking proline residues showed only the faster phase, as seen in Figure 9.7. We conclude that one or more proline residues are responsible for the slow phase of refolding under these conditions.

Role of His31 in the Folding Pathway

The pH dependence of the folding rate was monitored by fluorescence spectroscopy and by varying the pH of the refolding conditions. Under

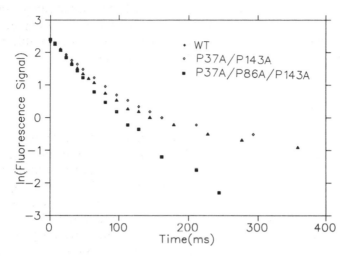

Figure 9.7 The fluorescence detected time-course of refolding in 1.5 M urea and 110 mM potassium phosphate, 12.5 mM potassium chloride, pH 6.8 and 23 °C, for the WT* and various proline to alanine replacements in T4 lysozyme. The data are presented as the natural logarithm of the change in fluorescence as a function of time. The linear relationship in the proline-free version of the protein indicates that proline isomerization plays a role in the low-amplitude, slower relaxation seen in the other versions of the protein that contain proline. The proline-free version of T4 lysozyme was provided by Hale Nicholson

these experimental conditions, the native, folded state is strongly favoured, so the observed relaxation time reflects the rate constant for folding. As shown in Figure 9.8, the observed first-order refolding rate is essentially pH independent in the range of pH values from 6 to 9. This is the range of pH values that spans the change in the pK_a value of His31 from 6.8 in the unfolded state to 9.1 in the folded state (Anderson *et al.*, 1990). The high pK_a value for His31 in the folded state is due to a strong salt bridge formed with Asp70 in the native state (see below). A formal interpretation of the lack of pH dependence in the folding rate constant over this range is that the pK_a value of His31 has the same value in the transition state as in the unfolded starting conditions. We therefore conclude that the native salt bridge formed between His31 and Asp70 has not yet been formed in the transition state for folding as judged by the tryptophan fluorescence signal. This simple experiment provides a direct method to probe the environment around ionizable groups in the transition state between the folded and unfolded states.

Amide Trapping Studies

Recently, two-dimensional NMR techniques combined with quenched flow measurements of hydrogen–deuterium exchange of amide protons

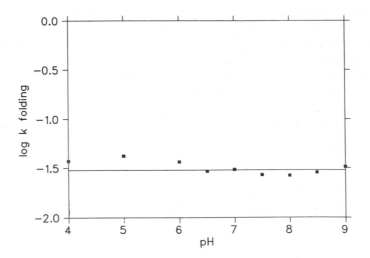

Figure 9.8 The pH dependence of the refolding rate for WT* T4 lysozyme at 23 °C and in 4.0 M urea

have been used to detect and characterize kinetic intermediates observed during protein folding. This approach can also provide some structural information about intermediates on the folding pathway. This method has been applied to several proteins including RNase A (Udgaonkar and Baldwin, 1988, 1990), egg-white lysozyme (Miranker *et al.*, 1991), cytochrome-c (Roder *et al.*, 1988), ubiquitin (Briggs and Roder, 1992) and barnase (Bycroft *et al.*, 1990).

Measurement of hydrogen–deuterium exchange rates of amide protons is a sensitive probe for the presence of structure in a fully or partially folded protein. The formation of hydrogen bonds such as those seen in α helices or β sheets dramatically reduces hydrogen–deuterium exchange rates of amide protons. In a typical pulse hydrogen exchange experiment (see Robertson and Baldwin, 1991), refolding of a deuterated protein is initiated in D_2O and a 'snapshot' of the protein is taken at different times after refolding by subjecting the sample to a proton labelling pulse of H_2O at high pH to rapidly label those amides which remain exposed at that particular time. After allowing the protein to refold under conditions where hydrogen exchange is slow, two-dimensional ^1H–^{15}N NMR is then used to determine which amide protons are protected at various times during the folding process (Lu and Dahlquist, 1992).

As a prerequisite to these studies, we characterized the hydrogen exchange properties of the unfolded state of WT* T4 lysozyme. In this experiment, the protein was unfolded in 8.0 M urea at pH* 7.5 and 20 °C in D_2O and allowed to stand to allow the amides to become fully deuterated. Exchange was initiated by diluting the solution 1:2 with the same buffer in H$_2$O. After various times of exchange in 67% H$_2$O, exchange was quenched by diluting the sample tenfold by injection into pH 3 buffer containing no urea. Under these conditions, unfolding and refolding is rapid compared to exchange. After folding was complete, the extent of deuterium labelling during the various times of incubation of the unfolded protein was determined by volume integration of the corresponding ^{1}H–^{15}N cross-peaks of the 80 individual amide sites used as probes. The observed time-courses for replacement of deuterium by hydrogen were all adequately described a single first-order decay. The results are summarized in Figure 9.9 as a plot of the first-order rate constant for exchange versus residue position for each of the 80 amide residues. As can be seen, there is very little difference in exchange rate constant as a function of position and the observed rates are essentially equal to those expected from unstructured model compounds. These data suggest that the unfolded state produced under these conditions has very little structure and that essentially all the amides are completely exposed to solvent. Similar slower exchange of valine and isoleucine residues in the unfolded state of RNase has been observed by Robertson and Baldwin (1991). Using model peptides, they have shown that fully exposed isoleucine residues have intrinsically slower hydrogen exchange than other residues. Using the results shown in Figure 9.9, we conclude that the unfolded state produced in 8 M urea at pH 7.5 has no significant

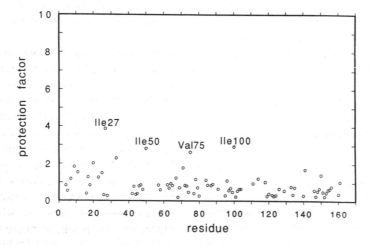

Figure 9.9   The amide exchange properties of the unfolded state of T4 lysozyme expressed as the calculated protection factor versus position at pH*2.4, 30 °C and in 3.0 M urea. The protection factors are calculated as the ratio of the observed first order rate constant for exchange for each residue as compared with a theoretical value which accounts for the pH* of the sample, the presence of urea and the effects of adjacent residues (see Lu and Dahlquist, 1992)

amount of stable secondary structure. Similar results have been obtained in 3.0 M urea at pH 2.7.

Using hydrogen exchange measurements of the unfolded state, we have shown that there is little residual structure present under the conditions we usually employ to unfold the protein. Fink (Goto *et al.*, 1990) has suggested that there may be conditions where T4 lysozyme exists as a 'molten globule state'. These conditions are at low pH, moderate urea concentrations and substantially higher salt concentration than we typically use. We have observed some changes in the NMR spectra of the protein in conditions which seem to correspond to the fully unfolded state. In one experiment, the protein was labelled with ^{15}N-phenylalanine and the ^{1}H–^{15}N HSMQC spectra were recorded at low pH and high urea, which gives a fully unfolded protein by most criteria. The spectra showed three resolved resonances in the ratio 1:1:3 for the five phenylalanine residues. This was not unexpected, since chemical shift of amides depends to some extent on the immediate near-neighbours in the sequence. However, as the temperature was raised, the resonances moved closer together. This suggests that there is fast exchange between a population of unfolded states and that raising the temperature may decrease the differences in average environment experienced by the different phenylalanine residues in the unfolded state.

During Folding Selected Regions of T4 Lysozyme are Protected from Solvent Exchange in 10 ms or Less

The time-scale for the pulsed hydrogen exchange-labelling experiments was chosen so that the rapid ($\tau = 120$ ms) phase of the refolding seen by fluorescence measurements (Figure 9.8) was monitored. Pulse-labelling experiments with WT* were performed at seven different refolding times between 8 ms and 400 ms. Unfolded protein in which the amide protons were exchanged for deuterium was allowed to refold in 50% D$_2$O at pH 5.1 for a variable length of time (between 8 ms and 400 ms) before it was exposed to a 41 ms proton-labelling pulse at pH 9.0. An exposed amide should exchange in about 1 ms under these conditions. Proton labelling was terminated at the end of the pulse by ejecting sample into quench buffer to bring the pH to 3.3. At this pH, amide proton exchange is halted but refolding can proceed to completion. After refolding was completed the protein samples were concentrated. The exchange was quantified from HSMQC ^{1}H–^{15}N spectra (Zuiderweg, 1990). A control experiment using the fully folded protein was performed to test the resistance of individual amide protons in native T4 lysozyme to hydrogen-exchange labelling. In this experiment the fully folded protein was subjected to the same pH 9.0 proton labelling pulse, quench and concentration followed by observation of the HSMQC spectrum. This experiment confirmed that proton label was completely excluded from all slowly exchanging amide sites which were used as probes for this study.

There is a substantial difference in the time-course of proton occupancy for different sites. This is seen more clearly in Figure 9.10 as a plot of the relative proton occupancy as a function of refolding time for the amides of Ile27, Asn68, Ile100 and Ala129. These four residues are representative of the range of behaviours seen for all probe residues. At the first time point, 8 ms, the various residues have become protected against exchange to different degree. Asn68 and Ala129 have properties similar to the majority of the probes. At 8 ms these residues are almost completely accessible to solvent and are strongly labelled by protons in the 40 ms pulse of pH 9.0 used. These residues become slowly protected against exchange with an apparent time constant of about 170 ms. They reflect regions which have not formed a structure capable of protecting those amides from exchange. The 170 ms folding of these residues is equal to the fast folding phase detected by fluorescence and reflects the 'global' formation of native structure.

Contrasting behaviour is represented by Ile100. The amides of this residue are completely protected against exchange by the high pH pulse, even at the earliest times. This result reflects the formation of some early structure which is capable of essentially complete protection of residue 100 against exchange by the high pH pulse.

The behaviour shown by Ile27 is intermediate. At early times, the high pH pulse results in labelling of about half of the molecules. To

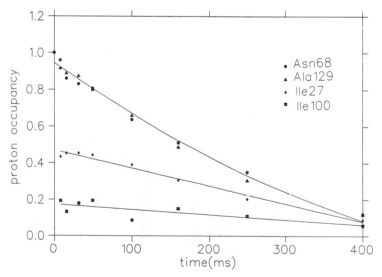

Figure 9.10 The proton occupancy of residues 68, 129, 27 and 100 observed in the time-course of refolding, as monitored by pulsed hydrogen exchange

characterize this unusual behaviour, the pH of the 40 ms labelling pulse was varied. The extent of labelling varied for different residues and varied with the pH of the pulse. For those residues that show protection from solvent exchange at early times, the proton occupancy observed increases as the pH of the probe pulse is increased. This pH dependence continues until the proton occupancy seen at early times approaches that of an unprotected residue such as Tyr88. These observations, combined with the absence of a pH dependence for the observed rate of folding to the native state, strongly suggest that the partial proton occupancies seen at early times for residues such as Ile27 are not the result of a bifurcation of the folding path. Rather, it appears that most, if not all, of the molecules form an intermediate or a population of intermediates that is characterized by significant protection from solvent exchange. This intermediate behaviour is observed at 25–30 residues with varying degrees of protection at early times.

This information is converted into a more useful form in Figure 9.11 as a plot of relative protection factor versus position. In this case the relative protection factor is calculated as the ratio of the base-catalysed rate constants of exchange for the intermediate to the base-catalysed rate constants for exchange of the unfolded state shown in Figure 9.9. These data clearly demonstrate that three regions are protected from amide exchange within 8 ms of folding. The calculated relative protection factors are at least 20 and vary to a maximum of 300 for the residues in three regions of the sequence, corresponding to residues 5–13 (helix A), those near residue 20 of the N-terminal domain β sheet, and residues 96–105 (helix E).

Our pulse-labelling experiments done at different pH values showed that several residues are substantially protected from solvent exchange

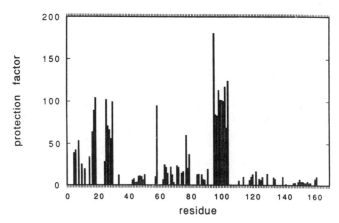

Figure 9.11 The ratio of the protection seen in the refolding intermediate as compared with the unfolded state after 32 ms of refolding at 20 °C in 1.5 M urea

in the intermediate. However, any given residue in the intermediate is much more susceptible to exchange than it is in the final native structure. Most protected residues contained in the β sheet in N-terminal domain and in helix A and helix E in C-terminal domain are protected about 100-fold relative to free amides. Protection factors of similar magnitude were observed in the partly folded apomyoglobin structure formed under equilibrium conditions by Hughson *et al.* (1990). Such behaviour is consistent with the formation of relatively isolated regions of secondary structure.

The region of the protein that shows the highest degree of protection from hydrogen exchange corresponds to residues 95–103. This region corresponds to helix E in the native structure of the protein. The crystal structure of the native state shows that this helix is the most buried. It is reasonable that this helix forms earliest during refolding. Helix A is in close contact with helix E and amide hydrogen exchange is also slow in helix A. It is possible that these early protected regions form a structure for the subsequent refolding of the entire protein.

Previous kinetic circular dichroism studies have indicated that secondary structure is formed before the final tertiary structure during refolding of α-lactalbumin (Kuwajima *et al.*, 1985; Gilmanshin and Ptitsyn, 1987), hen egg white lysozyme (Ikeguchi *et al.*, 1986), carbonic anhydrase B (McCoy *et al.*, 1980; Dolgikh *et al.*, 1984; Semisotnov *et al.*, 1987), ferrocytochrome-c and β-lactoglobulin (Kuwajima *et al.*, 1987). NMR trapping techniques applied to RNase A (Udgaonkar and Baldwin, 1988; 1990), horseheart cytochrome-c (Roder *et al.*, 1988), barnase (Bycroft *et al.*, 1990) and hen egg white lysozyme (Miranker *et al.*, 1991) have detected the presence of intermediates with a subset of native secondary structure. Thus, the behaviour seen for T4 lysozyme is qualitatively similar to that observed for several other proteins.

Our data suggest a preliminary pathway for the folding of T4 lysozyme under these conditions, as shown below. Here U, I and F refer to the fully unfolded state which lacks any stable secondary structure, the intermediate discussed above, and the final fully folded state, respectively. The rate-limiting step appears to involve the conversion of the intermediate into the final folded state. From the pH dependence of the observed folding rates we conclude that the H31–D70 salt bridge is formed after the rate-limiting step in the folding direction:

$$U \; \underset{}{\overset{}{\rightleftharpoons}} \; I \; \underset{}{\overset{}{\rightleftharpoons}} \; F$$

$$\text{fast} \qquad \text{slow}$$
$$(\tau < 10 \text{ ms}) \quad (\tau = 170 \text{ ms})$$

Preliminary evidence suggests that the intermediate may lie along the folding pathway as shown. We have measured the apparent folding rate constant for 20 mutants which affect the stability of the final folded state by varying amounts. These mutations were chosen because they are more or less uniformly distributed along the lysozyme sequence. The vast majority of these have no effect on the apparent rate constant for folding. Three show substantially slower rates of folding and all are mutations in one of the three protected regions of the intermediate. These mutations may destabilize the intermediate and, hence, slow down folding. In this view the rate-limiting step for folding involves conversion of the intermediate into the final transition state for folding. Of course, there may be many pathways and many transition states and these mutations may either remove certain pathways or may directly change the energetics of the lowest transition states.

4 Role of Ionizable Groups in Stabilization of the Folded State

The pH dependence of the thermodynamic stability of the folded state of T4 lysozyme suggests that interactions involving ionizable groups account for 12–15 kcal/mol of free energy stabilizing the folded state at neutral pH (see below). Dao-pin *et al.* (1991c) have shown that several basic residues in T4 lysozyme play virtually no role in stabilizing the folded state and can be replaced by acidic residues with little or no energy cost to the folded state. This observation suggests that most surface residues contribute little to the free energy of folding. Also, T4 lysozyme has an isoelectric point near 11, yet is maximally stable near pH 5.5. Together, these observations show that a few specific interactions can be very important to the overall folding energy. The roles of ionic groups in the energetics and structure of the folded state of proteins have not been extensively studied at the experimental level. We have begun a systematic examination of the roles of ionic interactions in the stability, structure and function of T4 lysozyme.

We have completed the first step in this analysis, describing the His31–Asp70 interaction. These residues form a highly energetic salt bridge that stabilizes the folded state relative to the unfolded state by 4–5 kcal/mol. In the folded state, the pK_a value of His31 is shifted to a value of 9.1 from a more normal value of 6.8 in the unfolded state. When Asp70 is replaced by asparagine, the histidine pK_a value returns to a value of about 6.8 in the folded state. This suggests that the interaction with aspartate dominates the pK_a value shift of His31. A corresponding lowering of the pK_a value for Asp70 also occurs.

We have been especially interested in characterizing all the electrostatic interactions between the individual carboxyl groups and their neighbours in the folded state. Figure 9.12 shows the ^{13}C NMR spectrum obtained from a sample of T4 lysozyme in which the side-chain carbonyl

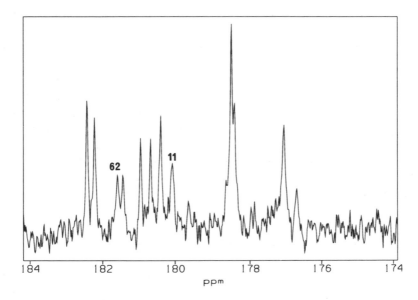

Figure 9.12 The ^{13}C spectrum of the glutamine and glutamate side-chain carboxyl groups of T4 lysozyme at pH 6.5 and 10 °C. The resonances corresponding to Glu62 and Glu11 are indicated

carbons of the eight glutamic acid and five glutamine residues have been enriched in ^{13}C by biosynthetic incorporation of labelled [δ-^{13}C]-glutamic acid. There are thirteen peaks and eight undergo the expected 3 ppm upfield changes in ^{13}C chemical shift, as pH is lowered. These eight peaks correspond to the eight glutamic acids in the protein. Two derivatives, E11Q and E62Q, in which a glutamic acid has been substituted with glutamine, were used to assign these two residues. Two peaks in the spectrum of the WT protein are systematically absent in the mutants from the group of titrating peaks at several pH values, providing the assignments shown in Figure 9.12. The remaining peaks have not been systematically assigned.

Figure 9.13 shows a ^{13}C NMR spectrum from a sample of T4 lysozyme with biosynthetic incorporation of ^{13}C at the side-chain γ-carboxyl groups of the ten aspartic acid residues, using an auxotrophic strain that does not convert aspartate to asparagine. The carboxyl of Asp70 is ionized under all accessible experimental conditions and has been assigned to the aspartic acid peak at 178.9 ppm by mutation (Anderson *et al.*, 1990). The remaining nine aspartate resonances were also assigned by mutation. In all but one case, a systematic absence of one carboxyl peak was observed in these labelled samples in which aspartate residues were replaced by glutamate, asparagine or cysteine residues.

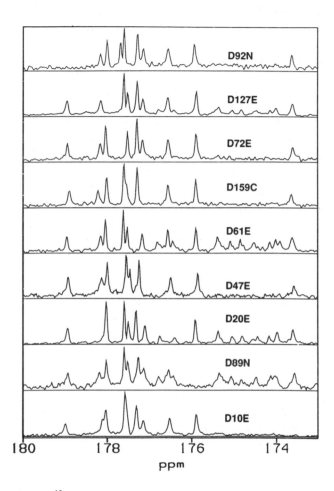

Figure 9.13 The ^{13}C spectra of the aspartate side-chain carboxyl resonances of T4 lysozyme at pH 6.5 and 10 °C. Each of the aspartate residues was replaced by a single residue as indicated. In some spectra, additional small peaks are observed. These peaks correspond to asparagine residues of known chemical shift (see Anderson *et al.*, 1990: Figure 5a). These arise from 'leakage' of one or both of the Asn A&B lesions that block the conversion of aspartate to asparagine in the host cells. A second complication is seen in the spectrum of derivative D159C (C54SC97S). This construction was produced in a different cysteine-free background from the other derivatives. The peak corresponding to Asp47 is not seen in the D47E derivative, as expected. However, a peak is present at 177.7 ppm in spectrum from D159C, which we tentatively assigned to Asp47. It is reasonable that different substitutions of residue 54 (serine for cysteine in this case) can affect the ionized chemical shift of Asp47, owing to the strong hydrogen bond between Asp47 and the amide of residue 54 observed in the high-resolution structure. From Weaver and Matthews (1987)

A second approach to assignment was to use correlations between carboxyl carbon chemical shifts and assigned α and β proton chemical shifts, using the HMBC (Bax and Summers, 1986; Bax *et al.*, 1988) spectrum shown in Figure 9.14. This experiment displays a crosspeak for the C^α and C^β protons that have coupling constants in the range of 10 Hz to the ^{13}C carboxyl group. Interestingly, the HMBC experiment failed to detect three aspartic acids entirely and was less successful at detecting beta protons. Correlations seen in the HMBC spectrum are consistent with the assignments for the remaining seven aspartic acids made by mutation as described above. Asp10, Asp47 and Asp70 were not detected. Presumably, these residues are rigidly held and the resulting increase in spin–spin relaxation effectively competes with the transfer of coherence. The complete assignments of aspartate resonances are shown in Figure 9.15.

The pH dependence of the chemical shifts of peaks corresponding to the ten aspartates and eight glutamates is shown in Figure 9.16 and 9.17, respectively. Smooth lines through data points represent best-fit lines of titration curves in which the pK_a value, ionized chemical shift and change in chemical shift upon protonation were the fit parameters. Aspartates 10 and 70 show no significant chemical shift as pH is varied. As discussed above, we have established that Asp70 is ionized at all pH values, owing to its interaction with His31 (Anderson *et al.*, 1990).

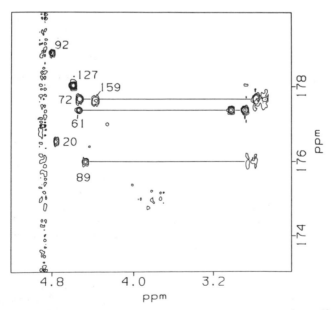

Figure 9.14 The HMBC spectra of aspartate side-chain carboxyl ^{13}C-labelled T4 lysozyme at pH 6.5 and 10 °C in D_2O. The lines show the C^α and C^β protons of the indicated aspartate residues

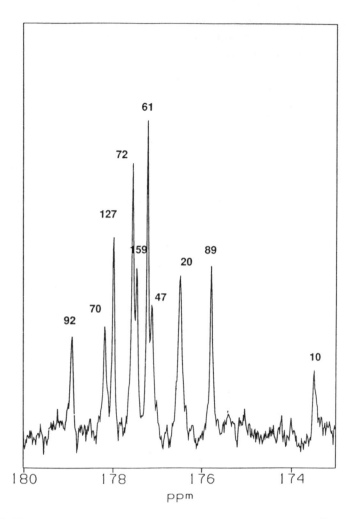

Figure 9.15 The assignments of the aspartate side-chain carboxyl ^{13}C resonances at pH 6.5 and 10 °C

Further experiments were necessary to determine the ionization state of Asp10. Two experiments effectively rule out the possibility that titration was not observed because of coincidence of the protonated and ionized chemical shift. First, replacement of two arginine residues, 145 and 148, with methionine, led to an absolute change in chemical shift of Asp10 but no apparent chemical shift titration between pH 7 and 2. Second, a difference titration between D10S and WT* revealed no differential proton uptake between pH 8 and 3. The above experiments indicate no titration occurs but could not establish whether the solvent inaccessible carboxyl was fully protonated or fully ionized over this pH

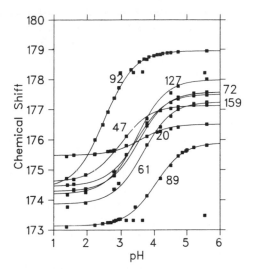

Figure 9.16 The pH dependence of the chemical shift of the aspartate side-chain carboxyl ^{13}C resonances of T4 lysozyme at 10 °C

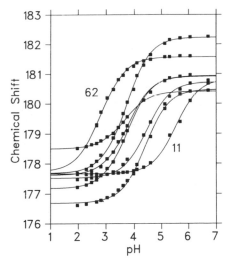

Figure 9.17 The pH dependence of the chemical shift of the glutamate side-chain carboxyl ^{13}C resonances of T4 lysozyme at 10 °C

range. The ^{13}C chemical shift of the sidechain carboxyl group of Asp10 is the most upfield of any of the aspartate resonances and is similar to that of a protonated carboxyl or an amide carbonyl group. While all thermodynamic evidence supports the view that Asp10 is a strong acid, the chemical shift suggests a relatively charge neutralized carboxyl for Asp10.

We have been able to demonstrate that Asp10 is unprotonated at acidic pH values by the following study. Qualitatively, below pH 2.0 only Asp70 and Asp10 have the possibility to be fully ionized, since all other residues have substantially larger pK_a values than 2. Upon unfolding, each could take up one proton at pH 2.0. Thus, the apparent dependence of the unfolding equilibrium constant on hydrogen ion concentration at pH 2.0 should reflect the ionization states of the acidic residues. We carefully measured the dependence of the unfolding equilibrium constant in the vicinity of pH 2.0. As shown in Figure 9.18, the C2 proton of His31 gives distinct resonances for the folded and unfolded states. These resonances are free of any overlap from other resonances when the amides are fully exchanged by deuterium and the reaction is observed in D_2O. The equilibrium constant for the folding reaction can be accurately measured by integrating the histidine C2 proton resonances of the folded and unfolded states. Using this approach, we measured the hydrogen ion dependence of the unfolding reaction for a variant of the normal enzyme and for the two mutants D70N and D10S.

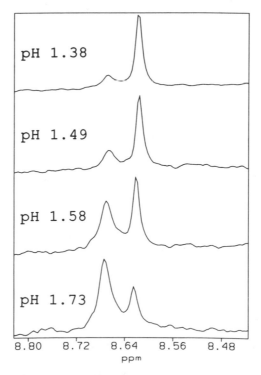

Figure 9.18 The pH dependence of the proton spectra of the C2 proton of His31 of the T4 lysozyme mutant M106A at 10 °C. The downfield resonance (8.67 ppm) corresponds to the folded state. The upfield resonance (8.61 ppm) corresponds to the unfolded state

We have had some difficulty in isolating reasonable amounts of several proteins mutated at position 10 highlighting its important role in stability. Mutant D10S has proven to be the easiest to isolate in reasonable yields. If both Asp10 and Asp70 are unprotonated at pH 2.0, we expect to see approximately second-order behaviour for the normal protein and first-order behaviour for the two single mutants. If Asp10 is protonated, both the normal enzyme and D10S should show the same first-order dependence on the unfolding equilibrium constant, while the equilibrium constant for D70N should show little or no dependence on hydrogen ion concentration.

The pH dependence of the stability of M106A, D10S and D70N is shown in Figure 9.19 as a plot of the logarithm of the equilibrium constant for folding as a function of pH. The mutant M106A was chosen as a control because it has about the same stability at pH 2 as the aspartate mutants. The environments of Asp10 and Asp70 are intact in this mutant but the hydrophobic core of the protein has been destabilized by the introduction of a cavity by replacement of the leucine sidechain by the methyl group of alanine. The dashed line shows the expected slope calculated for L99A, using the pK_a values of 0.5 for Asp10 and Asp70 and the values shown in Figure 9.20 for all the remaining acidic residues. The slopes are lower by approximately one unit for the mutants D10S and D70N. Thus, Asp10 and Asp70 each take up one proton when the protein unfolds at pH 2. This result establishes that both Asp10 and Asp70 are essentially fully ionized at pH 2.

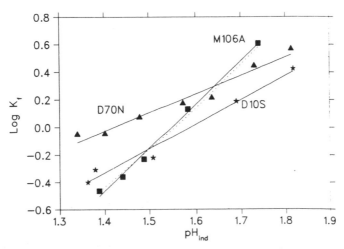

Figure 9.19 The pH dependence of the logarithm of the folding equilibrium constant K_f as a function of pH for the mutant M106A(■), D10S(★) and D70N(▲) versions of T4 lysozyme

The results of the pK_a value measurements are summarized in Figure 9.20 as a plot of the shift in pK_a value from the non-interacting value as a result of forming the folded state. The values for Asp10 and Asp70 are shown, using a pK_a value of 0.5 for each residue. As can be seen in the figure, Glu62 and Asp92 are also shifted substantially and many of the remaining acidic residues are slightly more acidic than the non-interacting value. Only Glu11 and Glu 108 are shifted to higher pK_a values as a result of their environments in the folded state.

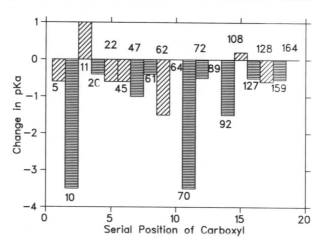

Figure 9.20 A summary of the shifts in pK_a value associated with the folded state of T4 lysozyme. Horizontal hatches indicate aspartate residues and diagonal hatches represent glutamate residues

pH-Dependent Amide Chemical Shifts

Although the amide group of an amino acid residue in a protein does not ionize, it is sensitive to environmental changes brought on by local ionization. Especially large amide titration chemical shifts have been observed for amides in direct hydrogen bonding interactions with carboxyl groups (Brown *et al.*, 1978). The assignments available for the backbone amides of wild-type protein make it possible to use individual amide resonances as titration reporting groups at specific locations in the protein.

Figure 9.21 shows a cartoon of the ^1H–^{15}N HSMQC spectrum obtained from a uniformly ^{15}N-labelled T4 lysozyme sample at pH 6.1. A series of 14 spectra were collected from samples of pH 6.6 to 1.9 and the chemical shifts of individual amides were followed as a function of pH. About a third of the resonances show fairly large shifts as pH is varied. While the individual peaks can be reasonably well tracked as a function of pH, there are several instances where resonances cross and

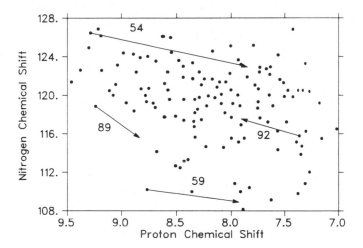

Figure 9.21 A summary of the pH dependence of the ^{1}H–^{15}N HSMQC of the backbone amide resonances of residues 54, 59, 89 and 92 of T4 lysozyme. The arrowheads indicate the position of the resonances at pH 1.9

the assignment to a particular residue becomes difficult. To confirm resonance assignments at low pH, two subsets (aspartate + asparagine and glutamate + glutamine + threonine + glycine + arginine) of amide resonances were specifically ^{15}N labelled and HSMQC spectra were collected at low pH (data not shown). The arrows on Figure 9.21 indicate the total change in chemical shift observed for the amide resonances of residues 54, 59, 89 and 92. The arrowhead indicates the chemical shifts at pH 1.9. Note that the arrows do not necessarily indicate the path followed by individual resonances during their apparent titration. These four resonances show fairly linear behaviour, indicating a simple two-state model for the protonation/deprotonation event.

Figure 9.22 shows the observed change in proton chemical shift between pH 6.56 and 1.92 for the assigned amide resonances of T4 lysozyme. The locations of carboxyl groups in the primary sequence are indicated by filled circles. Some amide titration chemical shifts are extremely large. In these cases, most notably residues 24, 54, 59, 93 and 95, examination of the X-ray derived T4 lysozyme structure reveals that these amides are involved in direct hydrogen bonds with carboxyl groups. In each case there is a large upfield shift in the amide resonance as pH of the solution is decreased. This is consistent with the view that these amides are involved in stronger or more polarized hydrogen bonds when the interacting carboxyl group is ionized. When the carboxyl group is protonated, the hydrogen bond may be broken or becomes less polarized, resulting in an upfield shift in the amide proton resonance. The details of some of these interactions will be discussed more fully

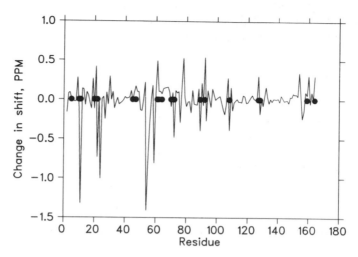

Figure 9.22 A summary of the pH dependence of the ^{1}H chemical shifts of the backbone amide resonances of T4 lysozyme. A negative change in chemical shift indicates a resonance that shifts upfield as pH is lowered

when we attempt to explain the observed perturbation in pK_a values of the carboxyl groups based on the crystallographic model of the structure.

Application of Mass Action Binding Model

Qualitatively, if the native structure of a protein causes a shift in pK_a value for a particular residue, the native structure must differentially stabilize the acidic form of the residue as compared to the conjugate base. For example, if the pK_a value of a carboxylic acid is lowered, the native structure must stabilize the carboxylate as compared with the protonated form. The coupling of thermodynamic functions requires that the protonated form of the acid must stabilize the folded state as compared with the unfolded state. This relationship is easily quantified. If there are m ionizable groups, the apparent value for the unfolding equilibrium constant at any pH, K_{app}, is given by

$$K_{app} = K \frac{\sum_{i=0}^{n} \phi_{\mu i} H^i}{\sum_{i=0}^{n} \phi_{fi} H^i}$$

where $\phi_{\mu i}$ and ϕ_{fi} represent the phenomenological overall (Adair) constants describing the binding of i protons to the unfolded and folded forms of the protein respectively. The microscopic equilibrium constant K represents the equilibrium between the fully unprotonated forms of the unfolded and folded states. For a single ionizable group this reduces to

$$K_{app} = K(1 + K_u H)/(1 + K_f H)$$

The value of K_{app} varies from the high pH value of K to a low pH value of $K(K_u/K_f)$. Thus the magnitude and direction of a shift in pK_a value defines the magnitude and direction of an ionizable group's contribution to the pH-dependent stability of a protein. When the folded environment results in lowering the pK_a value for an acidic residue, that residue is involved in the differential stabilization of the high-pH form of the protein to unfolding as compared with the low pH form. When the environment results in an increase in pK_a value, that group destabilizes the high-pH form of the folded state relative to the low pH form. Residues with pK_a values near the expected model values make little or no contribution to the pH-dependent stability of the protein.

As shown in Figure 9.20, all groups titrating between pH 1 and 10 have been assigned a pK_a value in T4 lysozyme. Two carboxyls, from Asp10 and Asp70, are strong acids and could not be protonated. The pK_a values of these carboxyl groups are almost certainly below 0.5 but this conservative value will be used here. We have applied the mass action binding model described above to these data and the results are summarized in Figure 9.23. Rather than attempt measurement of the pK_a value of each carboxyl group in the unfolded protein, we use model compound values of 4.4 and 3.9 for the pK_a values of unfolded aspartate

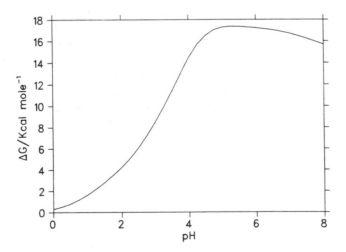

Figure 9.23 The calculated pH dependence of the free energy of unfolding, ΔG, as a function of pH, using the pK_a values from Figure 9.20

and glutamate residues. The pK_a values for His31 were determined in both the folded and unfolded states. We assume that our apparent pK_a values can be used to construct the Adair constants. This is equivalent to assuming that the carboxyl group ionizations are independent. While this is likely to be a good overall assumption, it is probably not true for a few sites.

As shown in the figure, the model suggests that the overall contribution of the ionizable groups is about 17 kcal/mol, well in excess of the net stability of the protein. The two interactions involving Asp10 and Asp70 account for about half of the free energy stabilizing the protein at neutral pH relative to low pH.

Structural Rationales for Electrostatic Perturbations

T4 lysozyme has been the subject of intensive crystallographic study. The protein's structure was first determined from the crystal available in the space group $P3_221$ with one molecule per asymmetric unit (Weaver and Matthews, 1987). Since then two additional crystal forms have given rise to six additional independent conformations of the molecule. The main difference between these conformations is a changed hinge bending angle relating the N-terminal and C-terminal subdomains as described by Faber and Matthews (1990) and by Dixon *et al.* (1992). These seven structures provide crystal contact-free views of all acidic residues, except Asp72. Furthermore, in some cases a multiplicity of views provides a qualitative measure of the uncertainty that may be present for side-chain placement.

It is helpful to separate the carboxyl groups of T4 lysozyme into four different groups based solely on their observed titration behaviour. In the first group, which includes Asp10 and Asp70, the interactions of the carboxyl group are so energetic that only a lower bound can be placed on each interaction's contribution to the protein's stability. In the second group, only Glu11 has a significantly higher pK_a value in the folded state than in the unfolded state. The third group comprises Asp47, Asp92 and Glu62. These three residues have pK_a values indicating a contribution of more than 1 kcal/mol to the stability of the protein. The 12 remaining carboxyl-bearing residues comprise the last group. Interactions involving these residues contribute less than 1 kcal/mol to the pH-dependent stability of the protein.

Aspartates 10 and 70

We have shown that Asp10 is effectively ionized in the folded protein under all accessible experimental conditions. The structural rationale for this property is revealed by the high-resolution structure of the protein as reproduced in Figure 9.24. This part of the structure is one of the most highly ordered in the protein with the lowest B factors found for any residues except the most buried residues in the protein. Aspartate 10 is the least solvent-exposed carboxyl group in the protein. There is no room for ordered water molecules to act as hydrogen bond donors for the carboxyl oxygen atoms. Using the solvent accessibility algorithm of Lee and Richards (1971) and a probe radius of 1.3 Å, the carboxyl group has an average solvent exposure of 16%. While Asp10 is not in contact with solvent, it is not very far away either. Only 4.4 Å separate Asp10 carboxyl oxygen and the nearest bound water which is contiguous with the solvent-filled cavities in the crystal. Although removal from solvent would be expected to raise the pK_a value of Asp10, this effect could be offset by interactions of the carboxylate with the nearby solvent (Sharp and Honig, 1990).

There is an array of specific polar interactions stabilizing the ionized form of Asp10. One of the carboxyl oxygens (OD2) provides the only possible hydrogen bonding interaction with the phenoxy moiety of Tyr161 and one of the side-chain amide hydrogens of Asn101. Asn101 is completely buried in the C-terminal core of the protein. The remaining carboxyl oxygen (OD1) is in close salt bridge interaction with the guanidinium group of Arg148. Hydrogen bonds can be made with both the secondary imino (2.8 Å) and one of the terminal imino groups (2.7 Å). A second guanidinium group, from residue 145, brings more

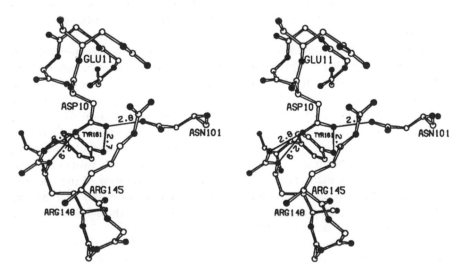

Figure 9.24 A stereo-pair drawing of the structure of T4 lysozyme in the vicinity of Asp70. Carbon atoms are shown as open circles and nitrogen or oxygen atoms are shown as filled circles. The lines connect various atoms with the distances given in Å. The drawing is based on the refined crystal structure of T4 lysozyme

potentially charge-bearing atoms to within 4 Å of OD2. The general details of these interactions change little in the various M6I conformations (Faber and Matthews, 1990). In one conformation, the guanidinium group of Arg145 may move towards Asp10 with the distance between potentially charge-bearing atoms shrinking to about 3.0 Å.

These interactions are apparently sufficient to lower the pK_a value of Asp10 drastically. The contribution to the pH-dependent stability of T4 lysozyme provided by the interactions involved in lowering the pK_a value is very large but undetermined. Using a conservative lower limit for the pK_a value of 0.5, the interaction is expected to provide at least 5 kcal/mol to the pH-dependent stability of the protein. Further attempts to establish a better estimate for the pK_a value of Asp10 (and 70) must await transfer to a more thermostable background to allow lower pH observation without the complexities involved with acid unfolding.

In a complex environment such as that surrounding the side-chain of Asp10, there is the possibility to delocalize charge density over all the polar residues of the complex hydrogen bonded network. The resonances corresponding to the Tyr161 phenolic OH proton and the ^{13}C resonance of the side-chain carboxyl of Asp10 provide some insight into that environment. As mentioned above, the carboxyl ^{13}C resonance of Asp10 is about 3 ppm upfield of the expected chemical shift of an ionized carboxyl group. This argues that much of the negative charge on the carboxyl group has been effectively removed. Similarly, the phenolic proton of Tyr161 has the most downfield proton resonance in the protein, 11.4 ppm (McIntosh *et al.*, 1990). In addition, this proton is in slow exchange with bulk water ($\tau > 1$ s) at neutral pH. These observations suggest that the Tyr161 phenolic OH bond may be unusually polarized, with some of the charge density serving to neutralize the charge on the side-chain carboxyl of Asp10.

Aspartic acid 70, like Asp10, has never been observed to protonate in the native state. As shown in Figure 9.25, Asp70 makes a salt bridge with histidine 31 with 'excellent' 2.7 Å hydrogen bonds between one carboxyl oxygen atom and Nδ1 of the histidine ring. A longer 3.2 Å distance is seen to the other oxygen atom of the carboxyl of Asp70. In contrast to the situation near Asp10, there are three water molecules (182, 265, 171) within hydrogen bonding distance to the Asp70 carboxyl oxygens. A fourth ordered water molecule (202) is 3.4 Å away. The salt bridge orientation is apparently maintained by van der Waals contacts with residues near the interacting groups. As with Asp10, we have not measured the pK_a value of Asp70, because it is lower than the lowest pH at which the predominant species is folded. A conservative upward bound of 0.5 leads to a stabilization of about 5 kcal/mol at a pH of 5.5.

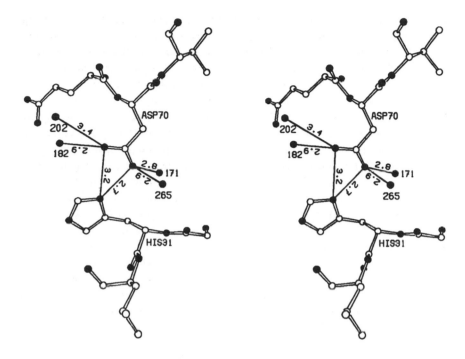

Figure 9.25 A stereo-pair drawing of the structure of T4 lysozyme in the vicinity of Asp10. Carbon atoms are shown as open circles and nitrogen or oxygen atoms are shown as filled circles. The lines connect various atoms with the distances given in Å. Four water molecules are indicated with the numbers 171, 182, 202 and 265. The drawing is based on the refined crystal structure of T4 lysozyme

The comparison of the environments and ^{13}C chemical shifts of Asp10 and Asp70 is fascinating. Both are rigidly held in place and both environments are retained in all the different crystal packing arrangements. Both are never observed to be protonated in the native state. Yet one is strongly solvated and the other is excluded from solvent. One has the most upfield ^{13}C chemical shift and the other is one of the two most downfield in ^{13}C shift. These observations suggest that simple qualitative arguments to explain the extreme acidity of both these residues are not likely to be satisfactory.

Catalytic Residues

The environment of Glu11 changes significantly with the hinge-bending motions observed in the $P2_12_12_1$ crystal form structures obtained from the T4 lysozyme mutant M6I (Faber and Matthews, 1990). The pK_a value measured for Glu11 is 5.4. Thus, it is the only carboxyl group which has the protonated form of the carboxyl group stabilized by the folded state. We would expect such a change to occur upon removal of the carboxyl group from solvent and hydrogen bonding interactions.

The wild-type structure and the three more closed conformations of the M6I structures share similar features in the locale of Glu11. A portion of that environment is shown in Figure 9.24. The glutamic acid carboxyl oxygen (OE1) is in reasonable hydrogen bonding distance to three different ordered waters (207, 204 and 229). These are contiguous with other bound waters which are then contiguous with solvent channels in the crystal. The other carboxyl oxygen (OE2) is in hydrogen-bonding distance to one bound water (256) and in close salt-bridging distance to the ε-NH of Arg145, with the potential charge-bearing atoms separated by 2.8 Å. The hydrogen-bonding geometry seems good, precluding conformational effects (Cybulski and Shreiner, 1989). Consequently, solvent exposure, hydrogen bonding and salt bridge interactions would be expected to lower the pK_a value of the carboxyl group below model compound values.

There is a very significant difference in the conformation of Glu11 seen in the most open conformation as compared with the most closed conformation from the M6I study. The carboxyl of Glu11 rotates away from a solvent-exposed state to take up a less solvent-exposed position facing up towards the C-terminal core of the protein. No water can access either carboxyl oxygen to make adequate hydrogen-bonding interactions. Only two possible hydrogen bonds can be made involving one carboxyl oxygen (OE1) and the ε and terminal amide groups of Arg145. Thus, in this most open conformation Glu11 is removed from a rich hydrogen-bonding environment to a position less than 3.5 Å from the backbone and side-chain carbonyl oxygens of the buried Asn101. This conformation is more consistent with the measured shift in the pK_a value of Glu11. The reason for the burial of Glu11 seems to be the requirement of Arg145 for hydrogen bonding interactions.

Aspartic acid 20, together with Glu11, compose the two catalytic carboxyls of T4 lysozyme. Substitution of Asp20 with asparagine nearly abolishes activity while substitution with alanine leads to measurable activity as observed also for the analogous residue in hen egg white lysozyme (Malcomb *et al.*, 1989). The pK_a value of Asp20 is 3.6, indicating a favourable interaction of about 0.5 kcal/mol. In the original crystal form of T4 lysozyme Asp20 is seen to provide hydrogen-bonding interactions with two waters through one carboxyl oxygen and hydrogen bonds with the amides of three residues, 22, 23 and 24. These residues make up a turn that interconnects the two beta strands of T4 lysozyme and the carboxyl group offers the hydrogen-bonding interactions necessary to stabilize this conformation. These amide groups report the titration of Asp20 through pH-dependent amide chemical shifts. No significant changes in these interactions are seen in the M6I conformations. These interactions would be expected to provide at least enough stabilization of the ionized carboxyl to explain the small perturbation seen. This interaction provides about 0.5 kcal/mol to the stability of the protein.

Glutamate 62 and Aspartates 47 and 92

Aspartic acid 47 has a pK_a value of 3.0 at 10 °C. The ideal geometry of the hydrogen bond between the amide of residue 54 and the carboxyl group of Asp47 was noted in the high-resolution structure of T4 lysozyme (Weaver and Matthews, 1987). Indeed, the amide resonances of residue 54 and 55 undergo large pH-dependent chemical shifts (1.5 ppm for residue 54) with apparent pK_a values of about 3.1. There are no well-ordered potential ion-pair partners available to produce this perturbation in pK_a value. Both these amide groups are involved in a hydrogen bond with the carboxyl group of Asp47. This feature is retained in the alternative M6I structures (Faber and Matthews, 1990). Substitution of Asp47 with alanine leads to a pH-dependent loss in stability of approximately 0.7 kcal/mol (Heinz et al., 1992). This is qualitatively consistent with expected stabilization to the folded protein of about 1.4 kcal/mol at 10 °C.

Glutamic acid 62 has a pK_a value of 2.8, indicating stabilization of about 2 kcal/mol of the protein at neutral pH. One carboxyl oxygen (OE2) is in excellent hydrogen-bonding geometry with water 259 and the guanidinium group of Arg52. The remaining oxygen (OE1) is in excellent hydrogen-bonding distance from amide of residue 59 and one of the terminal NH groups of Arg52. Both the amide resonance of residue 59 and a guanidinium group resonance of an unassigned arginine have pH-dependent chemical shifts that give apparent pK_a values nearly coincident with that reported by carbon measurements. The Glu62–Arg52 salt bridge lies between bulk solvent and a group of nonpolar residues including Ile58 and Val 57. The features of this interaction are retained in the alternate conformations of the M6I crystal form.

Aspartic acid 92 has a pK_a value of 2.5, determined from ^{13}C resonance shifts, indicating a contribution to stability of about 2 kcal/mol in going from acid to neutral pH. One oxygen of the carboxyl is in hydrogen-bonding distance to the amide hydrogens of residues 93 and 95. The amide resonances of these residues also show large pH dependent chemical shifts with apparent pK_a values of about 2.6. This interaction is due to a helix cap and has been the subject of a structural and thermodynamic study by Nicholson et al. (1991). The remaining oxygen of the carboxylate is in excellent hydrogen-bonding distance and orientation with the guanidinium group of Arg95 and a water molecule contiguous with solvent spaces. The guanidinium group of Arg95 itself makes hydrogen bond interactions with the backbone carbonyl groups of residues 153 and 154, two of the backbone carbonyl groups at the C-terminal end of helix H.

The Remaining Carboxyl-containing Residues

The grouping of these residues is based on the small shifts in apparent pK_a values from that expected in the non-interacting case. Some groups are very disordered in all or most of the crystal forms, while others (such as Asp20) have very precise interactions with other protein residues, suggesting relatively strong interactions. This group includes a number of acidic residues that appear to be in close proximity to basic residues but produce little or no shift in pK_a value as a result of this proximity. These include Glu5 (N terminus), Glu22 (Arg137), Glu45 (Lys48), Asp61 (helix cap), Asp72 (Lys76), Asp89 (Lys88), Asp127 (Arg154) and Asp159 (Lys162). We are particularly interested in the structural basis for the shift in pK_a values we see for some residues but not others. Both the obvious interactions with polar residues near the carboxyl of interest and the less obvious interactions associated with solvation surely play important roles in generating shifts in pK_a value. In some cases these potential salt bridge interactions are seen in only some of the crystal forms, suggesting that crystal packing forces are comparable to the weak electrostatic interactions between the groups. When we have measured a shift in pK_a value greater than one unit in solution, the salt bridge is seen in all crystal forms. This argues that the differential crystal packing forces are likely to be weaker than about 1.5 kcal/mol.

The Role of Charge–Charge Interactions in Protein Folding

We have addressed three questions in the work described above: What is the role of electrostatic interactions in stabilizing T4 lysozyme? What kind of structure provides electrostatic stabilization to the protein? What role do these interactions play in specifying the structure of T4 lysozyme?

Even using the modest estimates for the pK_a values of Asp10 and Asp70, mass action requires that electrostatic interactions provide enough stabilization to exceed the net stability of T4 lysozyme. This contribution to the stability of the protein is dominated by two highly energetic interactions. The interactions are so energetic that they lower the pK_a values of both aspartates close to the pK_a value of a strong acid.

In answering the first question it is important to remember that we have not studied the role of lysine and arginine residues in the stability of the protein. Basic residues involved in salt bridges will perturb the pK_a values of acids with which they interact and thus their contribution is indirectly accounted for. But basic residues interacting with concentrations of hydrogen-bond donors such as those found at the C-terminal ends of helices and elsewhere would perturb the pK_a value of the basic group but not provide a simple measure of the interaction. There are several cases in which arginine residues are found well ordered in excellent hydrogen-bonding geometry with hydrogen-bond-accepting groups. These interactions might be expected to contribute up to two or three kcal/mol of further stability for each interaction. At present the contribution of these interactions is being evaluated using site directed mutagenesis.

A further limitation is that we monitor apparent pK_a values only near the pH value where the residue is about equally protonated and unprotonated. This very clearly limits our ability to examine the effects on pK_a values of protonation at other residues. We see very little evidence for cooperative interactions between ionizable groups. In addition, it is possible that some pH-dependent conformational changes might occur. Although by no means conclusive, the observation that most amide resonances experienced very small changes in chemical shift in going from pH 6.6 to 1.9 is consistent with the view that there are no large-scale conformational transitions over this pH range.

The one case in which this may be especially important is the effect the ionization of Glu11 may have on the titration properties of Asp10. Protonation of Glu11 occurs at pH 5.4. Significantly below this pH value, Arg145 has lost its ion-pair partner. To what extent is the titration of Asp10 affected by earlier protonation of Glu11? Does Arg145 interact more strongly with Asp10 at low pH? One way of addressing this issue would have been to titrate each of the carboxyls in the absence of the other. Unfortunately, as demonstrated by the data shown in Figure 9.14, the pK_a value of these groups remains too low to measure in the mutants D10S and D70N.

These measurements raise several interesting points with respect to the catalytic mechanism. Glu11 is a catalytic residue and is analogous to Glu35 of hen egg white (HEW) lysozyme (Malcomb *et al.*, 1989). In both cases the glutamate carboxyl groups are more basic than normal in the absence of substrate. The catalytic glutamate residue is more important than the other catalytic carboxyl (Asp20 in T4, Asp52 in HEW). Is this decrease in acidity of the glutamates an essential feature of the catalytically important glutamates in the lysozyme family?

It is clear that placing an ionizable group in a non-polar environment will greatly change the pK_a value of that group. We have recently characterized two basic residues which have pK_a values shifted to more acidic values. In one mutant, Y25H, a histidine residue was placed in the hydrophobic environment of the N-terminal β sheet. This resulted in lowering the pK_a value of that histidine residue from the normal value of 6.8 to about 4.4, as judged by the pH dependence of the chemical shift of the C2 proton of the histidine residue. In another mutant, M102K (see Dao-pin *et al.*, 1991a), a lysine residue was placed in the hydrophobic core of the C-terminal domain. This resulted in a dramatic lowering in the pK_a value to a value of about 6.5 from its normal value of 10.5. Similar results have been obtained in Shortle's laboratory by placing a lysine residue in the core of staphylococcal nuclease (Stites *et al.*, 1991). Large increases in the pK_a values of acidic residues in hydrophobic environments have also been reported. Langsetmo *et al.* (1991) monitored an aspartate residue in thioredoxin which has a greatly increased pK_a value of about 7.5. Varadarajan *et al.* (1989) replaced Val68 of human myoglobin with aspartate. This placed the aspartate in a quite hydrophobic environment with a corresponding increase in pK_a value to about 8.9.

The most severe test of our understanding of charge–charge interactions is to construct a stabilizing interaction between oppositely charged groups. Our first attempts at the construction of designed interactions of this sort have been only partially successful. These include a potential histidine aspartate interaction in a rather hydrophobic environment which was constructed with the double mutant (W138H with Q105E). A hydrogen-bonded interaction between uncharged partners exists in the wild-type protein (Trp138 and Gln105) and we hoped to retain that interaction in the construction, using a charge pair. This construct had little interaction, as judged by the stabilities of the two single mutants and the double mutant. We also attempted to construct a lysine–aspartate interaction in the hydrophobic core (Dao-pin *et al.*, 1991a) designed to counteract the uncompensated charge of lysine in the mutant M102K by the mutation L133D. Again this gave only a modest stabilization of the mutation and the double mutant was still much less stable than the wild-type protein. Dao-pin *et al.* (1991b) also attempted to design salt bridge interactions in T4 lysozyme. The most successful of these showed a one unit increase in the pK_a value of an engineered histidine residue (S90H) when compensated by an engineered aspartate (Q122D). However, the salt bridge resulted in a net destabilization of the folded state, probably because of steric strain that was introduced into the structure. They also attempted to create salt bridges at the surface at six sites. These seemed to have very little overall effect on the stability of the folded state.

Clearly, the interactions found in the Asp70–His31 and in the Asp10–hydrogen bond and charge network interactions are not yet fully understood. We are making progress in understanding the basis of the highly energetic nature of these interactions, but our ability to construct such an interaction is still limited. This is partly due to the special combination of interactions needed to produce this effect. We can design pairwise interactions but the design of a complex network of interactions requires several mutational steps and the identification of a suitable site for the construction. We must develop more sophisticated algorithms to search for these suitable sites. With such methods in hand, it should be possible to design stabilizing interactions that could supply 5 kcal/mol to the pH dependent stability of any globular protein.

Acknowledgements

We thank Enoch Baldwin, Susan Baxter, Dirk Heinz, David Lowry, Andy Morton, Hale Nicholson, Cynthia Phillips, Amy Roth, Michael Strain, John Schellman and Brian Matthews for many discussions and helpful advice. We thank Jim Remington for construction of the stereopair drawings. This work has been supported by grants from the National Science Foundation (DMB8905322), National Institutes of Health (1S10RR04926) and the Markey Charitable Trust.

References

Alber, T. (1989). Mutational effects on protein stability. *Ann. Rev. Biochem.*, **58**, 765–798

Alber, T., Dao-pin, S., Wilson, K., Wozniak, J. A., Cook, S. P. and Matthews, B. W. (1987). Contributions of hydrogen bonds of Thr 157 to the thermodynamic stability of phage T4 lysozyme. *Nature*, **330**, 41–46

Anderson, D. E., Becktel, W. J. and Dahlquist, F. W. (1990). pH-induced denaturation of proteins: A single salt bridge contributes 3-5 kcal/mol to the free energy of folding of T4 lysozyme. *Biochemistry*, **29**, 2403–2408

Bax, A., Griffey, R. H. and Hawkins, B. L. (1983). Correlation of proton and nitrogen-15 chemical shifts by multiple quantum NMR. *J. Magn. Reson.*, **55**, 301–315

Bax, A., Sparks, S. W. and Torchia, D. A. (1988). Long range heteronuclear correlation. A powerful tool for the NMR analysis of medium-size proteins. *J. Am. Chem. Soc.*, **110**, 7926–7927

Bax, A. and Summers, M. F. (1986). Proton and carbon-13 assignments from sensitivity-enhanced detection of heteronuclear multiple-bond connectivity by 2D multiple quantum NMR. *J. Am. Chem. Soc.*, **108**, 2093–2094

Bowie, J. U., Reidhaar-Olson, J. F., Lim, W. A. and Sauer, R. T. (1990). Deciphering the message in protein sequences: tolerance to amino acid substitutions. *Science*, **247**, 1306–1310

Briggs, M. S. and Roder, H. (1992). Early hydrogen bonding events in the folding reaction of ubiquitin. *Proc. Natl Acad. Sci. USA*, **89**, 2017–2021

Brown, L. R., DeMaro, A., Richarz, R., Wagner, G. and Wüthrich, K. (1978). The influence of a single salt bridge on static and dynamic features of the globular solution conformation of the basic pancreatic trypsin inhibitor.

Proton and carbon-13 nuclear magnetic resonance studies of the native and transaminated inhibitor. *Eur. J. Biochem.*, **88**, 87–95

Bycroft, M., Matouschek, A., Kellis, J. T., Jr., Serrano, L. and Fersht, A. R (1990). Detection and characterization of a folding intermediate in barnase by NMR. *Nature*, **346**, 488–490

Cybulski, S. M. and Scheiner, S. (1989). Hydrogen bonding and proton transfers involving the carboxylate group. *J. Am. Chem. Soc.*, **111**, 23–31

Dao-pin, S., Anderson, D. E., Baase, W. A., Dahlquist, F. W. and Matthews, B. W. (1991a). Structural and thermodynamic consequences of burying a charged residue within the hydrophobic core of T4 lysozyme. *Biochemistry*, **30**, 11521–11529

Dao-pin, S., Sauer, U., Nicholson, H. and Matthews, B. W. (1991b). Contributions of engineered surface salt bridges to the stability of T4 lysozyme determined by directed mutagenesis. *Biochemistry*, **30**, 7142–7153

Dao-pin, S., Söderlind, E., Baase, W. A., Wozniak, J. A., Sauer, U. and Matthews, B. W. (1991c). Cumulative site-directed charge–change replacements in bacteriophage T4 lysozyme suggest that long-range electrostatic interactions contribute little to protein stability. *J. Mol. Biol.*, **221**, 873–887

Dill, K. A. (1990). Dominant forces in protein folding. *Biochemistry*, **29**, 7133–7155

Dixon, M. M., Nicholson, H., Shewchuck, L., Baase, W. A. and Matthews, B. W. (1992). The structure of a 'hinge-bending' T4 lysozyme mutation, Ile 3, Pro. *J. Mol. Biol.* (in press)

Dolgikh, D. A., Kolomiets, A. P., Bolotina, I. A. and Ptitsyn, O. B. (1984). 'Molten-globule' state accumulates in carbonic anhydrase folding. *FEBS Lett.*, **165**, 88–92

Eriksson, A. E., Baase, W. A., Wozniak, J. A. and Matthews, B. W. (1992a). A cavity-containing mutant of T4 lysozyme is stabilized by buried benzene. *Nature*, **355**, 371–373

Eriksson, A. E., Baase, W. A., Zhang, X.-J., Heinz, S. W., Blaber, M., Baldwin, E. P. and Matthews, B. W. (1992). Response of a protein structure to cavity-creating mutations and its relation to the hydrophobic effect. *Science*, **255**, 178–183

Faber, H. R. and Matthews, B. W. (1990). A mutant T4 lysozyme displays five different crystal conformations. *Nature*, **348**, 263–266

Gil'manshin, R. I. and Ptitsyn, O. B. (1987). An early intermediate of refolding α-lactalbumin forms within 20 ms. *FEBS Lett.*, **223**, 327–329

Goldenberg, D. P., Frieden, R. W., Haack, J. A. and Morrison, T. B. (1989). Mutational analysis of a protein-folding pathway. *Nature*, **338**, 127–132

Goto, Y., Calciano, L. J. and Fink, A. L. (1990). Acid-induced folding of proteins. *Proc. Natl Acad. Sci. USA*, **87**, 573–577

Griffey, R. H., Redfield, A. G., McIntosh, L. P., Oas, T. G. and Dahlquist, F. W. (1986). Assignment of proton amide resonances of T4 lysozyme using ^{13}C and ^{15}N multiple isotopic labelling. *J. Am. Chem. Soc.*, **108**, 6816–6817

Heinz, D. W., Baase, W. A. and Matthews, B. W. (1992). Folding and function of a T4 lysozyme containing 10 consecutive alanines illustrate the redundancy of information in an amino acid sequence. *Proc. Natl Acad. Sci. USA*, **89**, 3751–3755

Hughson, F. M., Wright, P. E. and Baldwin, R. L. (1990). Structural characterization of a partly folded apomyoglobin intermediate. *Science*, **249**, 1544–1548

Ikeguchi, M., Kuwajima, K., Mitani, M. and Sugai, S. (1986). Evidence for identity between the equilibrium unfolding intermediate and a transient folding: a comparative study of the folding reactions of α-lactalbumin and lysozyme. *Biochemistry*, **25**, 6965–6972

Karpusas, M., Baase, W. A., Matsumura, M. and Matthews, B. W. (1989). Hydrophobic packing in T4 lysozyme probed by cavity-filled mutants. *Proc. Natl Acad. Sci. USA*, **86**, 8237–8241

Kim, P. S. and Baldwin, R. L. (1990). Intermediates in the folding reactions of small proteins. *Ann. Rev. Biochem.*, **59**, 459–489

Kuwajima, K., Hiraoka, Y., Ikeguchi, M. and Sugai, S. (1985). Comparison of the transient folding intermediates in lysozyme and α-lactalbumin. *Biochemistry*, **24**, 874–881

Kuwajima, K., Yamaya, H., Miwa, S., Sugai, S. and Nagamura, T. (1987). Rapid formation of secondary structure framework in protein folding studied by stopped-flow circular dichroism. *FEBS Lett.*, **221**, 115–118

Langsetmo, K., Fuchs, J. A. and Woodward, C. (1991). The conserved, buried aspartic acid in oxidized Escherichia coli thioredoxin has a pK_a of 7.5. Its titration produces a related shift in global stability. *Biochemistry*, **30**, 7603–7609

Lee, B. and Richards, F. M. (1971). Interpretation of protein structures: estimation of static accessibility. *J. Mol. Biol.*, **55**, 379–400

Lim, W. A. and Sauer, R. T. (1989). Alternative packing arrangements in the hydrophobic core of lambda repressor. *Nature*, **339**, 31–36

Lu, J. and Dahlquist, F. W. (1992). Detection and characterization of an early folding intermediate of T4 lysozyme using pulsed hydrogen exchange and two-dimensional NMR. *Biochemistry*, **31**, 4749–4756

McCoy, L. F. J., Rowe, E. S. and Wong, K. P. (1980). Multiparameter kinetic study on the unfolding and refolding of bovine carbonic anhydrase B. *Biochemistry*, **19**, 4738–4743

McIntosh, L. P., Wand, A. J., Lowry, D. F., Redfield, A. G. and Dahlquist, F. W. (1990). Assignment of backbone ^{1}H and ^{15}N-NMR resonances of bacteriophage T4 lysozyme. *Biochemistry*, **29**, 6341–6362

Malcomb, B. A., Rosenberg, S., Corey, M. J., Allen, J. S., deBaetselier, A. and Kirsh, J. F. (1989). Site-directed mutagenesis of the catalytic residues Asp-52 and Glu-35 of chicken egg white lysozyme. *Proc. Natl Acad. Sci. USA*, **86**, 133–137

Matsumura, M. Becktel, W. J., Levitt, M. and Matthews, B. W. (1989). Stabilization of phage T4 lysozyme by engineered disulfide bonds. *Proc. Natl Acad. Sci. USA*, **86**, 6562–6566

Matsumura, M., Becktel, W. J. and Matthews, B. W. (1988). Hydrophobic stabilization in T4 lysozyme determined directly by multiple substitutions of Ile3. *Nature*, **334**, 406–410

Matsumura, M. and Matthews, B. W. (1989). Control of enzyme activity by an engineered disulfide bond. *Science*, **243**, 792–794

Matthews, B. W. and Remington, S. J. (1974). The three-dimensional structure of the lysozyme from bacteriophage T4. *Proc. Natl Acad. Sci. USA*, **71**, 4178–4182

Matthews, C. R. (1991). The mechanism of protein folding. *Curr. Opinion Struct. Biol.*, **1**, 28–35

Miranker, A., Radford, S. E., Karplus, M. and Dobson, C. M. (1991). Demonstration by NMR of folding domains in lysozyme. *Nature*, **349**, 633–636

Molday, R. S., Englander, S. W. and Kallen, R. G. (1972). Primary structure effects on peptide group hydrogen exchange. *Biochemistry*, **11**, 150–158

Muchmore, D. C., McIntosh, L. P., Russell, C. B., Anderson, E. E. and Dahlquist, F. W. (1990). Expression and ^{15}N labelling of proteins for proton and nitrogen-15 NMR. *Meth. Enzymol.*, **177**, 44–73

Nicholson, H., Anderson, D. E., Dao-pin, S. and Matthews, B. W. (1991). Analysis of the interaction between charged side chains and the α-helix dipole using designed thermostable mutants of phage T4 lysozyme. *Biochemistry*, **30**, 9816–9828

Nicholson, H., Becktel, W. J. and Matthews, B. W. (1988). Enhanced protein thermostability from designed mutations that interact with α-helix dipoles. *Nature*, **366**, 651–656

Pace, C. N., Laurents, D. V. and Thomson, J. A. (1990). pH dependence of the urea and guanidine hydrochloride denaturation of ribonuclease A and ribonuclease T1. *Biochemistry*, **29**, 2564–2572

Rennell, D., Bouvier, S. E., Hardy, L. W. and Poteete, A. (1991). Systematic mutation of bacteriophage T4 lysozyme. *J. Mol. Biol.*, **222**, 67–87

Robertson, A. D. and Baldwin, R. L. (1991). Hydrogen exchange in thermally denatured ribonuclease. *Biochemistry*, **30**, 9907–9914

Roder, H., Elove, G. A. and Englander, S. W. (1988). Structural characterization of folding intermediates in cytochrome c by hydrogen-exchange labelling and proton NMR. *Nature*, **335**, 700–704

Semisotnov, G. V., Rodionova, N. A., Kutyshenko, V. P., Ebert, B., Blanck, J. and Ptitsyn, O. B. (1987). Sequential mechanism of refolding of carbonic anhydrase B. *FEBS Lett.*, **224**, 9–13

Sharp, K. A. and Honig, B. (1990). Electrostatic interactions in macromolecules: theory and applications. *Ann. Rev. Biophys. Biophys. Chem.*, **19**, 301–332

Stites, W. E., Gittis, A. G., Lattman, E. E. and Shortle, D. (1991). In a staphylococcal nuclease mutant the side-chain of a lysine replacing valine 66 is fully buried in the hydrophobic core. *J. Mol. Biol.*, **221**, 7–14

Udgaonkar, J. B. and Baldwin, R. L. (1988). NMR evidence for an early framework intermediate on the folding pathway of ribonuclease A. *Nature*, **335**, 694–699

Udgaonkar, J. B. and Baldwin, R. L. (1990). Early folding intermediate of ribonuclease A. *Proc. Natl Acad. Sci. USA*, **87**, 8197–201

Varadarajan, R., Lambright, D. G. and Boxer, S. G. (1989). Electrostatic interactions in wild-type and mutant recombinant human myoglobins. *Biochemistry*, **28**, 3771–3781

Weaver, L. H. and Matthews, B. W. (1987). Structure of bacteriophage T4 lysozyme refined at 1.7 Å resolution. *J. Mol. Biol.*, **193**, 189–199

Zuiderweg, E. R. P. (1990). A proton-detected heteronuclear chemical-shift correlation experiment with improved resolution and sensitivity. *J. Magn. Reson.*, **86**, 346–357

Index